Dynamic Modelling and Simulation of Food Systems

Dynamic Modelling and Simulation of Food Systems

Editors

Carlos Vilas
Míriam R. García
Jose A. Egea

MDPI • Basel • Beijing • Wuhan • Barcelona • Belgrade • Manchester • Tokyo • Cluj • Tianjin

Editors

Carlos Vilas
IIM-CSIC (Spanish National Research Council)
Spain

Míriam R. García
IIM-CSIC (Spanish National Research Council)
Spain

Jose A. Egea
CEBAS-CSIC (Spanish National Research Council)
Spain

Editorial Office
MDPI
St. Alban-Anlage 66
4052 Basel, Switzerland

This is a reprint of articles from the Special Issue published online in the open access journal *Foods* (ISSN 2304-8158) (available at: https://www.mdpi.com/journal/foods/special_issues/dynamic_modelling_simulation_food_systems).

For citation purposes, cite each article independently as indicated on the article page online and as indicated below:

LastName, A.A.; LastName, B.B.; LastName, C.C. Article Title. *Journal Name* **Year**, *Volume Number*, Page Range.

ISBN 978-3-0365-6692-4 (Hbk)
ISBN 978-3-0365-6693-1 (PDF)

Contents

About the Editors

Carlos Vilas

Carlos Vilas, Ph.D., has been a Tenured Scientist at the IIM-CSIC (Vigo, Spain) since 2022. He received the B.Sc. degree in Chemistry in 2001 and his Ph.D. in Mathematics in 2008, both from University of Vigo. After that, in 2008, he started a postdoctoral research stay at the Automatic Control Laboratory of University of Mons (Belgium). In 2010, he moved back to the IIM-CSIC with a grant from the regional government (Xunta de Galicia). His research focuses on mathematical modelling, simulation, advanced control and optimization of processes, mainly from the biotechnological and food industries. In particular, he is interested in distributed process systems described by partial differential equations, in the evolution of quality and safety indicators in fish and shellfish, and in the development of digital twins for the processes of the food industry. He was also a teacher in Masters programs at different institutions, such as the University of Valladolid, University of Vigo, and Polytechnic University of Madrid.

Míriam R. García

Míriam R. García, Ph.D., is a Tenured Scientist at the BioSystems and BioProcess Engineering group (Bio2Eng) at the IIM-CSIC (Vigo, Spain). Her training is in chemistry (MSc) and applied mathematics (Ph.D. 2008 held at Bio2Eng) from the University of Vigo. After her PhD, she moved to Ireland to study the mathematics of systems biology for Parkinson's Disease at the Hamilton Institute (Maynooth University). After a five-year postdoctoral program, she came back to Bio2Eng and obtained a permanent position (2021) thanks to a new research line on modelling bacterial resistance to chemical stresses, like antimicrobials, which is of relevance to the food industry. Her research's main interest is to develop mathematical models in microbiology with sufficient predictive capabilities (i.e., not too complex or simple) to be used as tools in the food industry, together with proper optimization and control algorithms.

Jose A. Egea

Jose A. Egea, Ph.D., has been a Tenured Scientist at the CEBAS-CSIC (Murcia, Spain) since 2018. He received the M.Sc. degree in Chemical Engineering in 2003 (University of Murcia, Spain) and he developed his Ph.D Thesis on mathematical methods for complex process optimization at the Bioprocess Engineering Group at IIM-CSIC (Vigo, Spain), obtaining his Ph.D. degree by the University of Vigo in 2008 (Chemical Engineering Program). He was a member of the Department of Applied Mathematics and Statistics at the Technical University of Cartagena (Spain) from 2009 to 2018, as an Assistant Professor and later as an Associated Professor. Furthermore, he completed his research record with several pre-doctoral and post-doctoral research stays (e.g., KU Leuven, PUC Chile or EBI at UK) and the Master of Bioinformatics at the University of Murcia in 2015. In 2018, he obtained a permanent position at CEBAS-CSIC, joining the Plant Breeding Department. His research focuses on mathematical modelling, simulation, optimization and data analysis in agricultural and food sciences. He has scientific contributions in microbial inactivation in foods that include model development, optimal experimental design and optimization. His most recent research lines are related to modelling and analysis of phenology in *Prunus* species from an agro-meteorological point of view.

foods

MDPI

Editorial

Dynamic Modelling and Simulation of Food Systems: Recent Trends and Applications

Jose A. Egea [1], Míriam R. García [2] and Carlos Vilas [2,*]

1 Fruit Breeding Group, CEBAS-CSIC, Campus de Espinardo 25, 30100 Murcia, Spain
2 Biosystems and Bioprocess Engineering Group, IIM-CSIC, 36208 Vigo, Spain
* Correspondence: carlosvf@iim.csic.es; Tel.: +34-986-231-930

Several factors influence consumers' choices of food products. While price remains the main criterion, quality, pleasure, convenience, and health are also important driving factors in food market evolution. Food enterprises are making significant efforts to manufacture products that meet consumers' demands without compromising on safety standards. Additionally, the food industry also aims to improve the efficiency of transformation and conservation processes by minimizing energy consumption, process duration, and waste generation. However, foods are highly complex systems in which: (i) non-linear dynamics and interactions among different temporal and spatial scales must be considered; (ii) a wide range of physical phenomena (such as evaporation, mechanical changes, thawing, energy/mass transport, and color changes) occur; (iii) different food matrices (such as meat, vegetables, cereal, milk, and juices) with different microstructures and properties are involved; and (iv) the number of quality and safety indicators (such as bacteria, total volatile basic nitrogen, color, texture, odor, and sensory characteristics) is substantial. Mathematical modeling and simulation are key elements that allow us to gain a deeper understanding of food processes and enable the use of tools such as optimization and real-time control to improve their efficiency. This special issue aims to gather research on the development of dynamic mathematical models that describe the relevant factors in food processes from the perspectives of food safety (chemical or microbiological), food quality (organoleptic or nutritional), or resource consumption. Additionally, the development of model-based tools to improve food processes is also considered. This includes decision-making and optimization tools, the characterization of uncertainty/variability in model predictions, model simulation techniques, software sensors, and software development. The contributions published in this Special Issue can be grouped into two categories according to their main research topic: the evolution of safety and quality indicators in unprocessed food systems, and transformation and preservation processes.

Citation: Egea, J.A.; García, M.R.; Vilas, C. Dynamic Modelling and Simulation of Food Systems: Recent Trends and Applications. *Foods* **2023**, *12*, 557. https://doi.org/10.3390/foods12030557

Received: 10 January 2023
Accepted: 18 January 2023
Published: 27 January 2023

1. Evolution of Safety and Quality Indicators in Unprocessed Food Systems

The evolution of quality in food products is mainly dependent on microbial content, but also on other indicators such as nucleotide degradation; the formation of volatile nitrogenous bases or biogenic amines; and texture. In this Special Issue, we present four research articles on different aspects of bacterial growth or inactivation and a review paper analyzing the mathematical models in the literature that describe and predict food quality indicators.

1.1. Bacterial Dynamics

Microbial growth and inactivation rates are highly influenced by the food matrix. Therefore, matrix microstructure is a main factor to consider when deriving mathematical models that describe microbial dynamics in food systems. Verheyen and Van Impe [1] provide a comprehensive review of the models developed during the last two decades that study microstructure influence. Two types of model are identified: (i) macroscale secondary models including food microstructural factors, and (ii) microscale semi-mechanistic models.

The selection of the best approach depends on the particular application, the accuracy required, and the available computational power. The authors also identified current research trends: (i) the study of the effect of intrinsic factors on microbial dynamics, and (ii) the development of models considering the influence of food microstructure during non-thermal processes.

Some bacteria, such as *Carnobacterium maltaromaticum* CNCM I-3298, can be used in food biopreservation, flavor development processes, or in biological time–temperature integrators to track temperature variations during transport and storage. Puentes et al. [2] used the reaction scheme mechanism to derive an accurate mathematical model that describes the growth of *C. maltaromaticum* and the production of formic acid, acetic acid, lactic acid, and ethanol from trehalose. The surface response method was used to describe the relationships between the operating conditions (temperature and pH) and the specific growth and production rates. The authors also illustrated how the model can be used to compute the optimal operating conditions of the process (T and pH). Finally, they also proposed some interesting research directions such as incorporating the effects of other culture parameters or understanding the inhibition mechanisms of metabolites.

The efficiency of treatments to inactivate bacteria can be assessed by detecting and quantifying the sublethal injury of pathogenic microorganisms. However, existing methods of modeling the evolution of sublethal injury (SI) present several disadvantages related to the frequent occurrence of SI trends in these methods, which are, in part, artifacts. Akkermans et al. [3] proposed a new approach to modeling the evolution of SI during microbial inactivation that avoids unrealistic calculations. The method, based on the description of inactivation kinetics between subpopulations of healthy, sublethally injured, and dead cells, was designed to be used in combination with any existing microbial inactivation model. Log-linear inactivation, biphasic inactivation, and log-linear inactivation with tailing were used to validate the approach. The advantages of this approach make it suitable for describing SI during food processes.

Shewanella putrefaciens is one of the most important Specific Spoilage Organisms (SSOs) in fish products. Yi and Xie [4] focused on designing a nondestructive method, based on the use of an electronic nose, to describe the growth of *S. putrefaciens* during fish spoilage. Bacterial concentration was described using two classical primary models—Gompertz and logistic—whereas the dependence of growth rate and lag time on temperature was modeled using the square root model. The authors also derived a regression model based on the partial least squares method to correlate the electronic nose and electrical conductivity measurements with the spoilage potential of *S. putrefaciens*. Finally, gas chromatography/mass spectrometry was used to determine the characteristic volatile organic compounds of tuna inoculated with *S. putrefaciens*.

1.2. Other Quality Indicators

The development of methods to describe the evolution of other quality indicators has gained relevance in recent decades. García et al. [5] presented a comprehensive review of the different indicators used in the literature to assess the quality of fresh fish; the stress variables that affect the evolution of such indicators; and the mathematical models available to describe such evolution. The work also presented the main challenges currently faced in food quality modeling:

(i) There is a lack of mathematical models for some critical indicators, such as nutrients and odor.
(ii) There are many different model structures but a lack of proper comparisons between alternatives.
(iii) Uncertainty analysis of model parameters and bacterial load is usually missed.
(iv) Model validation is usually disregarded.
(v) The relationships between the shelf life and growth of SSOs are not well understood and are usually not described in dynamic modeling.

(vi) The potential of current models is not fully exploited towards their integration into software systems for online quality prediction.

2. Preservation and Transformation Processes

Preservation and transformation processes are of paramount importance in food systems. This Special Issue includes four articles focusing on different aspects of the fermentation process and three manuscripts that consider thermal processes.

2.1. Fermentation Processes

Mathematical models are useful tools to understand food systems, and combined with proper methodologies such as optimization, control, or scheduling, they enable the design of food processes and their operating conditions. Ritonja et al. [6] derived a fourth-order non-linear state-space model to explain the effect of temperature on the dynamics of CO_2 produced during milk fermentation. The structure of the proposed model is compact and simpler than other options in the literature, although it is able to represent experimental behavior. The authors also suggest that a non-linear adaptive control approach would be a reliable option to design a control law to force the process to follow the desired reference trajectory.

Fermentation is also used to ensure the safety and quality of foods, and to increase product shelf life. Predictive microbiology can be exploited to describe the growth and inactivation of bacteria as a function of the fermentation conditions. Racioppo et al. [7] used the Food Spoilage and Safety Predictor to model the effects of stress variables (such as temperature, pH, and salt) on the growth of lactic acid bacteria in fermented smoked fish. The maximal growth rate and the time taken to attain the critical threshold were modeled through a multiple regression procedure. This model was used to optimize the production of smoked fermented fish by combining the variables through a fractional design of experiments. The authors showed that the most critical factor in the fermentation process was liquid smoke, followed by temperature and salt.

Rapaport et al. [8] proposed a simple model that includes a maintenance term (giving rise to a variable yield) to describe the growth of yeast on nitrogen during the fermentation of wine. This maintenance term can explain a consumption of nitrogen that is not entirely converted into biomass. Additionally, the variable yield, that can be estimated from data, gives the approach the flexibility to suit different kinds of models or experimental data with a single common structure. The maintenance term encodes the underlying mechanisms of transporters and carbohydrate accumulation. The authors showed that this simple model can reproduce the experimental data and results of more sophisticated models, bringing new perspectives to the control of wine fermentation through the addition of nitrogen.

Dynamic models describing food processes must be accurate and reliable, but they must also be compatible with measurable variables in real industrial processes. Zamudio-Lara et al. [9] proposed two dynamic models of beer fermentation and performed parameter estimation, structural identifiability analysis, observability analysis, and cross-validation to assess the models' predictive capabilities. The proposed models were based on biomass dynamics and CO_2. A set of variables that should be monitored for each model to achieve complete observability was provided. The estimation procedure included some mathematical relationships to describe the thermal dependence of the kinetic parameters proposed, leading to a good prediction of the experimental data for both models. These new models enable measurement implementations in order to identify and quantify the process variables, thus improving process efficiency and controllability. These new models are good candidates for model-based process control in beer fermentation.

2.2. Thermal Processes

The analysis of multiple objectives is crucial when designing dynamic food processes. The different dynamics of the considered objectives may lead to counterintuitive conclusions. Peñalver-Soto et al. [10], the authors analyzed the microbial inactivation of *Geobacillus stearothermophilus* and acrylamide production in the thermal processing of pureed potato and prune juice, which may be present in infant formulations. The authors found that to ensure proper microbial inactivation and reduce acrylamide formation, high-temperature processes (with a short application time) are needed. This could be counterintuitive as acrylamide formation increases along with temperature. However, the sensitivities of the objectives to the process variables make the dynamics of acrylamide formation much slower than those of microbial inactivation at high temperatures. These results may facilitate the design of microbial inactivation thermal processes where acrylamide formation is an issue.

Quality parameters can be seriously affected when dynamic thermal processes are applied to foods. In some cases, high temperatures can produce a decline in some quality parameters while improving others. This is the case with fried potato chips, where higher temperatures improve yellowness and crunchiness (important indicators for consumer acceptance) but also accelerate the production of certain toxic compounds such as acrylamide. Peñalver-Soto et al. [11] presented a multi-objective optimization approach to simultaneously maximize yellowness and minimize acrylamide production in the potato-chip frying process. Their results showed that most of the solutions of the Pareto front led to levels of acrylamide above the maximum recommended by the European Food Safety Agency (EFSA). Low temperatures and high processing times should be used to avoid excess acrylamide. They also found that under mild processing conditions, there can be quasi-equivalent solutions (e.g., different processing conditions leading to the same relationship between yellowness and acrylamide) due to the sensitivities of the objectives to such conditions. Finally, parameter uncertainty and Pareto front uncertainties were higher at higher temperatures.

Innovations in the field of rapid heating technologies require foods' thermal properties to be determined accurately. Muniandy et al. [12] performed a study to determine the thermal conductivity of a model food using rapid heating. Two-dimensional heat transfer models based on finite differences were formulated, and experiments to monitor temperature were designed based on scaled sensitivity coefficients. The authors proposed three models for thermal conductivity—constant, linear, and re-parameterized linear—to improve identifiability, and obtained better estimates from the linear ones. To estimate the parameters with low errors, it was concluded that the constant temperature experiment should be conducted for at least 20 min, while the rapid heating experiment required only 30 s. The estimated trend of conductivity with temperature was more consistent with fatty foods in the rapid heating experiments. Additionally, the residual analysis for both types of experiment revealed that the parameter estimation in the rapid heating experiment was more reliable. Finally, prolonged exposure to temperature in the constant-temperature experiments could negatively impact the reliability of the estimated thermal properties due to changes in the food matrix.

In the face of climate change, it may advised that unused species of some crops should be recovered to ensure resistance against increasing pests and resilience against changing climate conditions. It is therefore important to determine their physicochemical properties and understand how they are affected by processing treatments.Sridhar et al. [13] determined the physicochemical properties of currant tomato (*Solanum pimpinellifolium*) and studied the effects of cold- and hot-break heat treatments on it. Color-related parameter values decreased significantly under all of the heat treatments. The apparent viscosity, lycopene, and total titratable acidity differed significantly between heat treatments (mostly at the highest temperatures). The change in the viscosity of tomato pulp and paste with temperature was modeled using Arrhenius. The findings of this research may strengthen the knowledge of process optimization designers, and thus, facilitate the development of currant tomato-based products.

Author Contributions: All authors contributed equally to this work. All authors have read and agreed to the published version of the manuscript.

Funding: We acknowledge receipt of the following grants: RTI2018-093560-J-I00, funded by Spanish Ministerio de Ciencia e Innovación MCIN/AEI/10.13039/501100011033 "ERDF—A way of making Europe"; Project Pescando4.0 (ref. 202178002), funded by NextGenerationEU within the framework of the Plan de Recuperación, Transformación y Resiliencia (PRTR) and managed by the Ministerio de Agricultura, Pesca y Alimentación through the RD 685/2021; and Project MODELSANO, funded by the Spanish MINECO (grant number PID2019-104212RB-I00/AEI/10.13039/501100011033).

Institutional Review Board Statement: Not applicable.

Informed Consent Statement: Not applicable.

Data Availability Statement: No data were created or used in this research.

Conflicts of Interest: The authors declare no conflict of interest.

References

1. Verheyen, D.; Van Impe, J.F.M. The Inclusion of the Food Microstructural Influence in Predictive Microbiology: State-of-the-Art. *Foods* **2021**, *10*, 2119. [CrossRef] [PubMed]
2. Puentes, C.; Girardeau, A.; Passot, S.; Fonseca, F.; Trelea, I.-C. Dynamic Modeling of *Carnobacterium maltaromaticum* CNCM I-3298 Growth and Metabolite Production and Model-Based Process Optimization. *Foods* **2021**, *10*, 1922. [CrossRef] [PubMed]
3. Akkermans, S.; Verheyen, D.; Smet, C.; Van Impe, J. A Population Balance Model to Describe the Evolution of Sublethal Injury. *Foods* **2021**, *10*, 1674. [CrossRef] [PubMed]
4. Yi, Z.; Xie, J. Prediction in the Dynamics and Spoilage of *Shewanella putrefaciens* in Bigeye Tuna (*Thunnus obesus*) by Gas Sensors Stored at Different Refrigeration Temperatures. *Foods* **2021**, *10*, 2132. [CrossRef] [PubMed]
5. García, M.R.; Ferez-Rubio, J.A.; Vilas, C. Assessment and Prediction of Fish Freshness Using Mathematical Modelling: A Review. *Foods* **2022**, *11*, 2312. [CrossRef] [PubMed]
6. Ritonja, J.; Goršek, A.; Pečar, D.; Petek, T.; Polajžer, B. Dynamic Modeling of the Impact of Temperature Changes on CO_2 Production during Milk Fermentation in Batch Bioreactors. *Foods* **2021**, *10*, 1809. [CrossRef] [PubMed]
7. Racioppo, A.; Campaniello, D.; Sinigaglia, M.; Bevilacqua, A.; Speranza, B.; Corbo, M.R. Use of Food Spoilage and Safety Predictor for an "A Priori" Modeling of the Growth of Lactic Acid Bacteria in Fermented Smoked Fish Products. *Foods* **2022**, *11*, 946. [CrossRef] [PubMed]
8. Rapaport, A.; David, R.; Dochain, D.; Harmand, J.; Nidelet, T. Consideration of Maintenance in Wine Fermentation Modeling. *Foods* **2022**, *11*, 1682. [CrossRef] [PubMed]
9. Lara, J.M.Z.; Dewasme, L.; Escoto, H.H.; Wouwer, A.V. Parameter Estimation of Dynamic Beer Fermentation Models. *Foods* **2022**, *11*, 3602. [CrossRef] [PubMed]
10. Peñalver-Soto, J.L.; Garre, A.; Aznar, A.; Fernández, P.S.; Egea, J.A. Dynamics of Microbial Inactivation and Acrylamide Production in High-Temperature Heat Treatments. *Foods* **2021**, *10*, 2535. [CrossRef] [PubMed]
11. Peñalver-Soto, J.L.; Muñoz-Guillermo, M.; Garre, A.; Iguaz, A.; Fernández, P.S.; Egea, J.A. Multiobjective Optimization of a Frying Process Balancing Acrylamide Formation and Quality: Solution Analysis and Uncertainty Propagation. *Foods* **2022**, *11*, 3689. [CrossRef] [PubMed]
12. Muniandy, A.; Benyathiar, P.; Mishra, D.K.; Ozadali, F. Dynamic Thermal Properties Estimation Using Sensitivity Coefficients for Rapid Heating Process. *Foods* **2021**, *10*, 1954. [CrossRef] [PubMed]
13. Sridhar, K.; Makroo, H.A.; Srivastava, B. Effect of Cold- and Hot-Break Heat Treatments on the Physicochemical Characteristics of Currant Tomato (*Solanum pimpinellifolium*) Pulp and Paste. *Foods* **2022**, *11*, 1730. [CrossRef] [PubMed]

foods

Review

The Inclusion of the Food Microstructural Influence in Predictive Microbiology: State-of-the-Art

Davy Verheyen [1,2,3] and Jan F. M. Van Impe [1,2,3,*]

1 BioTeC+, Chemical and Biochemical Process Technology and Control, Department of Chemical Engineering, KU Leuven, Gebroeders de Smetstraat 1, 9000 Ghent, Belgium; davy.verheyen@kuleuven.be
2 OPTEC, Optimization in Engineering Center-of-Excellence, KU Leuven, 3000 Leuven, Belgium
3 CPMF2, Flemish Cluster Predictive Microbiology in Foods—www.cpmf2.be, 9000 Ghent, Belgium
* Correspondence: jan.vanimpe@kuleuven.be

Abstract: Predictive microbiology has steadily evolved into one of the most important tools to assess and control the microbiological safety of food products. Predictive models were traditionally developed based on experiments in liquid laboratory media, meaning that food microstructural effects were not represented in these models. Since food microstructure is known to exert a significant effect on microbial growth and inactivation dynamics, the applicability of predictive models is limited if food microstructure is not taken into account. Over the last 10–20 years, researchers, therefore, developed a variety of models that do include certain food microstructural influences. This review provides an overview of the most notable microstructure-including models which were developed over the years, both for microbial growth and inactivation.

Keywords: predictive microbiology; food microstructure; food safety; mathematical models

Citation: Verheyen, D.; Van Impe, J.F.M. The Inclusion of the Food Microstructural Influence in Predictive Microbiology: State-of-the-Art. *Foods* **2021**, *10*, 2119. https://doi.org/10.3390/foods10092119

Academic Editors: Carlos Vilas, Míriam R. García and Jose A. Egea

Received: 30 July 2021
Accepted: 30 August 2021
Published: 8 September 2021

1. Introduction

Predictive microbiology encompasses the development of mathematical models to evaluate and predict the effect of environmental conditions (e.g., temperature, pH, CO_2 level, salt concentration, water activity) on the (growth, survival, and inactivation) behaviour of microorganisms in food (model) systems [1,2]. While some simple model types have been developed over the years, most predictive models are continuous dynamical models that consist of a set of ordinary differential equations (ODEs) [3]. The models are useful tools to assess and design processing, distribution, and storage operations to assure the microbiological safety and quality of food products [1,2]. To date, predictive models have been widely accepted by food producers, governments, and scientists as a sound scientific approach to accomplish legal food safety requirements [4]. The main advantages of predictive microbiology over traditional challenge tests are (i) an increased efficiency regarding financial costs, labour-intensiveness, and time, and (ii) the cumulative nature of the developed models [5].

Food structure, from natural or process-generated origins, is defined as the spatial arrangement of the structural elements of food products and their interactions [6,7]. Food structural elements can be interpreted at different scales, i.e., the molecular level (e.g., sugar, water, protein, and polysaccharide molecules), the nanoscale level (e.g., casein micelles), the microscale level (fat and water droplets in emulsions, granules, gel networks), and the macroscale level (e.g., air pockets, powders, foams) [7–9]. When investigating microbial behaviour, it is mainly the microscale level (i.e., food microstructure) which is of interest, with influencing aspects including physical constraints on the mobility of microorganisms, variations in oxygen availability, and nutrient diffusion related to the nature of the food matrix (i.e., viscous or gelled, rheological properties), and the presence of fat droplets inside the food matrix [10,11]. An important aspect of the microstructural influence on microbial dynamics is the effect on the growth morphology of microorganisms [12]. Depending on the specific microstructural complexity of foods, microorganisms can occur as single cells, small aggregates (i.e., radius < 1.5 μm), microcolonies (i.e., radius < 200 μm), macrocolonies (i.e., radius > 200 μm), and biofilms [13]. An example of this microstructural influence on bacterial growth morphology is illustrated in Figure 1.

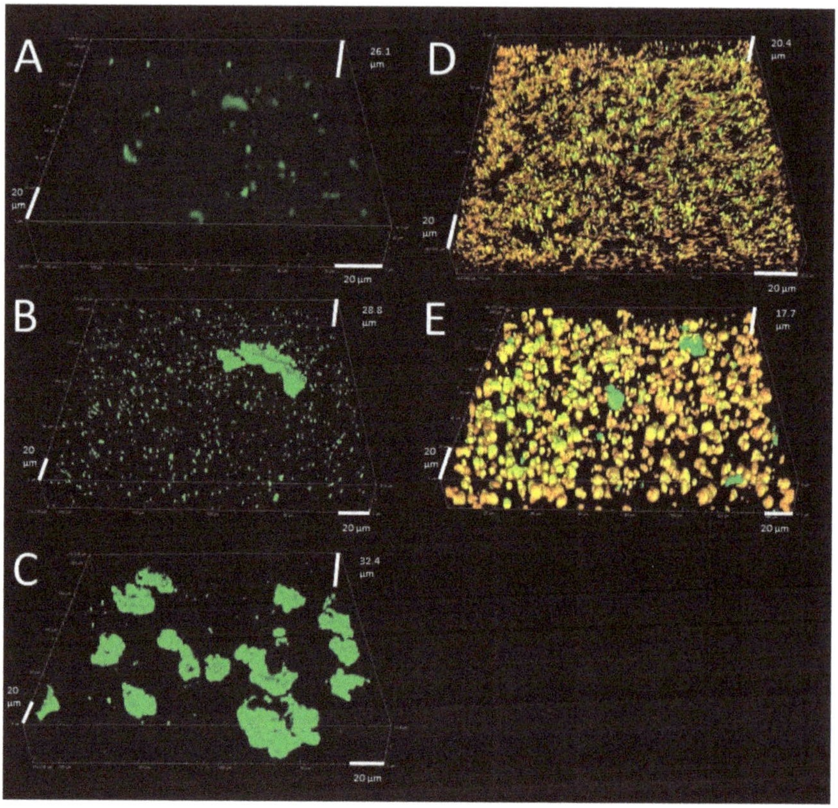

Figure 1. Confocal laser scanning microscopy images of growth of *Listeria monocytogenes* in food model systems with varying microstructural complexity (adapted from [13], with permission from American Society for Microbiology—Journals, 2021). Bacterial cells and fat droplets are shown in green and orange, respectively. The liquid/gelled aqueous phase was not stained and is, hence, shown by the absence of colour. The growth morphology of the bacteria clearly depends on the microstructural aspects of the food (model) system: (**A**) single cells, small aggregates and small microcolonies in a simple low-viscosity liquid system; (**B**) a large number of small aggregates and small microcolonies and some larger microcolonies in a liquid system with increased viscosity; (**C**) microcolonies of different sizes in an aqueous gel model system; (**D**) small aggregates and microcolonies growing in the spaces between fat droplets and around the fat droplets in an emulsion model system; (**E**) small aggregates and microcolonies growing in the spaces between the fat droplets and around the fat droplets in a gelled emulsion model system.

Traditionally, most predictive models are developed based on experiments in liquid laboratory media, in which case the effect of the food microstructure on the microbial behaviour, albeit a major influencing factor, is not taken into account [10,14]. The applicability of these models is, hence, limited to liquid foods with a relatively uniform distribution of nutrients and microorganisms occurring in a planktonic form [15,16]. In/on more structured foods (e.g., aqueous gels, emulsions, gelled emulsions), model predictions can be either fail-safe (i.e., predicting more growth or less inactivation than in reality) or fail-dangerous (i.e., predicting less growth or more inactivation than in reality) from a food safety point of view [10]. For microbial growth, most liquid-based models are considered fail-safe in/on more structured foods, while for microbial inactivation, most models are considered fail-dangerous [17–20]. Nevertheless, exceptions to the general trend have been reported for both microbial growth and inactivation in structured foods, meaning that including the food microstructural effect into predictive models would be beneficial for the overall accuracy and safety of predictive models [17,18,21,22].

During the 1990s and early 2000s, a few review papers brought to attention the lack of knowledge concerning the effect of the food microstructure on microbial dynamics in the context of predictive microbiology, also proposing modelling frameworks to address the issue in the future [10,23]. During the 20–30 years following on these pioneering works, however, the number of developed predictive models that incorporated food microstructural effects remained scarce. While some useful models have been reported, most of them mainly consisted of isolated efforts more focussed on specific applications (e.g., predicting the growth of microorganisms as a function of the gelatine concentration of the food [24]) rather than systematic modelling frameworks. Nevertheless, a significant amount of useful experimental and modelling-related approaches/concepts have been developed during these last decades. No relevant review papers on this topic have, however, been published in recent years, except for the review of Skandamis and Jeanson [25], focussing on the inclusion of the effect of the type of growth of the microorganisms (i.e., colonial vs. planktonic) into mathematical models for liquid, semi-liquid, and solid foods and food surfaces. Therefore, in the current review, the aim is to provide an extensive overview of the most promising food microstructure-including predictive models which were developed over the years, both for microbial growth and inactivation. For consistency and clarity purposes, the parameters and variables in the different presented models were occasionally renamed to remain uniform among the different examples.

2. Historical Overview on the Inclusion of Food Microstructure in Predictive Models

2.1. The Absence of Food Microstructure in the Early Days of Predictive Microbiology

Figure 2 depicts the evolution of the inclusion of food microstructural factors in predictive models over the years. In the early days of predictive microbiology, research did not specifically focus on the influence of food microstructure on microbial behaviour. One of the basic assumptions of predictive microbiology (which is still valid now) was that the growth or inactivation behaviour of microorganisms can be described/predicted based on a limited number of variables. Traditionally, researchers did not see food microstructure as one of those variables and focussed on factors such as temperature, pH, water activity (a_w), nutrient concentration, and the presence of preservatives [26]. Because of this reason, predictive models were mainly developed based on experiments in liquid laboratory media due to their ease of use during microbiological experiments [27].

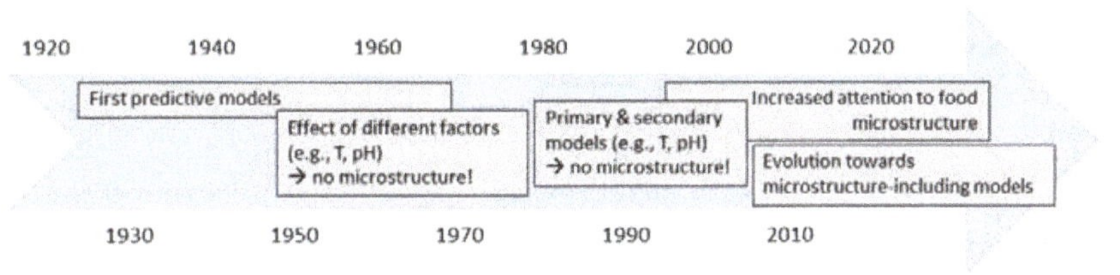

Figure 2. Timeline depicting the evolution on the inclusion of food microstructure into predictive models over the years.

Predictive growth models, both kinetic and probabilistic, started to appear in the 1960s and 1970s, describing the influence of factors such as storage temperature, salt concentration, and pH [28–30]. A renewed interest in predictive models arose in the 1980s, mainly due to a number of major food poisoning outbreaks and the consequent public (and political) awareness of the importance of food safety [1]. This renewed interest culminated in the identification of effective primary models (e.g., modified Gompertz model [31,32], Baranyi and Roberts model [33]), allowing the objective description of growth curves as mathematical equations. From then on, the development of secondary models describing the influence of important environmental factors on the parameters of the primary growth models was possible [26]. Such secondary models were incorporated into primary models to include the effect of those environmental factors on microbial growth into the predictions [34–37]. Over the years, secondary models became more sophisticated, describing the effect of different factors (e.g., temperature, pH, water activity, acid concentration) and the interaction between them on the growth behaviour of microorganisms [38–40]. However, research in the 1990s and early 2000s more and more brought the insight that such models could only adequately predict microbial growth in simple liquid food products. Microbial growth in/on more structured food products generally did not follow the predictions, with both fail-safe and fail-dangerous discrepancies being reported [10,41–43]. Extensive research on the effects of food microstructure on microbial growth was conducted in the following years, but models were only scarcely developed [17].

The most well-known early predictive model for the inactivation of microorganisms, although originating from before the introduction of the term predictive microbiology, was the *botulinum* cook of Esty and Meyer [44], describing a thermal process designed to kill 10^{12} spores of *Clostridium botulinum* type A [1]. This model was based on the Thermal Death Time (TDT) concept of Bigelow [45], which involves the use of D-values (i.e., the decimal reduction time, or the time necessary to accomplish a reduction in bacterial numbers of one log unit under isothermal conditions) and z-values (i.e., increase in temperature necessary to accomplish a 10-fold decrease in the D-value). Over the years, models based on this concept of D and z-values, such as the *botulinum* cook, have been used extensively in the food industry, with the canning industry being the most notable example [46]. However, the success of this approach in the canning industry is mostly due to overprocessing rather than due to modelling accuracy [47]. The models assume loglinear behaviour according to Equation (1) [48].

$$\log N = \log N_0 - \frac{t}{D_T} \tag{1}$$

with N, the cell density at time t; N_0, the initial cell density; D_T, the thermal reduction time (D-value). Similar to predictive growth models, secondary models describing the influence of important factors on the D-value of the primary inactivation model (Equation (1)) were developed. Arrhenius-type models represent one of the earliest methods to develop such secondary inactivation models (e.g., [49,50]). D-values can be expressed by means of the Arrhenius equation, as shown in Equation (2) [51].

$$k_{max} = \frac{2.303}{D_T} = A \cdot exp\left(\frac{-E}{R \cdot T}\right) \tag{2}$$

with k_{max}, the maximum specific inactivation rate; A, the frequency factor; E, the activation energy; R, the gas constant; T, the absolute temperature. This general Arrhenius equation can be extended with extra terms to, in addition to the influence of temperature, also include other factors such as pH and a_w [51,52], as for example already applied in the late 1970s by Davey et al. [53] for the effect of temperature and pH during thermal inactivation. The generalised form of such Arrhenius-type secondary models is represented by Equation (3) (based on [51]).

$$\ln k_{max} = a_0 + a_{1,1} \cdot V_1 + \cdots + a_{1,n} \cdot V_1^n + a_{2,1} \cdot V_2 + \cdots + a_{2,n} \cdot V_2^n + \cdots + a_{x,1} \cdot V_x + \cdots + a_{x,n} \cdot V_x^n \tag{3}$$

with n, the order of the secondary model; a_0–$a_{x,n}$, constants; V_1–V_x, environmental factors such as $1/T$, a_w or pH. Theoretically, secondary models, be it Arrhenius-type models or other model types, could include food microstructural factors, but this was not the case in the early days of predictive microbiology. Therefore, early models (i.e., including both primary and secondary models) were accurate when describing the inactivation behaviour of microorganisms in simple systems, but were inaccurate in complex food environments because the influence of food microstructure was not taken into account [46]. Again, model predictions could be fail-safe or fail-dangerous, depending on the specific situation [47,54,55]. In order to solve the possible inaccuracy of loglinear predictive inactivation models, new models were developed to deal with common non-loglinear inactivation trends, with the most notable models dating from after 1988 (e.g., [56–64]). Some of these models are able to implicitly include the microstructural effect by representing accurate inactivation dynamics in complexly structured food environments, e.g., via shoulder and/or tailing effects. However, a direct modelling of the food microstructural effect on microbial inactivation behaviour is not accomplished by the models if no food microstructural factors are included in the secondary models.

2.2. More Attention to Food Microstructure in the Last Decades

Following on the increasing number of studies that showed the significant influence of food microstructure on microbial dynamics, some concepts have been introduced to attempt to develop predictive models which take the food microstructural influence into account.

A relatively straightforward method to include the influence of food microstructure into predictive models is the development of models which are only valid for specific food products. In this case, microbial growth or inactivation experiments are conducted in/on the target food product. This methodology was occasionally already applied in the 1980s and 1990s, but has been widely used in recent literature. An extensive list of example studies exploiting this approach is provided in Table 1, both for microbial growth [65–86] and thermal inactivation [55,87–97], with a focus on early and recent examples. Interestingly, however, this predictive microbiology approach bears some similarities to the traditional challenge testing approach, in which microbial growth/inactivation experiments were also conducted directly in/on the food product of interest [5]. Hence, this approach could result in a large amount of models only suitable for a specific set of conditions (i.e., food product under certain environmental conditions), as also illustrated by the extensive (yet incomplete) list in Table 1. To the opinion of the authors, this food-specific approach, while certainly valuable, is not completely in line with the initial philosophy of predictive microbiology, which aims for "the accumulation of knowledge on microbial behaviour in foods" [1]. Additionally, and most relevant for this review, while those models inherently take the influence of food microstructure on microbial dynamics into account, they do not describe the influence of food microstructural factors on microbial behaviour [21].

Table 1. Examples of studies (ordered chronologically) in which predictive models only valid for specific food products were developed.

Year	Type	Microorganism(s)	Food Product	Ref.
1981	Growth	*Clostridium botulinum*	Pork slurry	[82]
1985	Growth	Various background microflora	Beef	[86]
1990	Growth	*Clostridium botulinum*	Fish filets	[66]
1992	Growth	*Salmonella* Typhimurium	Beef	[70]
1996	Growth	Various background microflora	Cut endive	[85]
1997	Inactivation	*Enterococcus faecium*	Bologna sausage	[55]
1998	Growth	Various background microflora	Sausage	[65]
1998	Growth	Various background microflora	Beef	[71]
1999	Inactivation	*Salmonella enteritidis*	Tarama salad	[93]
1999	Inactivation	*Enterobacter sakazakii*	Bovine whole milk	[94]
1999	Growth	*Salmonella* Typhimurium	Cooked chicken breast	[79]
1999	Growth	*Pseudomonas* spp. and *Shewanella putrefaciens*	Fresh bogue fish	[83]

Table 1. *Cont.*

Year	Type	Microorganism(s)	Food Product	Ref.
2003	Inactivation	*Staphylococcus aureus*	Surimi seafood sticks	[89]
2013	Inactivation	*Salmonella*	Ground chicken	[91]
2014	Inactivation	*Listeria monocytogenes*	Ground turkey	[92]
2016	Inactivation	*Salmonella*	Tree nuts	[96]
2018	Growth	*Bacillus cereus*	Cooked spinach	[73]
2018	Growth	*Bacillus cereus* (spores)	Cooked beans	[74]
2018	Growth	*Escherichia coli*	Mascarpone cheese	[76]
2018	Growth	*Weissella viridescens*	Vacuum-packaged ham	[78]
2018	Growth	*Escherichia coli*	Korean rice cake	[80]
2018	Inactivation	*Escherichia coli*	Ground chicken	[97]
2019	Growth	*Staphylococcus aureus*	Egg products	[68]
2019	Growth	*Vibrio parahaemolyticus*	Korean raw crab marinated in soy sauce	[69]
2019	Inactivation	*Listeria monocytogenes*	Gilthead sea bream fillets	[87]
2019	Growth	*Bacillus cereus*	Cooked rice	[72]
2019	Growth/inactivation	*Listeria monocytogenes*	Fish balls	[90]
2019	Growth	*Bacillus cereus* (spores)	Cooked pasta	[75]
2019	Growth	*Clostridium perfringens*	Roasted chicken and braised beef	[77]
2019	Growth	*Aeromonas hydrophila*	Lettuce	[81]
2019	Inactivation	*Salmonella*	Infant formula	[95]
2020	Growth	*Brochothrix thermosphacta, Leuconostoc gelidum* and *Pseudomonas* spp.	Minced pork	[67]
2020	Inactivation	*Salmonella* Thompson	Iceberg lettuce	[88]
2020	Growth	*Salmonella* Reading and lactic acid bacteria	Iceberg lettuce	[84]

An alternative approach to include the effect of food microstructure into predictive models is conducting experiments in artificial food model systems which simulate, to a certain extent, the microstructure of the product. In general, artificial food model systems are advantageous due to (i) their use being more simple and less labour-intensive than real food products, (ii) the absence of background microflora, (iii) repeatability of experimental results, (vi) the possibility to alter factors independently, and (v) the straightforward transferability of findings to other food products [15,21,98]. Wilson et al. [10] defined six categories of food architectures which could all be represented by artificial food model systems, i.e., liquids, oil-in-water emulsions, water-in-oil emulsions, aqueous gels, gelled emulsions, and surfaces. The use of liquid food model systems (e.g., meat broth to simulate meat products) was already a common practice in the early days of predictive microbiology, mimicking mostly the composition, pH, and water activity of the food products of interest [28,54,99–102]. In the early 2000s, a large number of research groups underwent a paradigm shift towards the use of more structured model systems. Artificial model systems of different structures are frequently used to study the effect of certain influencing factors on microbial growth [11,15,103–108] or inactivation [109–111]. Seldomly, predictive models are developed based on experiments in those more structure model systems. In this regard, two approaches can generally be identified, i.e., (A) using one model system to model (or include into the model) the effect of certain environmental or food intrinsic factors (e.g., oxygen diffusion and heat transfer in structured foods) (e.g., [98,112–115]) and (B) using model systems varying in certain compositional or microstructural factors to include the effect of those factors in relevant secondary models (e.g., [24,27,116,117]). Both approaches A and B can be regarded as systematic approaches to develop predictive models which take food microstructure into account, although not all models developed according to these approaches would explicitly model the microstructural effect. For instance, one could develop a model which describes the influence of the salt content on microbial growth in structured products. This hypothetical model would be valid in structured products, but would not directly express the influence of any microstructural factor. The following sections provide an overview of existing models which include food microstructural effects, for microbial growth and inactivation, respectively.

3. Growth Models Incorporating Food Microstructure

To the best knowledge of the authors, all existing growth models which include food microstructural effects can be divided into two categories. The first category encompasses the introduction of macroscale secondary models describing the influence of food microstructural factors on microbial growth, while the second category encompasses semi-mechanistic microscale models which take the local (structured) environment of the bacterial cell(s) into account. Table 2 provides an overview of the most relevant models of the two categories. A more detailed description of the respective models can be found in the following sections.

Table 2. Overview of the most relevant growth models incorporating food microstructure.

Macroscale Secondary Models			
Model Description	**Microstructural Factors**	**Non-Microstructural Factors**	**Ref.**
Listeria innocua and *Lactococcus lactis* growth (mono- and co-culture) in a gelled system	Gelatine concentration	Undissociated lactic acid concentration, pH, physiological state of the cells (for lag phase)	[24]
Aspergillus carbonarius growth in broth	Gelatine concentration	Temperature, water activity, physiological state of the cells (for lag phase)	[118]
Salmonella Typhimurium growth in broth	Gelatine concentration	Water activity, pH, physiological state of the cells (for lag phase)	[27]
Semi-Mechanistic Microscale Models			
Model Description	**Included Environmental Factors**		**Ref.**
Mixed population growth model for homogeneous food products, with 2-dimensional space dependency	Food structure (via firmness of the food), biomass transport (via diffusion)		[119]
Listeria innocua growth in solid or paste foods	Dissolved oxygen concentration (and diffusion), biomass transport (via diffusion)		[98,115]
Escherichia coli growth in a (3D)-structured leafy product during handling and storage	Temperature/heat transfer, biomass transfer (via diffusion), leafy structure (via inter-leaf contact points and entrapped air pockets)		[120]
Listeria monocytogenes growth on the surface of smear soft cheese and vacuum-packed cold-smoked salmon	Local pH, local water activity, temperature, structural environment (e.g., hollows, crests)		[121,122]

3.1. Macroscale Secondary Models Including Food Microstructural Factors

The traditional approach which has been employed to model the effect of food microstructure on microbial growth dynamics is the development of secondary models based on food microstructural factors of food products. This approach can be regarded as a macroscopic simplification of the real microbial behaviour. In essence, heterogeneity (i.e., both concerning the bacterial distribution and the food environment) is ignored and an average macroscopic relation is assumed. Due to its relative straightforwardness and limited required computing power, this secondary modelling approach has been dominant in predictive microbiology over the years. A remarkable observation is that, in all of the relevant studies focussing on secondary models for the food microstructural influence on bacterial growth, gelatine concentration was used as a variable to include the influence of product rheology on the maximum specific growth rate μ_{max} in the Baranyi and Roberts growth model [24,27,118].

Antwi et al. [24] developed a predictive model for the growth of *Listeria innocua* and *Lactococcus lactis* (mono- and coculture), quantifying the influence of a gelatine gel matrix. For this purpose, the model of Vereecken et al. [123], describing bacterial growth in function of undissociated lactic acid concentration and pH, was fit to growth data in model systems containing different gelatine concentrations. After their model optimisation procedure, Antwi et al. [24] obtained the model represented by Equations (4) and (5).

$$\frac{dN}{dt} = \frac{Q}{1+Q} \cdot \mu_{max}(G_c) \cdot \mu_{LaH,H}([LaH], [H^+]) \cdot N \tag{4}$$

$$\mu_{max}(G_c) = a_0 + a_1 \cdot exp(-a_2 \cdot G_c) \tag{5}$$

with Gc, the gelatine concentration; Q, the physiological state of the cells; $\mu_{LaH,H}([LaH], [H^+])$, the factor bringing the inhibition of growth into the model via the undissociated lactic acid concentration and the pH; a_0, a_1, and a_2 constant factors. Compared to the traditional Baranyi and Roberts [33] growth model, this model uses a more mechanistically inspired coupled-ODE method to explain growth inhibition, i.e., via the effect of the undissociated lactic acid concentration, the pH, and the gel strength. The authors concluded that the model satisfactorily predicted the effect of the gelatine concentration on the lactic acid dissociation and pH evolution and, in turn, also on the growth of the target microorganisms (in mono- and coculture). A moderate decrease in the growth rate was observed with an increasing gelatine concentration for both *Listeria innocua* and *Lactococcus lactis*, possibly explained by the increasing medium solidness imposing additional stresses on the cells.

Kapetanakou et al. [118] developed a model for the combined effect of water activity, temperature, and gelatine concentration on the growth of the fungus *Aspergillus carbonarius* in food model systems. A two-step procedure was employed during the model development, as the authors first fitted the Baranyi and Roberts [33] model to the growth data, and then fitted a polynomial secondary model (Equation (6)) to the square root of the maximum specific growth rates μ_{max}.

$$\sqrt{\mu_{max}} = a_0 + a_1 \cdot T + a_2 \cdot G_c + a_3 \cdot a_w + a_4 \cdot T \cdot G_c + a_5 \cdot T \cdot a_w + a_6 \cdot G_c \cdot a_w + a_7 \cdot T^2 + a_8 \cdot {G_c}^2 + a_9 \cdot a_w^2 \tag{6}$$

with $a_0, a_1, a_2, \ldots$, and a_9 the constants to be estimated; T, the temperature; G_c, the gelatine concentration; a_w, the water activity. The developed secondary model showed that the addition of gelatine caused a large decrease in $\sqrt{\mu_{max}}$, but that the structural influence was less pronounced at lower a_w and T. The model was also validated in three commercial products, i.e., custard, marmalade, and jelly. While the model predictions agreed well with growth data on custard and marmalade, they agreed poorly with the observed data on jelly.

Theys et al. [27] developed a model for the growth of *Salmonella* Typhimurium in function of pH, a_w, and gelatine concentration in a broth model system. Similar to the two previous studies, the authors incorporated a secondary model into the primary growth model of Baranyi and Roberts [33]. This secondary model, based on the model of Ross et al. [124], is presented by Equation (7).

$$\sqrt{\mu_{max}} = a_0 \cdot \sqrt{a_w - a_{w,min}} \cdot \sqrt{1 - 10^{(pH_{min} - pH)}} \cdot \sqrt{\frac{G_c \cdot \mu_{liq} + a_1 \cdot \mu_{sol}}{(a_1 + G_c) \cdot \mu_{sol}}} \tag{7}$$

with a_0 and a_1 constants; $a_{w,min}$ and pH_{min}, the theoretical minimal values of a_w and pH below which no growth occurs; μ_{liq}, the maximum specific growth rate in liquid media (i.e., gelatine concentration equal to zero); μ_{sol}, the maximum specific growth rate in the strongest possible solid medium (i.e., theoretical infinite gelatine concentration); G_c, the gelatine concentration. In brief, the structural factor becomes equal to zero if G_c equals zero, while the factor becomes equal to the ratio of μ_{sol} and μ_{liq} if G_c is infinitely high. Based on the developed model, it was observed that the curves of the secondary model were much flatter with respect to the effect of pH and a_w for the 1% and 5% gelatine concentrations than for 0% gelatine. Hence, the decrease in μ_{max} at a higher gelatine concentration was large at a mild pH and a_w conditions, while the μ_{max} decrease with an increasing gelatine concentration was much smaller at a stressful pH and a_w conditions.

In general, the three aforementioned gelatine concentration models, although exhibiting different model structures, managed to accurately describe microbial growth in the respective studied food model systems. However, those models are only valid for microbial growth in gelatine-containing food products, evidently limiting their general applicability. In order to address this issue, Aspridou et al. [103] recommended the use of a single uniform rheological parameter to describe the structure of food matrices in predictive models. The most suitable rheological parameters to include in predictive models need to be determined in future research, but might also be specific to the different food product categories. For example, the structure of more liquid food products could be described by the viscosity parameters of the Power law model of Reiner [125], while the structure of visco-elastic and solid products could be described by the storage modulus G', the loss modulus G'', or the loss tangent $\tan \delta$ [11,103]. It should, however, also be taken into account that rheological properties of food products can be time-dependent and that the handling of the food product (e.g., stirring, shaking) could exert a more significant influence on microbial growth than the rheological properties [27].

In addition to food product rheology, other food intrinsic factors related to food microstructure could also be incorporated into secondary models (e.g., fat droplet size, food matrix fat content), although more dedicated research towards the effect of those factors on microbial growth should first be conducted [126]. An extra complexity for these kinds of models, however, is related to the determination of the most suitable model structure. One could raise the question whether the food microstructural influence should be added to factors that describe the influence of other (traditional) factors on microbial growth (e.g., temperature, pH), or, whether (an) additional food microstructural factor(s) should be added to existing models. In the latter case, it would theoretically also be possible that these additional food microstructural factors are dependent on other variables which are already represented by their own factor in the traditional model (e.g., a pH-dependent influence of the gelling agent concentration on microbial growth, while a pH factor is already present in the model) [11,27]. Moreover, certain variables that are included in the microstructural factor could be dependent on other variables (e.g., a temperature-dependent rheology in a rheology-related factor). While none of these options can a priori be defined as the sole correct choice, the resulting possible complexity should be taken into account when selecting the macroscale secondary modelling approach for certain complex applications.

3.2. Semi-Mechanistic Microscale Models

While the secondary modelling approach described in Section 3.1 "Macroscale Secondary Models including Food Microstructural Factors" can result in accurate microbial growth predictions, it does not provide and/or require an extensive fundamental knowledge of the food microstructural influence on microbial dynamics. When looking at the rheology-based examples in the previous section, the microbial behaviour which is not explained by the influence of traditional factors (e.g., temperature, pH, water activity) is explained by a "black box" rheology-based factor which is not based on any physical phenomena. In reality, the influence of food rheology on microbial behaviour consists of different interacting effects, e.g., oxygen and metabolite diffusion, the mechanical distribution of water, the chemical redistribution of organic acids, and physical constraints on the mobility of microorganisms [6].

A more mechanistic modelling approach can be applied by modelling the interaction of microorganisms with the local environment based on physical phenomena. In this regard, the elementary model structure for microbial evolution is provided by Equation (8) [127].

$$\frac{\partial N(x,y,z,t)}{\partial t} = \mu(local\ environment) \cdot N(x,y,z,t) \tag{8}$$

with $N(x,y,z,t)$, the local cell density and μ, the specific growth rate of the microorganisms. As shown in the equation, the growth rate of the cells is dependent on the local environment, which comprises a plethora of different factors, e.g., temperature, pH, substrate concentration, metabolite concentration, and interactions with other microorganisms (i.e., same or competing species) [127]. Accurate modelling of the microbial behaviour, hence, requires an accurate modelling of the local environment of the cells. Hereby, it is important to include all factors which exert a significant influence of the microorganisms, e.g., oxygen and metabolite diffusion, the surrounding microorganisms (same species and other species), the local temperature, and local physicochemical conditions such as pH and a_w. In this regard, partial differential equations (PDEs), describing changes in variables as a function of space and time, are often introduced into the predictive models [3]. This methodology also allows the inclusion of changes in environmental factors over time in a straightforward way. This can be useful to include the time-dependent nature of food structure into models; it is, for example, known that the rheological properties of food products can change over time [27]. Taking the local conditions into account leads to the development of microscale models, rather than the macroscale models (i.e., focussing on total cell populations and macroscopic food properties) discussed in the previous section. Since microscale models need to deal with a high level of detail (e.g., spatial and microbial heterogeneity), they have a high complexity, possibly leading to significant computational costs [128]. It should also be noted that no fully mechanistic models for microbial dynamics exist to date, since (i) the current microbiological knowledge is too limited for fully mechanistic relations and (ii) empiric relations are sometimes used to describe some environmental conditions in order to save computational efforts. Consequently, most predictive models are semi-mechanistic, meaning that they contain some mechanistic information in their structure and physically measurable parameters [128].

Dens and Van Impe [119] proposed a general modelling approach to take spatial heterogeneity in structured foods into account. While the model was solely based on model simulations and, hence, not based on experiments in structured model systems, the study was included in this review because of its importance for models developed in later studies. In brief, the authors extended a previously developed mixed population growth model for homogeneous food products, unifying the growth model of Baranyi and Roberts [33] and the Lotka–Volterra model [129] for two-species competition (i.e., in this case, *Escherichia coli* and *Lactobacillus plantarum*), and introducing a (two-dimensional) space-dependency into the model. The general model structure is shown by Equations (9) and (10), with the respective first terms representing bacterial growth via the combined Baranyi and Roberts [33] and Lotka–Volterra [129] model, and the respective second terms representing the biomass transport via diffusion.

$$\frac{\partial N_1(x,y,t)}{\partial t} = \mu_1(x,y,t)\cdot N_1(x,y,t) + D\cdot\nabla^2 N_1(x,y,t) \tag{9}$$

$$\frac{\partial N_2(x,y,t)}{\partial t} = \mu_2(x,y,t)\cdot N_2(x,y,t) + D\cdot\nabla^2 N_2(x,y,t) \tag{10}$$

The complete form of these model equations is represented by Equations (11) and (12).

$$\frac{\partial N_1(x,y,t)}{\partial t} = \mu_{max,1}\cdot\frac{Q_1(x,y,t)}{1+Q_1(x,y,t)}\cdot\frac{N_1(x,\,y,t)}{N_{max,1}}\cdot(N_{max,1} - N_1(x,y,t) - \alpha_{1,2}\cdot N_2(x,y,t)) + D\cdot\left(\frac{\partial^2 N_1(x,y,t)}{\partial x^2} + \frac{\partial^2 N_1(x,y,t)}{\partial y^2}\right) \tag{11}$$

$$\frac{\partial N_2(x,y,t)}{\partial t} = \mu_{max,2}\cdot\frac{Q_2(x,y,t)}{1+Q_2(x,y,t)}\cdot\frac{N_2(x,\,y,t)}{N_{max,2}}\cdot(N_{max,2} - N_2(x,y,t) - \alpha_{2,1}\cdot N_1(x,y,t)) + D\cdot\left(\frac{\partial^2 N_2(x,y,t)}{\partial x^2} + \frac{\partial^2 N_2(x,y,t)}{\partial y^2}\right) \tag{12}$$

with N_1 and N_2, the cell densities of the two bacterial species; Q_1 and Q_2, the internal physiological state of both species used to describe the lag phase; $\mu_{max,1}$ and $\mu_{max,2}$, the maximum specific growth rate of both species; $N_{max,1}$ and $N_{max,2}$, the maximum population density of both species when grown in monoculture; $\alpha_{1,2}$ and $\alpha_{2,1}$, the interaction coefficients measuring the effects of species one on species two and vice versa; D, the cell diffusivity; ∇^2, the diffusive operator. For their simulations, the authors assumed a 10 by 10 cm agar gel, 1 mm thick, to allow a two-dimensional model. Space was considered as a grid of lattice sites, with each site being assumed homogeneous. The biomass mass transfer was taken into account via the diffusion law, while the food structure was included via factor D, a measure for the firmness of the food. Factor D was assumed infinite for very fluid foods and zero for solid foods in which no movement of microorganisms is possible. In their concluding remarks, Dens and Van Impe [119] state that, while their model simulations should not be regarded as accurate, the main message of their work is that an extended model structure taking space into account is necessary to model microbial growth in structured food environments. In later years, other authors have indeed used this approach to model microbial growth in structured environments, as shown by the two following examples of Noriega et al. [98,115] and De Bonis and Ruocco [120].

Noriega et al. [98,115] developed a predictive model for *L. innocua* growth in solid or paste foods, taking into account oxygen diffusion limitations. Three different dissolved oxygen concentrations were investigated in solidified broth systems, i.e., (i) aerobic conditions (7.6–8.0 mg/L), (ii) hypoxic conditions (0.2–2.6 mg/L), and (iii) anoxic conditions (<0.01 mg/L). In brief, the used modelling approach was a combination of the logistic Riccati equation for microbial growth (Equation (13) [130]), on the one hand, and oxygen (Equation (14)) and substrate (Equation (15)) mass balances, on the other.

$$\frac{dN}{dt} = K \cdot N \cdot (1 - \tau \cdot N) \tag{13}$$

$$\frac{\partial N}{\partial t} = D \cdot \frac{\partial^2 N}{\partial z^2} + \frac{dN}{dt} \tag{14}$$

$$\frac{\partial C_{O_2}}{\partial t} = D_{O_2} \cdot \frac{\partial^2 C_{O_2}}{\partial z^2} + \frac{dC_{O_2}}{dt} \tag{15}$$

with D, the cell diffusivity; D_{O_2}, the oxygen diffusion rate; C_{O_2}, the oxygen concentration; K and τ, kinetic parameters obtained in liquid media, with τ being a function of the oxygen concentration. The authors concluded that the used approach of combining kinetic parameters as a function of oxygen concentration, obtained in liquid medium with the assumption of oxygen as a limiting substrate for cell growth, resulted in accurate model predictions for structured media.

De Bonis and Ruocco [120] developed a mathematical 3D model of a structured leafy product to simulate *Escherichia coli* growth in fresh iceberg lettuce during handling and storage. In brief, the model combined general first-law equations for heat transfer (Equation (16)), biomass kinetics and transfer (Equation (17)), primary bacterial growth kinetics (Equation (18) [131]), and secondary temperature-dependent bacterial growth kinetics (Equation (19) [132]). Structural features of the food were taken into account by a multiplicity of inter-leaf contact points and insulating air pockets, which influence microbial growth.

$$\rho \cdot c_P \cdot \frac{\partial T}{\partial t} = k_S \cdot \nabla^2 T \tag{16}$$

$$\frac{\partial N}{\partial t} = D \cdot \nabla^2 N + \frac{dN}{dt} \tag{17}$$

$$\frac{dN}{dt} = \frac{Q}{1+Q} \cdot \mu_{max} \cdot \left(1 - \frac{N}{N_{max}}\right) \cdot N \tag{18}$$

$$\sqrt{\mu_{max}} = b \cdot (T - T_{min}) \tag{19}$$

with ρ, the substrate density; c_p, the substrate specific heat capacity; T, the temperature; k_S, the substrate conductivity; D, the cell diffusivity in the substrate; μ_{max}, the maximum specific growth rate; Q, the physiological state of the cells; N_{max}, the maximum cell density; b, the kinetic parameter of the Ratkowsky equation; T_{min}, the reference temperature. The predicted thermal profiles were validated in real iceberg lettuce samples, showing that the model predicted temperature evolution nicely at all considered depths. The microbial behaviour was not experimentally validated, but depended on general data on microbial generation and diffusion. A value for bacterial (constant temperature) diffusivity was obtained from literature for this purpose. In general, the main advantage of the model of De Bonis and Ruocco [120] is that it is a general engineering tool which stems from the integration of partial differential equations that describe heat and mass transfer. The model allows the prediction of local and volume-averaged bacterial cell growth with proper accuracy, and both in function of the initial contamination and the operating thermal regime of the product.

The previously discussed approaches can be classified as grid-based or biomass-based models (BbM), as the cell density in a small volume unit of the food product was used as the basic unit to model microbial growth. A further step towards more accurate models would be to include direct intercellular reactions by using individual cells as the basic model units in individual-based models (IbMs) [128]. IbMs provide realistic bacterial dynamics and can be designed to include accurate descriptions of complex micro-structures and environments [133]. Over the last two decades, IbMs have shown increasing potential for the modelling of microbial behaviour due to the development of specialised software [134]. Examples of IbM software tools for predictive microbiology include BacSim [135], INDISIM [136], MICRODIMS [137,138], BSim [133], and iDynoMiCS [139]. The potential of IbMs to predict microbial dynamics in complex systems (i.e., complex environment and/or complex microflora), even including the complex behaviour in microbial biofilms, is a major advantage for modelling applications in structured food products [140]. While an in-depth explanation of IbMs lies outside the scope of this study, it is worth mentioning the most relevant example of an IbM approach applied to microbial growth on (i.e., surface growth) structured food products.

Ferrier et al. [121] and Augustin et al. [122] developed an IBM approach to describe the behaviour of a small number of *Listeria monocytogenes* cells contaminating the surface of smear soft cheese and vacuum-packed cold-smoked salmon. Microscale models describing the local pH and a_W over the food surface were constructed based on microelectrode measurements. These models were combined with the IBM approach to simulate the stochastic growth of the bacteria on the product; simulations were also validated on real cheese and salmon samples. On the one hand, the authors concluded that, for no-growth or poor-growth situations (i.e., a small number of cells), the accuracy of their coupled IBM approach surpassed the classical macroscale approach. On the other hand, the results of the two approaches were similar when assessing the impact of changes in control measures influencing the growth of the bacteria. Therefore, the IbM approach was mainly useful to predict single-cell growth probability of foodborne pathogens contaminating food with a small number of cells. Nevertheless, more microenvironmental factors, as well as the interaction of *L. monocytogenes* with background microflora, should be added to the model in order to further increase its accuracy.

4. Inactivation Models Incorporating Food Microstructure

Due to the long history of thermal processing as a means of food preservation in the food industry, most inactivation models that incorporate the food microstructural influence are thermal inactivation models, which are, hence, the main focus of this section. Although predictive inactivation models have been developed for novel non-thermal technologies, these are in most cases solely based on experiments in liquid laboratory media (e.g., for high pressure processing [141]).

Similar to the microbial growth models, all reported thermal inactivation models which include food microstructure can be classified into two categories, i.e., (i) macroscale secondary models including food microstructural factors and (ii) semi-mechanistic microscale models. Table 3 provides an overview of the most relevant models of the two categories. A more detailed description of the respective models can be found in the following sections.

Table 3. Overview of the most relevant thermal inactivation models incorporating food microstructure.

Macroscale Secondary Models			
Model Description	**Microstructural Factors**	**Non-Microstructural Factors**	**Ref.**
Listeria monocytogenes inactivation in homogenised milk model systems	Fat content	Temperature, pH	[142]
Salmonella inactivation in whey protein powder model systems	Water mobility	Temperature, water activity	[143]
Salmonella inactivation in whey protein–peanut oil powders model systems	Fat content	Temperature, water activity	[144]
Semi-Mechanistic Microscale Models			
Model Description	**Included Environmental Factors**		**Ref.**
Escherichia coli K12 thermal inactivation (microwave) in calcium alginate gels	Local temperature (via microwave dielectric heating and heat transfer, taking thermophysical properties of the gels into account)		[113]
Escherichia coli K12 thermal inactivation in pre-packed ground beef in water baths	Fluid flow in water bath, local temperature (via heat transfer, taking thermophysical properties of the ground beef into account)		[145]

4.1. Macroscale Secondary Models Including Food Microstructural Factors

To the best knowledge of the authors, the only existing macroscopic thermal inactivation models that incorporate food microstructural factors are based on fat content and water mobility. Fat content is often categorised as a food compositional factor, but since fat content also determines the microstructural characteristics of emulsion-type foods, the classification of fat content as a food microstructural factor is justified in some cases. While some models were developed specifically for certain food products types (e.g., for poultry with different fat content [19]), this section is focussed on inactivation models developed in artificial food model systems, because this approach fits more into the general predictive microbiology mindset.

Chhabra et al. [142] were among the first to develop a fat content-based inactivation model using food model systems, based on homogenised milk. They developed a model for the inactivation of *L. monocytogenes* in function of the fat content, pH, and processing temperature. The developed model was a modified Gompertz equation, as shown in Equation (20), including parameters S_L, k_{max}, and N_{red}, characterising the shoulder phase, the maximum inactivation rate, and the overall change in the number of survivors, respectively. These parameters depended on the fat content, pH, and processing temperature, as indicated in Equations (21)–(23).

$$\log N = N_{red} \cdot e^{-e^{(S_L + k_{max} \cdot t)}} - N_{red} \cdot e^{-e^{(S_L)}} \tag{20}$$

$$S_L = a_0 + a_1 \cdot C_F + a_2 \cdot pH + a_3 \cdot T + a_{1,2} \cdot C_F \cdot pH + a_{1,3} \cdot C_F \cdot T + a_{2,3} \cdot pH \cdot T + a_{1,2,3} \cdot C_F \cdot pH \cdot T \tag{21}$$

$$k_{max} = b_0 + b_1 \cdot C_F + b_2 \cdot pH + b_3 \cdot T + b_{1,2} \cdot C_F \cdot pH + b_{1,3} \cdot C_F \cdot T + b_{2,3} \cdot pH \cdot T + b_{1,2,3} \cdot C_F \cdot pH \cdot T \tag{22}$$

$$N_{red} = c_0 + c_1 \cdot C_F + c_2 \cdot pH + c_3 \cdot T + c_{1,2} \cdot C_F \cdot pH + c_{1,3} \cdot C_F \cdot T + c_{2,3} \cdot pH \cdot T + c_{1,2,3} \cdot C_F \cdot pH \cdot T \tag{23}$$

with C_F, the fat content parameter; $a_0 - c_{1,2,3}$, constant factors. Theoretically, the model could have also been extended to include terms of higher order. It should, however, be noted that not all parameters were deemed significant. For instance, the shoulder region of the inactivation was only affected by the pH, while the death rate was only affected by the temperature and fat content. It was also shown that, as temperature increased, there was a decrease in heat resistance due to the presence of milkfat. However, modified Gompertz models for inactivation are, apart from some mathematical limitations, characterised by two major modelling problems, i.e., (i) $N(t = 0)$ is not equal to $N(0)$ in the static version of the model and (ii) there is no explicit dependency on $N(0)$ in the dynamic version of the model, which should be avoided [61,146].

Santillana Farakos et al. [143] used whey protein powder model systems to develop a predictive model for the thermal inactivation of *Salmonella* in low-moisture foods in function of temperature, a_w, and water mobility. The water mobility of the different model systems was acquired by a pH adjustment and heat denaturation, and equilibrated to a_w levels between 0.19 ± 0.03 and 0.54 ± 0.02. The specific water mobility values were determined by means of wide-line proton-NMR (Nuclear Magnetic Resonance). Four different models were fitted to the inactivation data, but the Weibull model (Equation (24) [147]) was selected for secondary modelling, because it best described the data over all temperatures.

$$\log N = \log N_0 - \left(\frac{t}{\delta}\right)^{\beta} \quad (24)$$

with N_0, the initial cell population; δ, the scale parameter representing the treatment time (in min) required to for the first decimal log reduction in the cell population; β, the shape factor value. In order to develop secondary models, the significance of the temperature, a_w, and water mobility on $\log \delta$ and $\log \beta$ was assessed. The temperature was deemed to be a significant influencing factor on both Weibull parameters, while a_w was only deemed to be a significant influencing factor on $\log \delta$. The water mobility, however, being the only investigated food microstructural factor, did not exert a significant influence on any of the parameters. Hence, the developed model, as shown in Equations (25) and (26), did not explicitly incorporate the influence of food microstructure.

$$\log \delta = -0.10T - 4.34a_w + 9.91 \quad (25)$$

$$\log \beta = -0.006T \quad (26)$$

Nevertheless, the model achieved acceptable predictions in different real food products, i.e., low-fat cocoa powder, low-fat peanut meal, non-fat dry milk, wheat flour, and whey protein.

More recently, Trimble et al. [144] adapted the aforementioned model of Santillana Farakos et al. [143] to include the influence of fat content on *Salmonella* inactivation in low-water-activity foods, using whey protein–peanut oil powders as model systems. For this purpose, they eliminated the correlation between δ and β by using a fixed value of $\beta = 0.3644$. This resulted in a secondary model expressing a significant influence of the temperature, fat content (C_F), and water activity on $\log \delta$, as shown in Equation (27).

$$\log \delta = 22.90 - 0.167T + 0.051C_F - 4.38a_w \quad (27)$$

The adapted model was validated in fat-containing low-a_w foods, i.e., toasted oats, animal crackers, chia seed powder, and natural peanut butter. A slight underestimation of thermal inactivation by the model was reported, but the model was still deemed successful for the prediction of *Salmonella* survival in low-a_w foods. In their final remarks, Trimble et al. [144] suggested that the model could still be expanded to include a wider fat content and/or temperature range and a dependency on fat type. In general, it should, however, be mentioned that the Weibull model is characterized by two disadvantages. First of all, the frequency distribution of the viability of bacterial cells, the concept on which the Weibull model is based, is difficult to interpret and to validate experimentally [148]. Secondly, the Weibull model lacks a suitable differential model form. Due to the frequent use of ODEs and PDEs in predictive modelling frameworks that include food microstructure, the developed model strategy may, hence, have limited applicability for more complex microbial inactivation cases.

4.2. Semi-Mechanistic Microscale Models

An approach which has been extensively used to implicitly incorporate food microstructure into predictive thermal inactivation models is linking the bacterial inactivation model to (space- and time-dependent) heat and mass transfer models. This methodology is especially suitable for thermal inactivation because structural food properties tend to change during thermal processing, e.g., the viscosity of liquid foods which decreases with an increasing temperature or certain foods which further solidify during frying [149,150]. Zanoni et al. [55] were among the first to adopt this approach, although for experiments carried out in a real food product (i.e., bologna sausage). They combined the Whiting et al. [151] inactivation model with a heat and mass transfer model validated for bologna sausage cooking. A similar approach was later on conducted in food model systems by researchers from the School of Chemical Engineering (University of Birmingham) and Institute of Food Research (Reading Laboratory) (e.g., [112,114]). These authors conducted thermal inactivation experiments in agar cylinders and modelled both the heat transfer and thermal inactivation kinetics, the latter by means of experimentally obtained D and z-values. These studies marked an important step in the thermal inactivation model development because the used approach could be used with food products of different geometry or thermal conductivity, and for different bacteria.

Over the last decade, computer modelling techniques for heat and mass transfer have become more and more common when modelling treatments of different traditional and novel thermal processing technologies, e.g., beverage pasteurization [152], agitated retort heating [153], microwave heating [154], continuous deep frying [155], and radio frequency heating [156]. In brief, such techniques involve solving the heat and mass transfer equations with applied initial and boundary conditions using either (i) theoretical numerical finite difference and finite elements solutions or (ii) a computational fluid dynamics (CFD) approach [157]. While such models would be suitable to be coupled to microbial thermal inactivation models, this approach has only been scarcely employed for these industry-relevant processes [145].

Hamoud-Agha et al. [113] investigated the thermal inactivation of *Escherichia coli* K12 in calcium alginate gels during microwave processing. The inactivation model of Geeraerd et al. [61], including a Bigelow-type temperature dependency of the inactivation rate, was coupled to heat transfer and Maxwell's equations into a 3D finite elements model under dynamic heating conditions. By providing space-dependent predictions, the model was able to handle the thermal heterogeneity inherent to microwave treatments and the resulting differences in inactivation efficiency between the different locations within samples. Consequently, Hamoud-Agha et al. [113] demonstrated the reliability of the coupled modelling approach which links microbial inactivation models to heat transfer models.

A similar approach was used by Albuquerque et al. [145] for the thermal inactivation of *Escherichia coli* K12 in pre-packed ground beef in water baths programmed to deliver different heating rates to the product. The authors coupled a 3D-CFD and heat transfer finite elements model to the inactivation model of Geeraerd et al. [61], including a Bigelow-type temperature dependent inactivation rate. Even though the large heating rates caused large temperature gradients and heterogeneous inactivation distributions over the samples, there was still satisfactory agreement between model predictions and experimental data. Moreover, and most relevant for this review, the model was able to handle the typical microstructural complexity of the real food product under study.

The studies of Hamoud-Agha et al. [113] and Albuquerque et al. [145], hence, demonstrate that coupling microbial thermal inactivation models to heat and mass transfer models is a promising, and probably the most optimal, approach to develop microbial (thermal) inactivation models which take the food microstructural influence on microbial inactivation into account. The fact that the other possible method to include the food microstructural effect into predictive models (i.e., macroscale secondary models) has so far solely relied on less optimal model types for inactivation, such as the modified Gompertz and

Weibull model, strengthens this conclusion. Similar to microbial growth, the accuracy of the microscale methods could be further improved by including an IbM approach. The inactivation of small cell populations is often characterised by a high variability in inactivation behaviour, originating from individual cell heterogeneity [158,159]. However, an appropriate theoretical IbM approach for modelling of the variability in individual cell heterogeneity during the inactivation process has not been developed thus far [160].

5. Conclusions

While the influence of food microstructure on microbial dynamics was for the largest part neglected in the early days of predictive microbiology, significant progress on the subject has been achieved during the last two decades. Both for microbial growth and thermal inactivation, two general model types can be distinguished in the scientific literature, i.e., (i) macroscale secondary models including food microstructural factors and (ii) microscale semi-mechanistic models. These model types have benefited from the introduction of advanced mathematical modelling techniques and the increased usage of artificial food model systems to collect experimental data (i.e., rather than real food products).

In general, both the macroscopic secondary models and the semi-mechanistic microscale models have shown their potential as predictive modelling tools/frameworks when predicting microbial growth and inactivation in/on structured food products. The selection of one approach over another, or a combination of the two approaches, should ideally depend on the specific application, the required accuracy of the model, and the available computing power. Specifically for the macroscopic secondary models, other food intrinsic factors related to food microstructure could also be taken into account in addition to food product rheology (growth) and fat content (thermal inactivation), although more dedicated research towards the effect of those additional factors on microbial dynamics should first be conducted. Moreover, a lot of progress is still to be made for microbial inactivation via non-thermal technologies, as models including the food microstructural influence on microbial inactivation for those technologies are to date virtually non-existent. Specifically for the microscale semi-mechanistic models, the development of models taking more aspects of the complex microstructural and microbial environment into account forms an interesting research opportunity.

Author Contributions: Conceptualization, D.V. and J.F.M.V.I.; methodology, D.V.; software, D.V.; validation, D.V.; formal analysis, D.V.; investigation, D.V.; resources, J.F.M.V.I.; data curation, D.V.; writing—original draft preparation, D.V.; writing—review and editing, D.V. and J.F.M.V.I.; visualization, D.V.; supervision, J.F.M.V.I.; project administration, D.V. and J.F.M.V.I.; funding acquisition, D.V. and J.F.M.V.I. All authors have read and agreed to the published version of the manuscript.

Funding: This work was funded by the KU Leuven Research Fund through project C24/18/046, by the Research Foundation Flanders (FWO) through project G0B4121N, and by the EU H2020 research and innovation program under the Marie Skłodowska-Curie grant agreement No. 956126. Author Davy Verheyen was funded by the Research Foundation Flanders (FWO), grant number 1254421N.

Institutional Review Board Statement: Not applicable.

Informed Consent Statement: Not applicable.

Data Availability Statement: No new data were created or analysed in this study. Data sharing is not applicable to this article.

Conflicts of Interest: The authors declare no conflict of interest.

References

1. McMeekin, T.A.; Olley, J.; Ratkowsky, D.A.; Ross, T. Predictive microbiology: Towards the interface and beyond. *Int. J. Food Microbiol.* **2002**, *73*, 395–407. [CrossRef]
2. Stavropoulou, E.; Bezirtzoglou, E. Predictive modeling of microbial behavior in food. *Foods* **2019**, *8*, 654. [CrossRef]
3. Lopatkin, A.J.; Collins, J.J. Predictive biology: Modelling, understanding and harnessing microbial complexity. *Nat. Rev. Microbiol.* **2020**, *18*, 507–520. [CrossRef] [PubMed]

4. Pérez-Rodríguez, F.; Carrasco, E.; Pradhan, A.K.; Sant'Ana, A.S.; Valdramidis, V.P.; Valero, A. Special issue on 10th international conference of predictive modelling in foods: Towards a new paradigm in predictive microbiology. *Int. J. Food Microbiol.* **2019**, *291*, 65–66. [CrossRef]
5. McDonald, K.; Sun, D.-W. Predictive food microbiology for the meat industry: A review. *Int. J. Food Microbiol.* **1999**, *52*, 1–27. [CrossRef]
6. Heertje, I. Structure and function of food products: A review. *Food Struct.* **2014**, *1*, 3–23. [CrossRef]
7. Aguilera, J.M. Why food microstructure? *J. Food Eng.* **2005**, *67*, 3–11. [CrossRef]
8. Bhopatkar, D.; Hamaker, B.R.; Campanella, O.H. Micro to macro level structures of food materials. In *Food Materials Science and Engineering*; Bhandari, B., Roos, Y.H., Eds.; Blackwell Publishing Ltd.: Hoboken, NJ, USA, 2012; pp. 26–51.
9. Ubbink, J.; Burbridge, A.; Mezzenga, R. Food structure and functionality: A soft matter perspective. *Soft Matter* **2008**, *4*, 1569–1581. [CrossRef] [PubMed]
10. Wilson, P.D.G.; Brocklehurst, T.F.; Arino, S.; Thualt, D.; Jakobsen, M.; Lange, M.; Farkas, J.; Wimpenny, J.W.T.; Van Impe, J.F. Modelling microbial growth in structured foods: Towards a unified approach. *Int. J. Food Microbiol.* **2002**, *73*, 275–289. [CrossRef]
11. Verheyen, D.; Bolívar, A.; Pérez-Rodríguez, F.; Baka, M.; Skåra, T.; Van Impe, J.F. Effect of food microstructure on growth dynamics of *Listeria monocytogenes* in fish-based model systems. *Int. J. Food Microbiol.* **2018**, *283*, 7–13. [CrossRef]
12. Wimpenny, J.W.T.; Leistner, L.; Thomas, L.V.; Mitchell, A.J.; Katsaras, K.; Peetz, P. Submerged bacterial colonies within food and model systems: Their growth, distribution and interactions. *Int. J. Food Microbiol.* **1995**, *28*, 299–315. [CrossRef]
13. Verheyen, D.; Xu, X.M.; Govaert, M.; Baka, M.; Van Impe, J.F. Food microstructure and fat content affect growth morphology, growth kinetics, and preferred phase for cell growth of *Listeria monocytogenes* in fish-based model systems. *Appl. Environ. Microbiol.* **2019**, *85*, e00707-19. [CrossRef]
14. Mertens, L.; Geeraerd, A.H.; Dang, T.D.T.; Vermeulen, A.; Serneels, K.; Van Derlinden, E.; Cappuyns, A.M.; Moldenaers, P.; Debevere, J.; Devlieghere, F.; et al. Design of an experimental viscoelastic food model system for studying *Zygosaccharomyces bailii* spoilage in acidic sauces. *Appl. Environ. Microbiol.* **2009**, *75*, 7060–7069. [CrossRef] [PubMed]
15. Baka, M.; Noriega, E.; Van Langendonck, K.; Van Impe, J.F. Influence of food intrinsic complexity on *Listeria monocytogenes* growth in/on vacuum-packed model systems at suboptimal temperatures. *Int. J. Food Microbiol.* **2016**, *235*, 17–27. [CrossRef]
16. Pérez-Rodríguez, F.; Valero, A. *Predictive Microbiology in Foods*; SpringerBriefs in Food, Health, and Nutrition; Springer: New York, NY, USA, 2013.
17. Theys, T. Modelling the (Boundaries of) Microbial Growth in Structured Media: Effect of pH, Water Activity and Gelatin on the Growth of *Salmonella* Typhimurium. Ph.D. Thesis, KU Leuven, Leuven, Belgium, 2009.
18. Velliou, E.G.; Noriega, E.; Van Derlinden, E.; Mertens, L.; Boons, K.; Geeraerd, A.H.; Devlieghere, F.; Van Impe, J.F. The effect of colony formation on the heat inactivation dynamics of *Escherichia coli* K12 and *Salmonella* typhimurium. *Food Res. Int.* **2013**, *54*, 1746–1752. [CrossRef]
19. Juneja, V.K.; Eblen, B.S.; Marks, H.M. Modeling non-linear survival curves to calculate thermal inactivation of *Salmonella* in poultry of different fat levels. *Int. J. Food Microbiol.* **2001**, *70*, 37–51. [CrossRef]
20. Murphy, R.Y.; Marks, B.P.; Johnson, E.R.; Johnson, M.G. Thermal inactivation kinetics of *Salmonella* and *Listeria* in ground chicken breast meat and liquid medium. *J. Food Sci.* **2000**, *65*, 706–710. [CrossRef]
21. Verheyen, D. Micro- and Macroscopic Investigation of the Food Microstructural Influence on Microbial Dynamics: Case Study in/on Fish Products. Ph.D. Thesis, KU Leuven, Leuven, Belgium, 2020.
22. Smet, C.; Noriega, E.; Van Mierlo, J.; Valdramidis, V.P.; Van Impe, J.F. Influence of the growth morphology on the behaviour of *Salmonella* Typhimurium and *Listeria monocytogenes* under osmotic stress. *Food Res. Int.* **2015**, *77*, 515–526. [CrossRef]
23. Robins, M.M.; Wilson, P.D.G. Food structure and microbial growth. *Trends Food Sci. Technol.* **1994**, *5*, 289–293. [CrossRef]
24. Antwi, M.; Bernaerts, K.; Van Impe, J.F.; Geeraerd, A.H. Modelling the combined effects of structured food model system and lactic acid on *Listeria innocua* and *Lactococcus lactis* growth in mono- and coculture. *Int. J. Food Microbiol.* **2007**, *120*, 71–84. [CrossRef]
25. Skandamis, P.N.; Jeanson, S. Colonial vs. Planktonic type of growth: Mathematical modelling of microbial dynamics on surfaces and in liquid, semi-liquid and solid foods. *Front. Microbiol.* **2015**, *6*, 1178. [CrossRef]
26. Buchanan, R.L. Predictive food microbiology. *Trends Food Sci. Technol.* **1993**, *4*, 6–11. [CrossRef]
27. Theys, T.E.; Geeraerd, A.H.; Verhulst, A.; Poot, K.; Van Bree, I.; Devlieghere, F.; Moldenaers, P.; Wilson, D.; Brocklehurst, T.; Van Impe, J.F. Effect of pH, water activity and gel micro-structure, including oxygen profiles and rheological characterization, on the growth kinetics of *Salmonella* Typhimurium. *Int. J. Food Microbiol.* **2008**, *128*, 67–77. [CrossRef] [PubMed]
28. Genigeorgis, C.; Martin, S.; Franti, C.E.; Riemann, H. Initiation of Staphylococcal growth in laboratory media. *Appl. Microbiol.* **1971**, *21*, 934–939. [CrossRef] [PubMed]
29. Nixon, P.A. Temperature integration as a means of assessing storage conditions. In *Report on Quality in Fish Products, Seminar No. 3*; Fishing Industry Board: Wellington, New Zealand, 1971; pp. 34–44.
30. Spencer, R.; Baines, C.R. The effect of temperature on the spoilage of wet fish: I. Storage at constant temperature between −1 °C and 25 °C. *Food Technol. Champ.* **1964**, *18*, 769–772.
31. Gompertz, B. On the nature of the function expressive of the law of human mortality, and on a new mode of determining the value of life contingencies. *Philos. Trans. R. Soc. Lond.* **1825**, *115*, 513–585.

32. Zwietering, M.H.; Jongenburger, I.; Rombouts, F.M.; van't Riet, K. Modeling of the bacterial growth curve. *Appl. Environ. Microbiol.* **1990**, *56*, 1875–1881. [CrossRef] [PubMed]
33. Baranyi, J.; Roberts, T.A. A dynamic approach to predicting bacterial growth in food. *Int. J. Food Microbiol.* **1994**, *23*, 277–294. [CrossRef]
34. Bhaduri, S.; Turner-Jones, C.O.; Buchanan, R.L.; Phillips, J.G. Response surface model of the effect of pH, sodium chloride and sodium nitrite on growth of *Yersinia enterocolitica* at low temperatures. *Int. J. Food Microbiol.* **1994**, *23*, 333–343. [CrossRef]
35. George, S.M.; Richardson, L.C.C.; Peck, M.W. Predictive models of the effect of temperature, pH and acetic and lactic acid on the growth of *Listeria monocytogenes*. *Int. J. Food Microbiol.* **1996**, *32*, 73–90. [CrossRef]
36. Ng, T.M.; Schaffner, D.W. Mathematical models for the effects of pH, temperature, and sodium chloride on the growth of *Bacillus stearothermophilus* in salty carrots. *Appl. Environ. Microbiol.* **1997**, *63*, 1237–1243. [CrossRef] [PubMed]
37. Sutherland, J.P.; Bayliss, A.J.; Braxton, D.S. Predictive modelling of growth of *Escherichia coli* O157:H7: The effects of temperature, pH and sodium chloride. *Int. J. Food Microbiol.* **1995**, *25*, 29–49. [CrossRef]
38. Augustin, J.-C.; Carlier, V. Modelling the growth rate of *Listeria monocytogenes* with a multiplicative type model including interactions between environmental factors. *Int. J. Food Microbiol.* **2000**, *56*, 53–70. [CrossRef]
39. Le Marc, Y.; Huchet, V.; Bourgeois, C.M.; Guyonnet, J.P.; Mafart, P.; Thuault, D. Modelling the growth kinetics of *Listeria* as a function of temperature, pH and organic acid concentration. *Int. J. Food Microbiol.* **2002**, *73*, 219–237. [CrossRef]
40. Panagou, E.Z.; Skandamis, P.N.; Nychas, G.-J.E. Modelling the combined effect of temperature, pH and a_w on the growth rate of *Monascus ruber*, a heat-resistant fungus isolated from green table olives. *J. Appl. Microbiol.* **2003**, *94*, 146–156. [CrossRef]
41. Little, C.L.; Knøchel, S. Growth and survival of *Yersinia enterocolotica*, *Salmonella* and *Bacillus cereus* in Brie stored at 4, 8 and 20 °C. *Int. J. Food Microbiol.* **1994**, *24*, 137–145. [CrossRef]
42. Meldrum, R.J.; Brocklehurst, T.F.; Wilson, D.R.; Wilson, P.D.G. The effects of cell immobilization, pH and sucrose on the growth of *Listeria monocytogenes* Scott A at 10 °C. *Food Microbiol.* **2003**, *20*, 97–103. [CrossRef]
43. Ongeng, D.; Ryckeboer, J.; Vermeulen, A.; Devlieghere, F. The effect of micro-architectural structure of cabbage substratum and or background bacterial flora on the growth of *Listeria monocytogenes*. *Int. J. Food Microbiol.* **2007**, *119*, 291–299. [CrossRef]
44. Esty, J.R.; Meyer, K.F. The heat resistance of the spore of *B. botulinus* and allied anaerobes XI. *J. Infect. Dis.* **1922**, *31*, 650–663. [CrossRef]
45. Bigelow, W.D. The logarithmic nature of thermal death time curves. *J. Infect. Dis.* **1921**, *29*, 528–536. [CrossRef]
46. Bevilacqua, A.; Speranza, B.; Sinigaglia, M.; Corbo, M.R. A focus on the death kinetics in predictive microbiology: Benefits and limits of the most important models and some tools dealing with their application in foods. *Foods* **2015**, *4*, 565–580. [CrossRef] [PubMed]
47. Desriac, N.; Vergos, M.; Achberger, V.; Coroller, L.; Couvert, O. Predicting heat process efficiency in thermal processes when bacterial inactivation is not log-linear. *Int. J. Food Microbiol.* **2019**, *290*, 36–41. [CrossRef] [PubMed]
48. Ball, C.O.; Olson, F.C.W. *Sterilization in Food Technology: Theory, Practice and Calculation*; McGraw-Hill: New York, NY, USA, 1957.
49. Garrett, E.R. Prediction of stability in pharmaceutical preparation II. Vitamin stability in liquid multivitamin preparations. *J. Am. Pharm. Assoc.* **1956**, *45*, 171–178. [CrossRef] [PubMed]
50. Levine, S. Determination of the thermal death rate of bacteria. *Food Res.* **1956**, *21*, 295–301. [CrossRef]
51. Davey, K.R. Linear-Arrhenius models for bacterial growth and death and vitamin denaturations. *J. Ind. Microbiol.* **1993**, *12*, 172–179. [CrossRef]
52. Cerf, O.; Davey, K.R.; Sadoudi, A.K. Thermal inactivation of bacteria—A new predictive model for the combined effect of three environmental factors: Temperature, pH and water activity. *Food Res. Int.* **1996**, *29*, 219–226. [CrossRef]
53. Davey, K.R.; Lin, S.H.; Wood, D.G. The effect of pH on continuous high-temperature/short-time sterilization of liquid. *Am. Inst. Chem. Eng. J.* **1978**, *3*, 537–540. [CrossRef]
54. Blackburn, C.d.W.; Curtis, L.M.; Humpheson, L.; Billon, C.; McClure, P.J. Development of thermal inactivation models for *Salmonella enteritidis* and *Escherichia coli* O157:H7 with temperature, pH and NaCl as controlling factors. *Int. J. Food Microbiol.* **1997**, *38*, 31–44. [CrossRef]
55. Zanoni, B.; Peri, C.; Garzaroli, C. A dynamic mathematical model of the thermal inactivation of *Enterococcus faecium* during Bologna Sausage Cooking. *Lebensm. Wiss. Technol.* **1997**, *30*, 727–734. [CrossRef]
56. Albert, I.; Mafart, P. A modified Weibull model for bacterial inactivation. *Int. J. Food Microbiol.* **2005**, *100*, 197–211. [CrossRef]
57. Baranyi, J.; Jones, A.; Walker, C.; Kaloti, A.; Robinson, T.P.; Mackey, B.M. A combined model for growth and subsequent thermal inactivation of *Brochothrix thermosphacta*. *Appl. Environ. Microbiol.* **1996**, *62*, 1029–1035. [CrossRef] [PubMed]
58. Casolari, A. Microbial death. In *Physiological Models in Microbiology 2*; Bazin, M.J., Prosser, J.I., Eds.; CRC Press: Boca Raton, FL, USA, 2009; pp. 1–44.
59. Chiruta, J.; Davey, K.R.; Thomas, C.J. Combined effect of temperature and pH on microbial death in continuous pasteurisation of liquids. In *Engineering and Food at ICEF7*; Jowitt, R., Ed.; Sheffield Academic Press: Sheffield, UK, 1997; pp. A109–A112.
60. Daugthry, B.J.; Davey, K.R.; Thomas, C.J.; Verbyla, A.P. Food processing–A new model for the thermal destruction of contaminating bacteria. In *Engineering and Food at ICEF7*; Jowitt, R., Ed.; Sheffield Academic Press: Sheffield, UK, 1997; pp. A113–A116.
61. Geeraerd, A.H.; Herremans, C.H.; Van Impe, J.F. Structural model requirements to describe microbial inactivation during a mild heat treatment. *Int. J. Food Microbiol.* **2000**, *59*, 185–209. [CrossRef]

62. Sapru, V.; Teixeira, A.A.; Smerage, G.H.; Lindsay, J.A. Predicting thermophilic spore population dynamics for UHT sterilization processes. *J. Food Sci.* **1992**, *575*, 1248–1252. [CrossRef]
63. Whiting, R.C. Modeling bacterial survival in unfavorable environments. *J. Ind. Micro* **1993**, *12*, 240–246. [CrossRef]
64. Xiong, R.; Xie, G.; Edmondson, A.E.; Sheard, M.A. A mathematical model for bacterial inactivation. *Int. J. Food Microbiol.* **1999**, *46*, 45–55. [CrossRef]
65. Aggelis, G.; Samelis, J.; Metaxopoulos, J. A novel modelling approach for predicting microbial growth in a raw cured meat product stored at 3 °C and at 12 °C in air. *Int. J. Food Microbiol.* **1998**, *43*, 39–52. [CrossRef]
66. Baker, D.A.; Genigeorgis, C. Predicting the safe storage of fresh fish under modified atmospheres with respect to *Clostridium botulinum* toxigenesis by modeling length of the lag phase of growth. *J. Food Prot.* **1990**, *53*, 131–140. [CrossRef]
67. Cauchie, E.; Delhalle, L.; Baré, G.; Tahiri, A.; Taminiau, B.; Korsak, N.; Burteau, S.; Fall, P.A.; Farnir, F.; Daube, G. Modeling the growth and interaction between *Brochothrix thermosphacta*, *Pseudomonas* spp., and *Leuconostoc gelidum* in minced pork samples. *Front. Microbiol.* **2020**, *11*, 639. [CrossRef]
68. Choi, W.-S.; Son, N.; Cho, J.-I.; Joo, I.-S.; Han, J.-A.; Kwak, H.-S.; Hong, J.-H.; Suh, S.H. Predictive model of *Staphylococcus aureus* growth on egg products. *Food Sci. Biotechnol.* **2019**, *28*, 913–922. [CrossRef]
69. Chung, K.-H.; Park, M.S.; Kim, H.-Y.; Bahk, G.J. Growth prediction and time–temperature criteria model of *Vibrio parahaemolyticus* on traditional Korean raw crab marinated in soy sauce (*ganjang-gejang*) at different storage temperatures. *Food Control* **2019**, *98*, 187–193. [CrossRef]
70. Dickson, J.S.; Siragusa, G.R.; Wray, J.E., Jr. Predicting the growth of *Salmonella* typhimurium on beef by using the temperature function integration technique. *Appl. Environ. Microbiol.* **1992**, *58*, 3482–3487. [CrossRef] [PubMed]
71. Giannuzzi, L.; Pinotti, A.; Zaritzky, N. Mathematical modelling of microbial growth in packaged refrigerated beef stored at different temperatures. *Int. J. Food Microbiol.* **1998**, *39*, 101–110. [CrossRef]
72. Hwang, C.-A.; Huang, L. Growth and survival of *Bacillus cereus* from spores in cooked rice–One-step dynamic analysis and predictive modelling. *Food Control* **2019**, *96*, 403–409. [CrossRef]
73. Hyun, J.-E.; Yoon, J.-H.; Lee, S.-Y. Response surface modeling for the inactivation of *Bacillus cereus* on cooked spinach by natural antimicrobial at various temperatures. *J. Food Saf.* **2018**, *38*, e12484. [CrossRef]
74. Juneja, V.K.; Mishra, A.; Pradhan, A.K. Dynamic predictive model for growth of *Bacillus cereus* from spores in cooked beans. *J. Food Prot.* **2018**, *81*, 308–315. [CrossRef] [PubMed]
75. Juneja, V.K.; Golden, C.E.; Mishra, A.; Harrison, M.A.; Mohr, T.B. Predictive model for growth of *Bacillus cereus* at temperature applicable to cooling of cooked pasta. *J. Food Sci.* **2019**, *84*, 590–598. [CrossRef]
76. Kowalik, J.; Lobacz, A.; Zulewska, J.; Dec, B. Analysis and mathematical modelling of the behaviour of *Escherichia coli* in the mascarpone cheese during cold storage. *Int. J. Food Sci. Technol.* **2018**, *53*, 1541–1548. [CrossRef]
77. Li, M.; Huang, L.; Zhu, Y.; Wei, Q. Growth of *Clostridium perfringens* in roasted chicken and braised beef during cooling–One-step dynamics analysis and modelling. *Food Control* **2019**, *106*, 106739. [CrossRef]
78. Longhi, D.A.; da Silva, N.B.; Martins, W.F.; Carciofi, B.A.M.; de Aragão, G.M.F.; Laurindo, J.B. Optimal experimental design to model spoilage bacteria growth in vacuum-packaged ham. *J. Food Eng.* **2018**, *216*, 20–26. [CrossRef]
79. Oscar, T.P. Response surface models for effects of temperature and previous growth sodium chloride on growth kinetics of *Salmonella* typhimurium on cooked chicken breast. *J. Food Prot.* **1999**, *62*, 1470–1474. [CrossRef]
80. Park, S.Y.; Ha, S.-D. Predictive growth model of the effects of temperature on the growth kinetics of generic *Escherichia coli* in the Korean traditional rice cake product "Garaetteok". *J. Food Sci. Technol.* **2018**, *55*, 506–512. [CrossRef]
81. Park, S.Y.; Choi, S.-Y.; Ha, S.-D. Predictive modeling for the growth of *Aeromonas hydrophila* on lettuce as a function of combined storage temperature and relative humidty. *Foodborne Pathog. Dis.* **2019**, *16*, 376–383. [CrossRef] [PubMed]
82. Roberts, T.A.; Gibson, A.M.; Robinson, A. Prediction of toxin production by *Clostridium botulinum* in pasteurized pork slurry. *J. Food Technol.* **1981**, *16*, 337–355. [CrossRef]
83. Taoukis, P.S.; Koutsoumanis, K.; Nychas, G.J.E. Use of time-temperature integrators and predictive modelling for shelf life control of chilled fish under dynamic storage conditions. *Int. J. Food Microbiol.* **1999**, *53*, 21–31. [CrossRef]
84. Tarlak, F.; Johannessen, G.; Villegas, I.B.; Bolívar, A.; Posada-Izquierdo, G.D.; Pérez-Rodríguez, F. Modelling of the behaviour of *Salmonella enterica serovar* Reading on commercial fresh-cut iceberg lettuce stored at different temperatures. *Foods* **2020**, *9*, 946. [CrossRef]
85. Vankerschaver, K.; Willocx, F.; Smout, C.; Hendrickx, M.; Tobback, P. The influence of temperature and gas mixtures on the growth of the intrinsic micro-organisms on cut endive: Predictive versus actual growth. *Food Microbiol.* **1996**, *13*, 427–440. [CrossRef]
86. Zamora, M.C.; Zaritzky, N.E. Modeling of microbial growth in refrigerated packaged beef. *J. Food Sci.* **1985**, *50*, 1003–1006. [CrossRef]
87. Costa, J.C.C.P.; Bover-Cid, S.; Bolívar, A.; Zurera, G.; Pérez-Rodríguez, F. Modelling the interaction of the sakacin-producing *Lactobacillus sakei* CTC494 and *Listeria monocytogenes* in filleted gilthead sea bream (*Sparus aurata*) under modified atmosphere packaging at isothermal and non-isothermal conditions. *Int. J. Food Microbiol.* **2019**, *297*, 72–84. [CrossRef]
88. Cuggino, S.G.; Bascón-Villegas, I.; Rincón, F.; Pérez, M.A.; Posada-Izquierdo, G.; Marugán, J.; Carro, C.P.; Pérez-Rodríguez, F. Modelling the combined effect of choline, benzyl isothiocyanate, exposure time and cut size on the reduction of *Salmonella* in fresh-cut lettuce during washing process. *Food Microbiol.* **2020**, *86*, 103346. [CrossRef]

89. Jaczynski, J.; Park, J.W. Predictive models for microbial inactivation and texture degradation in surimi seafood during thermal processing. *J. Food Sci.* **2003**, *68*, 1025–1030. [CrossRef]
90. Jia, Z.; Li, C.; Fang, T.; Chen, J. Predictive modeling of the effect of ε-polylysine hydrochloride on growth and thermal inactivation of *Listeria monocytogenes* in fish balls. *J. Food Sci.* **2019**, *84*, 127–132. [CrossRef] [PubMed]
91. Juneja, V.K.; Gonzales-Barron, U.; Butler, F.; Yadav, A.S.; Friedman, M. Predictive thermal inactivation model for the combined effect of temperature, cinnamaldehyde and carvacrol on starvation-stressed multiple *Salmonella* serotypes in ground chicken. *Int. J. Food Microbiol.* **2013**, *265*, 184–199. [CrossRef] [PubMed]
92. Juneja, V.K.; Garcia-Dávila, J.; Lopez-Romero, J.C.; Pena-Ramos, E.A.; Camou, J.P.; Valenzuela-Melendres, M. Modeling the effects of temperature, sodium chloride, and green tea and their interactions on the thermal inactivation of *Listeria monocytogenes* in Turkey. *J. Food Prot.* **2014**, *77*, 1696–1702. [CrossRef] [PubMed]
93. Koutsoumanis, K.; Lambropoulou, K.; Nychas, G.-J.E. A predictive model for the non-thermal inactivation of *Salmonella enteritidis* in a food model system supplemented with a natural antimicrobial. *Int. J. Food Microbiol.* **1999**, *49*, 63–74. [CrossRef]
94. Nazarowec-White, M.; McKellar, R.C.; Piyasena, P. Predictive modelling of *Enterobacter sakazakii* inactivation in bovine milk during high-temperature short-time pasteurization. *Food Res. Int.* **1999**, *32*, 375–379. [CrossRef]
95. Portela, J.B.; Coimbra, P.T.; Cappato, L.P.; Alvarenga, V.O.; Oliveira, R.B.A.; Pereira, K.S.; Azeredo, D.R.P.; Sant'Ana, A.S.; Nascimento, J.S.; Cruz, A.G. Predictive model for inactivation of *Salmonella* in infant formula during microwave heating processing. *Food Control* **2019**, *104*, 308–312. [CrossRef]
96. Santillana Farakos, S.M.; Pouillot, R.; Anderson, N.; Johnson, R.; Son, I.; Van Doren, J. Modeling the survival kinetics of *Salmonella* in tree nuts for use in risk assessment. *Int. J. Food Microbiol.* **2016**, *227*, 41–50. [CrossRef]
97. Sheen, S.; Huang, C.-Y.; Ramos, R.; Chien, S.-Y.; Scullen, J.; Sommers, C. Lethality prediction for *Escherichia Coli* O157:H7 and uropathogenic *E. coli* in ground chicken treated with high pressure processing and trans-cinnamaldehyde. *J. Food Sci.* **2018**, *83*, 740–749. [CrossRef]
98. Noriega, E.; Laca, A.; Díaz, M. Modelling of diffusion-limited growth to predict *Listeria* distribution in structured model foods. *J. Food Eng.* **2008**, *87*, 247–256. [CrossRef]
99. Broughall, J.M.; Anslow, P.A.; Kilsby, D.C. Hazard analysis applied to microbial growth in foods: Development of mathematical models describing the effect of water activity. *J. Appl. Bacteriol.* **1983**, *55*, 101–110. [CrossRef]
100. Devlieghere, F.; Geeraerd, A.H.; Versyck, K.J.; Vandewaetere, J.; Van Impe, J.; Debevere, J. Growth of *Listeria monocytogenes* in modified atmosphere packed cooked meat products: A predictive model. *Food Microbiol.* **2001**, *18*, 53–66. [CrossRef]
101. Gibson, A.M.; Bratchell, N.; Roberts, T.A. Predicting microbial growth: Growth responses of *salmonellae* in a laboratory medium as affected by pH, sodium chloride and storage temperature. *Int. J. Food Microbiol.* **1988**, *6*, 155–178. [CrossRef]
102. Juneja, V.K.; Marmer, B.S.; Phillips, J.G.; Miller, A.J. Influence of the intrinsic properties of food on thermal inactivation of spores of nonproteolytic *Clostridium botulinum*: Development of a predictive model. *J. Food Saf.* **1995**, *15*, 349–364. [CrossRef]
103. Aspridou, Z.; Moschakis, T.; Biliaderis, C.G.; Koutsoumanis, K.P. Effect of the substrate's microstructure on the growth of *Listeria monocytogenes*. *Food Res. Int.* **2014**, *64*, 683–691. [CrossRef] [PubMed]
104. Boons, K.; Van Derlinden, E.; Mertens, L.; Peeters, V.; Van Impe, J.F. Effect of immobilization and salt concentration on the growth dynamics of *Escherichia coli* K12 and *Salmonella* typhimurium. *J. Food Sci.* **2013**, *78*, 567–574. [CrossRef]
105. Costello, K.M.; Gutierrez-Merino, J.; Bussemaker, M.; Ramaioli, M.; Baka, M.; Van Impe, J.F.; Velliou, E.G. Modelling the microbial dynamics and antimicrobial resistance development of *Listeria* in viscoelastic food model systems of various structural complexities. *Int. J. Food Microbiol.* **2018**, *286*, 15–30. [CrossRef]
106. Kabanova, N.; Stulova, I.; Vilu, R. Microcalorimetric study of the growth of bacterial colonies of *Lactococcus lactis* IL1403 in agar gels. *Food Microbiol.* **2012**, *29*, 67–79. [CrossRef]
107. Prachaiyo, P.; McLandsborough, L.A. Oil-in-water emulsion as a model system to study the growth of *Escherichia coli* O157:H7 in a heterogeneous food system. *J. Food Sci.* **2003**, *68*, 1018–1024. [CrossRef]
108. Zalazar, A.L.; Gliemmo, M.F.; Campos, C.A. Effect of stabilizers, oil level and structure on the growth of *Zygosaccharomyces bailii* and on physical stability of model systems simulating acid sauces. *Food Res. Int.* **2016**, *85*, 200–208. [CrossRef]
109. Castro, M.P.; Rojas, A.M.; Campos, C.A.; Gerschenson, L.N. Effect of preservatives, tween 20, oil content and emulsion structure on the survival of *Lactobacillus fructivorans* in model salad dressings. *LWT Food Sci. Technol.* **2009**, *42*, 1428–1434. [CrossRef]
110. Verheyen, D.; Baka, M.; Akkermans, S.; Skåra, T.; Van Impe, J.F. Effect of microstructure and initial cell conditions on thermal inactivation kinetics and sublethal injury of *Listeria monocytogenes* in fish-based food model systems. *Food Microbiol.* **2019**, *84*, 103267. [CrossRef] [PubMed]
111. Verheyen, D.; Govaert, M.; Seow, T.K.; Ruvina, J.; Mukherjee, V.; Baka, M.; Skåra, T.; Van Impe, J.F. The complex effect of food matrix fat content on thermal inactivation of *Listeria monocytogenes*: Case study in emulsion and gelled emulsion model systems. *Front. Microbiol.* **2020**, *10*, 3149. [CrossRef] [PubMed]
112. Bellara, S.R.; Fryer, P.J.; McFarlane, C.M.; Thomas, C.R.; Hocking, P.M.; Mackey, B.M. Visualization and modelling of the thermal inactivation of bacteria in a model food. *Appl. Environ. Microbiol.* **1999**, *65*, 3095–3099. [CrossRef] [PubMed]
113. Hamoud-Agha, M.M.; Curet, S.; Simonin, H.; Boillereaux, L. Microwave inactivation of *Escherichia coli* K12 CIP 54.117 in a gel medium: Experimental and numerical study. *J. Food Eng.* **2013**, *116*, 315–323. [CrossRef]

114. Mackey, B.M.; Kelly, A.F.; Colvin, J.A.; Robbins, P.T.; Fryer, P.J. Predicting the thermal inactivation of bacteria in a solid matrix: Simulation studies on the relative effects of microbial thermal resistance parameters and process conditions. *Int. J. Food Microbiol.* **2006**, *107*, 295–303. [CrossRef] [PubMed]
115. Noriega, E.; Laca, A.; Díaz, M. Modelling of diffusion-limited growth for food safety in simulated cheeses. *Food Bioprod. Process.* **2008**, *86*, 122–129. [CrossRef]
116. Mertens, L.; Van Derlinden, E.; Dang, T.D.T.; Cappuyns, A.M.; Vermeulen, A.; Debevere, J.; Moldenaers, P.; Devlieghere, F.; Geeraerd, A.H.; Van Impe, J.F. On the critical evaluation of growth/no growth assessment of *Zygosaccharomyces bailii* with optical density measurements: Liquid versus structured media. *Food Microbiol.* **2011**, *28*, 736–745. [CrossRef]
117. Ter Steeg, P.F.; Otten, G.D.; Alderliesten, M.; De Weijer, R.; Naaktgeboren, G.; Bijl, J.; Vasbinder, A.J.; Kershof, I.; Van Duijvendijk, A.M. Modelling the effects of (green) antifungals, droplet size distribution and temperature on mould outgrowth in water-in-oil emulsions. *Int. J. Food Microbiol.* **2001**, *67*, 227–239. [CrossRef]
118. Kapetanakou, A.E.; Ampavi, A.; Yanniotis, S.; Drosinos, E.H.; Skandamis, P.N. Development of a model describing the effect of temperature, water activity and (gel) structure on growth and ochratoxin A production by *Aspergillus carbonarius* in vitro and evaluation in food matrices of different viscosity. *Food Microbiol.* **2011**, *28*, 727–735. [CrossRef]
119. Dens, E.J.; Van Impe, J.F. On the need for another type of predictive model in structured foods. *Int. J. Food Microbiol.* **2001**, *64*, 247–260. [CrossRef]
120. De Bonis, M.V.; Ruocco, G. A heat and mass transfer perspective of microbial behavior modelling in a structured vegetable food. *Int. J. Food Eng.* **2016**, *190*, 72–79. [CrossRef]
121. Ferrier, R.; Hezard, B.; Lintz, A.; Stahl, V.; Augustin, J.-C. Combining individual-based modelling and food microenvironment descriptions to predict the growth of *Listeria monocytogenes* on smear soft cheese. *Appl. Environ. Microbiol.* **2013**, *19*, 5870–5881. [CrossRef]
122. Augustin, J.-C.; Ferrier, R.; Hezard, B.; Lintz, A.; Stahl, V. Comparison of individual-based modeling and population approaches for prediction of foodborne pathogens growth. *Food Microbiol.* **2015**, *45*, 205–215. [CrossRef] [PubMed]
123. Vereecken, K.M.; Devlieghere, F.; Bockstaele, A.; Debevere, J.; Van Impe, J.F. A model for lactic acid-induced inhibition of *Yersinia enterocolitica* in mono- and coculture with *Lactobacillus sakei*. *Food Microbiol.* **2003**, *20*, 701–713. [CrossRef]
124. Ross, T.; Ratkowsky, D.A.; Mellefont, L.A.; McMeekin, T.A. Modelling the effects of temperature, water activity, pH and lactic acid concentration on the growth rate of *Escherichia coli*. *Int. J. Food Microbiol.* **2003**, *82*, 33–43. [CrossRef]
125. Reiner, M. Über die strömung einer elastichen flüssigkeit durch eine kapillare. *Kolloid, Z.* **1926**, *39*, 80–87. [CrossRef]
126. Verheyen, D.; Bolívar, A.; Pérez-Rodríguez, F.; Baka, M.; Skåra, T.; Van Impe, J.F. Isolating the effect of fat content on *Listeria monocytogenes* growth dynamics in fish-based emulsion and gelled emulsion systems. *Food Control* **2020**, *108*, 106874. [CrossRef]
127. Van Impe, J.F.; Poschet, F.; Geeraerd, A.H.; Vereecken, K.M. Towards a novel class of predictive microbial growth models. *Int. J. Food Microbiol.* **2005**, *100*, 97–105. [CrossRef]
128. Tack, I. Metabolic Differentiation in Microbial Colonies and Biofilms: A Multiscale Modelling Approach. Ph.D. Thesis, KU Leuven, Leuven, Belgium, 2016.
129. Ayala, F.J.; Gilpin, M.J.; Ehrenfield, J.G. Competition between species: Theoretical models and experimental results. *Theor. Pop Biol.* **1973**, *4*, 331–356. [CrossRef]
130. Bailey, J.E.; Ollis, D.F. *Biochemical Engineering Fundamentals*, 2nd ed.; McGraw-Hill: New York, NY, USA, 1986.
131. Baranyi, J.; Robinson, T.P.; Kaloti, A.; Mackey, B.M. Predicting the growth of *Brochothrix thermosphacta* at changing temperature. *Int. J. Food Microbiol.* **1995**, *27*, 61–75. [CrossRef]
132. Ratkowsky, D.A.; Olley, J.; McMeekin, T.A.; Ball, A. Relationship between temperature and growth rate of bacterial cultures. *J. Bacteriol.* **1982**, *149*, 1–5. [CrossRef] [PubMed]
133. Gorochowski, T.E.; Matyjaszkiewicz, A.; Todd, T.; Oak, N.; Kowalska, K.; Reid, S.; Tsaneva-Atanasova, K.T.; Savery, N.J.; Grierson, C.S.; di Bernardo, M. BSim: An agent-based tool for modeling bacterial populations in systems and synthetic biology. *PLoS ONE* **2012**, *7*, e42790. [CrossRef] [PubMed]
134. González-Cabaleiro, R.; Mitchell, A.M.; Smith, W.; Wipat, A.; Ofiteru, I.D. Heterogeneity in pure microbial systems: Experimental measurements and modeling. *Front. Microbiol.* **2017**, *8*, 1813. [CrossRef] [PubMed]
135. Kreft, J.-U.; Booth, G.; Wimpenny, J.W.T. BacSim, a simulator for individual-based modelling of bacterial colony growth. *Microbiology* **1998**, *144*, 3275–3287. [CrossRef]
136. Ginovart, M.; López, D.; Valls, J. INDISIM, an individual-based discrete simulation model to study bacterial cultures. *J. Theor. Biol.* **2002**, *214*, 305–319. [CrossRef]
137. Verhulst, A.J.; Cappuyns, A.M.; Van Derlinden, E.; Bernaerts, K.; Van Impe, J.F. Analysis of the lag phase to exponential growth transition by incorporating inoculum characteristics. *Food Microbiol.* **2011**, *28*, 656–666. [CrossRef]
138. Tack, I.L.M.M.; Nimmegeers, P.; Akkermans, S.; Hashem, I.; Van Impe, J.F.M. Simulation of *Escherichia coli* dynamics in biofilms and submerged colonies with an individual-based model including metabolic network information. *Front. Microbiol.* **2017**, *8*, 2509. [CrossRef]
139. Lardon, L.A.; Merkey, B.V.; Martins, S.; Dötsch, A.; Picioreanu, C.; Kreft, J.-U.; Smets, B.F. iDynoMiCS: Next-generation individual-based modelling of biofilms. *Environ. Microbiol.* **2011**, *13*, 2416–2434. [CrossRef]
140. Hellweger, F.L.; Clegg, R.J.; Clark, J.R.; Plugge, C.M.; Kreft, J.U. Advancing microbial sciences by individual-based modelling. *Nat. Rev. Microbiol.* **2016**, *14*, 461–471. [CrossRef]

141. Possas, A.; Pérez-Rodríguez, F.; Valero, A.; Rincón, F.; García-Gimeno, R.M. Mathematical approach for the *Listeria monocytogenes* inactivation during high hydrostatic pressure processing of a simulated meat medium. *Innov. Food Sci. Emerg. Technol.* **2018**, *47*, 271–278. [CrossRef]
142. Chhabra, A.T.; Carter, W.H.; Linton, R.H.; Cousin, M.A. A predictive model to determine the effects of pH, milkfat, and temperature on thermal inactivation of *Listeria monocytogenes*. *J. Food Prot.* **1999**, *62*, 1143–1149. [CrossRef]
143. Santillana Farakos, S.M.; Frank, J.F.; Schaffner, D.W. Modeling the influence of temperature, water activity and water mobility on the persistence of *Salmonella* in low-moisture foods. *Int. J. Food Microbiol.* **2013**, *166*, 280–293. [CrossRef]
144. Trimble, L.M.; Frank, J.F.; Schaffner, D.W. Modification of a predictive model to include the influence of fat content on *Salmonella* inactivation in low-water-activity foods. *J. Food Prot.* **2020**, *83*, 801–815. [CrossRef]
145. Albuquerque, C.D.D.; Curet, S.; Boillereaux, L. A 3D-CFD-heat-transfer-based model for the microbial inactivation of pasteurized food products. *Innov. Food Sci. Emerg. Technol.* **2019**, *54*, 172–181. [CrossRef]
146. Gil, M.M.; Miller, F.A.; Brandão, T.R.S.; Silva, C.L.M. On the use of the Gompertz model to predict microbial thermal inactivation under isothermal and non-isothermal conditions. *Food Eng. Rev.* **2011**, *3*, 17–25. [CrossRef]
147. Mafart, P.; Couvert, O.; Gaillard, S.; Leguerinel, I. On calculating sterility in thermal preservation methods: Application of the Weibull frequency distribution model. *Int. J. Food Microbiol.* **2002**, *72*, 107–113. [CrossRef]
148. Huang, L. Thermal inactivation of *Listeria monocytogenes* in ground beef under isothermal and dynamic temperature conditions. *J. Food Eng.* **2009**, *90*, 380–387. [CrossRef]
149. Chen, H.-H.; Kang, H.-Y.; Chen, S.-D. The effects of ingredients and water content on the rheological properties of batters and physical properties of crusts in fried foods. *J. Food Eng.* **2008**, *88*, 45–54. [CrossRef]
150. Rao, M.A. *Rheology of Fluid and Semifluid Foods: Principles and Applications*; Springer Science+Business Media, LLC: New York, NY, USA, 2007.
151. Whiting, R.C.; Sackitey, S.; Calderone, S.; Morely, K.; Phillips, J.G. Model for the survival of *Staphylococcus aureus* in nongrowth environments. *Int. J. Food Microbiol.* **1996**, *31*, 231–243. [CrossRef]
152. Bhuvaneswari, E.; Anandharamakrishnan, C. Heat transfer analysis of pasteurization of bottled beer in a tunnel pasteurizer using computational fluid dynamics. *Innov. Food Sci. Emerg. Technol.* **2014**, *23*, 156–163. [CrossRef]
153. Erdogdu, F.; Tutar, M.; Sarghini, F.; Skipnes, D. Effects of viscosity and agitation rate on temperature and flow field in cans during reciprocal agitation. *J. Food Eng.* **2017**, *213*, 76–88. [CrossRef]
154. Topcam, H.; Karatas, O.; Erol, B.; Erdogdu, F. Effect of rotation on temperature uniformity of microwave processed low-high viscosity liquids: A computational study with experimental validation. *Innov. Food Sci. Emerg. Technol.* **2020**, *60*, 102306. [CrossRef]
155. Wu, H.; Karayiannis, T.G.; Tassou, S.A. A two-dimensional frying model for the investigation and optimisation of continuous industrial frying systems. *Appl. Therm. Eng.* **2013**, *51*, 926–936. [CrossRef]
156. Bedane, T.F.; Erdogdu, F.; Lyng, J.G.; Marra, F. Effects of geometry and orientation of food products on heating uniformity during radio frequency heating. *Food Bioprod. Process.* **2021**, *125*, 149–160. [CrossRef]
157. Erdogdu, F.; Karatas, O.; Sarghini, F. A short update on heat transfer modelling for computational food processing in conventional and innovative processing. *Curr. Opin. Food Sci.* **2018**, *23*, 113–119. [CrossRef]
158. Aspridou, A.; Koutsoumanis, K.P. Individual cell heterogeneity as variability source in population dynamics of microbial inactivation. *Food Microbiol.* **2015**, *45*, 216–221. [CrossRef] [PubMed]
159. Abe, H.; Koyama, K.; Kawamura, S.; Koseki, S. Stochastic evaluation of *Salmonella enterica* lethality during thermal inactivation. *Int. J. Food Microbiol.* **2018**, *285*, 129–135. [CrossRef]
160. Koseki, S.; Koyama, K.; Abe, H. Recent advances in predictive microbiology: Theory and application of conversion from population dynamics to individual cell heterogeneity during inactivation process. *Curr. Opin. Food Sci.* **2021**, *42*, 60–67. [CrossRef]

foods

MDPI

Article

Dynamic Modeling of *Carnobacterium maltaromaticum* CNCM I-3298 Growth and Metabolite Production and Model-Based Process Optimization

Cristian Puentes [1,2], Amélie Girardeau [1], Stephanie Passot [1], Fernanda Fonseca [1] and Ioan-Cristian Trelea [1,*]

[1] INRAE, AgroParisTech, UMR SayFood, Université Paris-Saclay, F-78850 Thiverval-Grignon, France; cristian.puentes@centralesupelec.fr (C.P.); amelie.girardeau@inrae.fr (A.G.); stephanie.passot@inrae.fr (S.P.); fernanda.fonseca@inrae.fr (F.F.)

[2] CentraleSupélec, LGPM, Université Paris-Saclay, F-91192 Gif-sur-Yvette, France

* Correspondence: ioan-cristian.trelea@inrae.fr; Tel.: +33-13-081-5490

Abstract: *Carnobacterium maltaromaticum* is a species of lactic acid bacteria found in dairy, meat, and fish, with technological properties useful in food biopreservation and flavor development. In more recent years, it has also proven to be a key element of biological time–temperature integrators for tracking temperature variations experienced by perishable foods along the cold-chain. A dynamic model for the growth of *C. maltaromaticum* CNCM I-3298 and production of four metabolites (formic acid, acetic acid, lactic acid, and ethanol) from trehalose in batch culture was developed using the reaction scheme formalism. The dependence of the specific growth and production rates as well as the product inhibition parameters on the operating conditions were described by the response surface method. The parameters of the model were calibrated from eight experiments, covering a broad spectrum of culture conditions (temperatures between 20 and 37 °C; pH between 6.0 and 9.5). The model was validated against another set of eight independent experiments performed under different conditions selected in the same range. The model correctly predicted the growth kinetics of *C. maltaromaticum* CNCM I-3298 as well as the dynamics of the carbon source conversion, with a mean relative error of 10% for biomass and 14% for trehalose and the metabolites. The paper illustrates that the proposed model is a valuable tool for optimizing the culture of *C. maltaromaticum* CNCM I-3298 by determining operating conditions that favor the production of biomass or selected metabolites. Model-based optimization may thus reduce the number of experiments and substantially speed up the process development, with potential applications in food technology for producing starters and improving the yield and productivity of the fermentation of sugars into metabolites of industrial interest.

Keywords: *Carnobacterium maltaromaticum*; modeling; microbial growth; optimization; fermentation

Citation: Puentes, C.; Girardeau, A.; Passot, S.; Fonseca, F.; Trelea, I.-C. Dynamic Modeling of *Carnobacterium maltaromaticum* CNCM I-3298 Growth and Metabolite Production and Model-Based Process Optimization. *Foods* **2021**, *10*, 1922. https://doi.org/10.3390/foods10081922

Academic Editor: Ursula Andrea Gonzales-Barron

Received: 16 July 2021
Accepted: 17 August 2021
Published: 19 August 2021

1. Introduction

Carnobacterium maltaromaticum is a psychotropic species of lactic acid bacteria widely found in food such as dairy products, fish, and meat. It is a Gram-positive, facultative anaerobic bacterium, able to grow at alkaline pH (up to 9.6) [1,2].

In the food industry, *C. maltaromaticum* has potential applications related to health protection and organoleptic properties. These include the biopreservation of food, by inhibiting the growth of foodborne pathogens such as *Listeria* sp. in cold conditions, and the development of flavor in ripened cheese varieties [2–4].

This lactic acid bacterium may also be used as a biological indicator in time–temperature integrators (TTI): 'smart-labels' that monitor the time–temperature history of chilled products throughout the cold-chain [5,6]. Concentrates of the strain CNCM I-3298 have been selected as inoculum for TopCryo® labels, the only biological TTI that has been taken to market to date. A pH decline of the label medium, associated with bacterial growth and acid production, produces an irreversible color change from green to red as an indication to the consumer about the spoilage of the food to which the TTI is attached [7].

In these applications, *C. maltaromaticum* concentrates produced by fermentation may be used alone or in association with other microorganisms. Some experimental studies on *C. maltaromaticum* fermentation under different culture conditions have been reported in the literature [3,4,6–14]. The effect of temperature and pH on the acidifying activity was evaluated and modelled by Girardeau et al. [7]. However, there is a lack of knowledge on the characterization and optimization of Carnobacteria growth and production of various metabolites such as acids or flavor compounds in a bioreactor.

Carnobacteria are considered to be homofermentative lactic acid bacteria that produce lactic acid from glucose, with pyruvate as a central metabolic intermediate (via the Embden–Meyerhof pathway) [15–17]. However, pyruvate may be alternatively converted to acetate, ethanol, formate, and CO_2 [16,18] under anaerobic conditions and substrate limitation [19], arising for example at the end of fermentation [20]. The production of organic acids by Carnobacteria is also strain-dependent [8,16,21]. A recent study reported that lactic, formic, and acetic acids are key organic acids produced by *C. maltaromaticum* in a meat juice medium [22], indicating that this microorganism has the enzymatic machinery to perform mixed-acid fermentation (Figure 1).

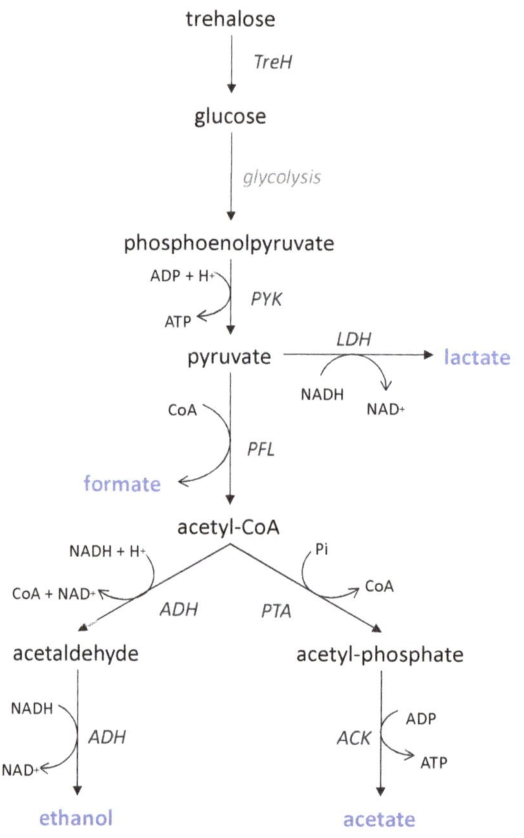

Figure 1. Mixed-acid fermentation pathway likely used by *C. maltaromaticum* to ferment trehalose. End products are shown in blue. ACK, acetate kinase; ADH, acetaldehyde dehydrogenase; LDH, lactate dehydrogenase; PFL, pyruvate formate lyase; PTA, phosphate acetyltransferase; PYK, pyruvate kinase; TreH, neutral trehalose. Adapted from [19–23].

For optimization purposes, modeling has proven to be a powerful tool, enabling the exploration of a wider range of operating conditions while minimizing cost, compared with the experimental approach [24–29]. To our knowledge, the only dynamic model dealing with *C. maltaromaticum* strains has been published by Ellouze et al. [6]. That research was oriented towards a biological TTI setting associated with a sausage-like packaging instead of a bioreactor and taking into account lactic acid as the single metabolite.

The aim of this study was thus to develop and validate a dynamic model predicting the impact of fermentation conditions (temperature and pH) on the growth and bioconversion fermentation dynamics of *C. maltaromaticum* CNCM I-3298 using trehalose as a carbon source and considering the four main identified metabolites: formic acid, acetic acid, lactic acid, and ethanol. This study was conducted as part of a research project on the production and conservation *of C. maltaromaticum* concentrates. In that context, the growth of *C. maltaromaticum* was tested in different sugars: glucose, maltose, mannitol, and trehalose, with similar growth rates. Trehalose was chosen in this study because this molecule is known for its ability to protect cells during bacterial stabilization processes (freeze-drying in particular). Therefore, the residual trehalose (not consumed during fermentation) could be used as cryoprotectant after production of bacterial concentrates.

The model development involved four major steps, presented in Section 3: derivation of the main governing equations based on the known mixed-acid fermentation pathway, mass balances, and kinetic rate expressions (Section 3.1); parameter identification for each fermentation experiment (Section 3.2); construction of response surfaces of the calibrated parameters as a function of temperature and pH (Section 3.3); and final validation of the complete model. The resulting model is shown to be a useful tool in determining the optimal conditions for producing bacterium concentrates in bioreactors and for assessing the productivity of the bioconversion fermentation of sugars into metabolites of potential industrial interest (Section 4.4).

2. Materials and Methods

Data used to calibrate and validate the model were partially reported in a previous study, in which a modified central composite experimental design was carried out to study the effect of operating conditions on the technological properties of *C. maltaromaticum* CNCM I-3298 [7]. Sixteen lab-scale fermentations (hereafter named F01 to F16) were performed using a wide range of regulated operating conditions (Figure 2): temperature between 20 and 37 °C and pH between 6.0 and 9.5.

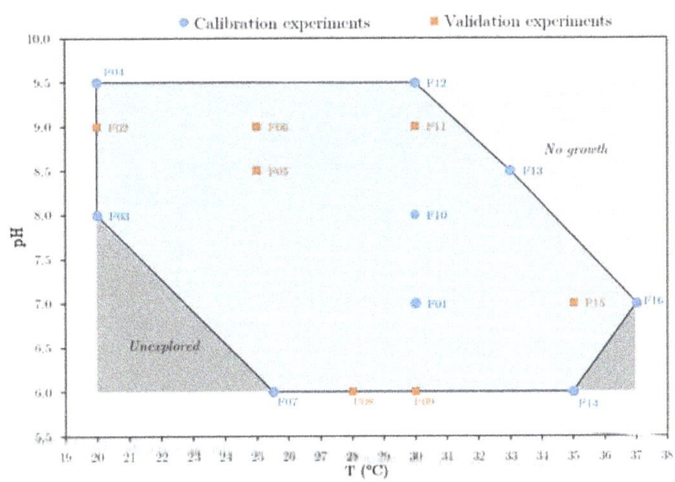

Figure 2. Operating conditions of *C. maltaromaticum* CNCM I-3298 fermentation experiments.

Fermentation durations varied between 20 h and 45 h, and the initial conditions were: for biomass (X_0) 0.077 $mol_C \cdot L^{-1}$, trehalose (S_0) between 0.091 $mol \cdot L^{-1}$ and 0.107 $mol \cdot L^{-1}$, and medium volume (V_0) 3.5 L.

The main fermentation settings and the kinetic measurements are reported below.

2.1. *Fermentation*

2.1.1. Culture Medium and Bacterial Strain

The fermentation medium was composed of the following ingredients for 1 kg of final solution: 40 g of trehalose (Treha™; Tokyo Japan); 10 g of proteose peptone (Oxoid; Waltham, MA, USA); 5 g of yeast extract (Humeau; La-Chapelle-sur-Erdre, France); 5 g of Tween 80 (VWR; Leuven, Belgium); 0.41 g of $MgSO_4$ (Merck; Darmstadt, Germany); 0.056 g $MnSO_4$ (Merck; Darmstadt, Germany); and water to reach a total of 1 kg of solution. All medium components were sterilized together at 121 °C for 20 min. Fermentations were carried out on *C. maltaromaticum* CNCM I-3298 pre-cultures. Pre-cultures were prepared by inoculating 10 mL of sterilized fermentation medium with 100 µL of *C. maltaromaticum* CNCM I-3298 stock culture and were incubated for 13 to 16 h at 30 °C. An amount of 1 mL of the resulting culture was transferred into 50 mL of fresh medium and then incubated again for 11 h under the same conditions. The resulting culture was then used to inoculate the bioreactor. Inoculation was performed at an initial concentration of approximately 10^7 CFU mL^{-1}.

2.1.2. Bioreactor and Parameter Control

The bioreactor (Minifors, Infors HT, Bottmingen, Switzerland) had a total volume of 5 L and was equipped with a heat mantle and a cryostat for temperature control. It contained 3.5 L of fermentation medium, inoculated with an initial cell concentration of approximately 10^7 $CFU \cdot mL^{-1}$. Initial pH was adjusted to the desired value with 5 M NaOH or 0.01 M H_2SO_4 solutions. During fermentation, pH was controlled to the desired setpoint for each investigated condition (Figure 2) by automatic addition of 5 M NaOH. Culture homogenization was performed with an agitation device set at 150 rpm. Temperature was set according to the investigated operating conditions mentioned above (Figure 2).

2.2. *Kinetic Measurements*

2.2.1. Cell Growth

Cell growth was monitored using an infrared probe (Excell210, CellD, Roquemaure, France) continuously measuring absorbance at 880 nm and storing data every minute. The absorbance data were calibrated in dry weight. Dry cell weight was determined by filtering 10 mL of bacterial suspension (straight out of the bioreactor) through a 0.20 µm polyethersulfone membrane (Supor®, PALL Biotech, Saint-Germain-en-Laye, France). The filter was then dried for 24 h at 80 °C. Measurements were obtained in triplicate. Mass concentrations were finally converted to mol_C L^{-1} (carbon-mol of biomass per liter), assuming the simplified unit-carbon biomass formula $CH_{1.8}O_{0.5}$ [30].

2.2.2. Total Acid Production

Total acid production was determined according to the volume of NaOH solution injected into the bioreactor to maintain a constant pH. The pH was regulated/controlled to set values using the IRIS NT V5 software (Infors, AG, Bottmingen, Switzerland).

2.2.3. Substrate Consumption and Metabolite Production

Trehalose consumption and metabolite production were determined using high-performance liquid chromatography (HPLC, Waters Associates, Millipore; Molsheim, France). HPLC was performed on culture media samples of a few mL, aseptically retrieved from the bioreactor at different times during fermentation and filtered through 0.22 µm pores (Sartorius stedim, Biotech; Göttigen, Germany). Analyses were made using a cation exchange column (Aminex Ion Exclusion HPX-87 300 × 7.8 mm, Bio-Rad, Richmond, VA,

USA) at 35 °C. Mobile phase was 0.0005 M H_2SO_4, and flow rate was set at 0.6 mL·min^{-1} (LC-6A pump, Shimadzu, Courtaboeuf, France).

HPLC analysis showed that *C. maltaromaticum* CNCM I-3298 produced not only lactic acid but also formic acid, acetic acid, and ethanol in variable proportions according to the fermentation conditions.

3. Dynamic Model

The mathematical model was a set of ordinary differential equations implemented in MATLAB R2018b (the MathWorks Inc. Natick, MA, USA). Model parameters and response surface coefficients were identified by nonlinear regression analysis using the Statistic and Machine Learning Toolbox of MATLAB.

3.1. Model Formulation

The dynamic model developed in this study combined biochemical knowledge about the metabolism of the selected bacterium and mass balances of the main compounds: substrate, biomass, and identified metabolites. Expressions of specific growth and metabolite production rates included substrate limitation, product inhibition phenomena, and time lags due to microbial metabolism adaptation [31]. The surface response method was used to express the empiric dependence of some model parameters on operating conditions. The model assumed the bioreactor was perfectly stirred and there were no differences between individual cells. It was thus unsegregated and zero-dimensional, predicting average spatial concentrations [32].

Seven state variables were considered: six volume concentrations (biomass [X], trehalose [S], formic acid [F], acetic acid [A], lactic acid [L], and ethanol [E], Figure 1) and the culture medium volume (V). This latter variable varied continuously with the addition of base (NaOH) for pH control but also changed in a discrete way due to periodic sampling for biological and chemical analysis.

Mass balances for the considered metabolites resulted in the following set of differential equations:

$$\frac{d[X]}{dt} = \mu_X[X] - \frac{Q}{V}[X] \quad (1)$$

$$\frac{d[F]}{dt} = \pi_F[X] - \frac{Q}{V}[F] \quad (2)$$

$$\frac{d[A]}{dt} = \pi_A[X] - \frac{Q}{V}[A] \quad (3)$$

$$\frac{d[L]}{dt} = \pi_L[X] - \frac{Q}{V}[L] \quad (4)$$

$$\frac{d[E]}{dt} = \pi_E[X] - \frac{Q}{V}[E] \quad (5)$$

$$\frac{d[A_T]}{dt} = \frac{d[F]}{dt} + \frac{d[A]}{dt} + \frac{d[L]}{dt} \quad (6)$$

$$\frac{d[S]}{dt} = -\left(\frac{\mu_X}{Y_{X/S}} + \frac{\pi_F}{Y_{F/S}} + \frac{\pi_A}{Y_{A/S}} + \frac{\pi_L}{Y_{L/S}} + \frac{\pi_E}{Y_{E/S}}\right)[X] - \frac{Q}{V}[S] \quad (7)$$

$$\frac{dV}{dt} = Q \quad (8)$$

Here, μ_X is the specific growth rate (h^{-1}); π_F, π_A, π_L, and π_E are the specific production rates of four metabolites (h^{-1}); and $Y_{X/S}$, $Y_{F/S}$, $Y_{A/S}$, $Y_{L/S}$, and $Y_{E/S}$ are the yield of biomass and metabolites with respect to the substrate (mol.mol^{-1}). Q is the experimentally measured rate of NaOH solution (L.h^{-1}) added for pH control throughout fermentation.

In Equation (6), [A_T] is the total acid concentration, defined as the sum of formic, acetic, and lactic acid concentrations. These compounds are assumed to be mainly responsible for the pH change of the liquid medium.

Specific growth and production rates were defined using the Monod law to account for substrate limitation, modified with product inhibition and enzymatic activation factors [33–35]:

$$\mu_X = \mu_{max,X} I_X E_A \frac{[S]}{K_{SX} + [S]} \tag{9}$$

$$\pi_m = \pi_{max,m} I_m E_A \frac{[S]}{K_{Sm} + [S]} \quad m = F,\ A,\ L,\ E \tag{10}$$

In these equations, I_X and I_m are inhibition factors that depend on the inhibitor concentration. They vary between 1 and 0. Inhibition increases with the inhibitor concentration, and its effect on the specific rate is maximal when the corresponding factor is 0. In this model, progressive inhibition factors of the following form were used [36,37]:

$$I_X = \frac{1}{1 + \left(\frac{C_I}{K_{IX}}\right)^n} \tag{11}$$

$$I_m = \frac{1}{1 + \left(\frac{C_I}{K_{Im}}\right)^p} \quad m = F,\ A,\ L,\ E \tag{12}$$

K_{IX} and K_{Im} represent characteristic concentrations of the inhibitors (mol L^{-1}) such that the corresponding rates (μ_X and π_m) are reduced by a factor of 2 compared with the absence of inhibitor, n and p are shape factors, and C_I is the concentration of the inhibitor. Since all the metabolites were produced in similar proportions and no biochemical knowledge about their relative inhibiting nature was available, C_I was simply defined as the sum of the four metabolite concentrations:

$$C_I = [F] + [A] + [L] + [E] \tag{13}$$

To illustrate the role of the shape factor n, Figure 3a depicts the evolution of I_X with C_I for different n values and a lag-time of 5 h. A more or less sharp change in the inhibition factor occurs around the characteristic inhibitor concentration, $C_I = K_{IX}$. The significance of the shape factor p is similar.

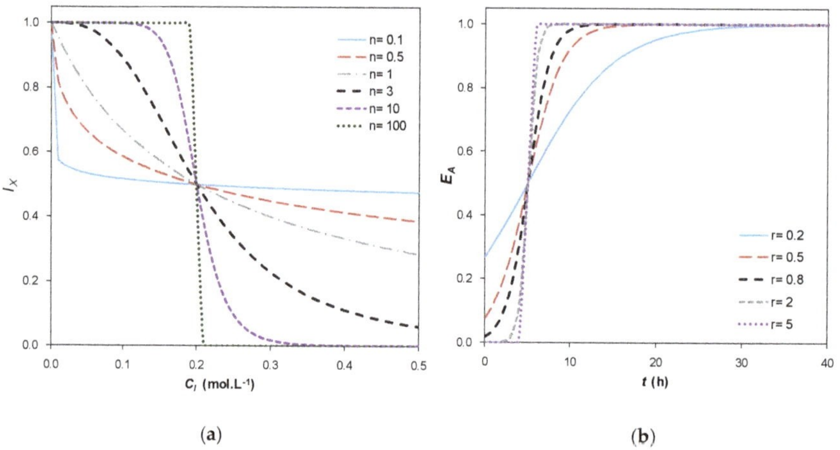

Figure 3. (**a**) Example of inhibition factor I_X as a function of C_I for different n values and K_{IX} = 0.2 mol L^{-1}. (**b**) Example of enzymatic activation factor E_A as a function of t for different r values, and t_{lag} = 5 h.

The enzymatic adaptation factor E_A is an empirical representation of the lag time, a period of adaptation to the culture environment where the microorganism produces new enzymatic machinery [38–40]. Based on the shape of experimental data, the following equation was proposed:

$$E_A = \frac{1}{1 + \exp\left(-r\left(t - t_{lag}\right)\right)} \tag{14}$$

where t_{lag} (h) is the lag time experimentally observed. Figure 3b shows that E_A is an increasing function of time, tending to 1 when $t \gg t_{lag}$. In analogy with n, r is a shape factor that describes the gradual transition from the lag phase to the active phase of growth. A higher value of r implies a steeper change of E_A around $t = t_{lag}$.

To illustrate the features of the proposed model, a representation of the dimensionless specific growth and production rates (μ/μ_{max} and π/π_{max}) over time is depicted in Figure 4. The dynamic behavior of both variables is similar given the similarity of Equations (1)–(5). The specific rates achieve a maximum value in the active growth phase, and they are zero when $t \ll t_{lag}$ and when the substrate is depleted. The shape of the curve is defined by three factors: in the increasing region (0 to 10 h in Figure 4), the dominant effect is enzyme activation E_A (Equation (14)); in the slowly decreasing region (10 to 20 h), the rate is controlled by inhibition (Equation (11) or (12)), whereas in the sharply decreasing region (20 to 22 h) it is controlled by substrate limitation, corresponding to the Monod-like factor in Equation (9) or (10).

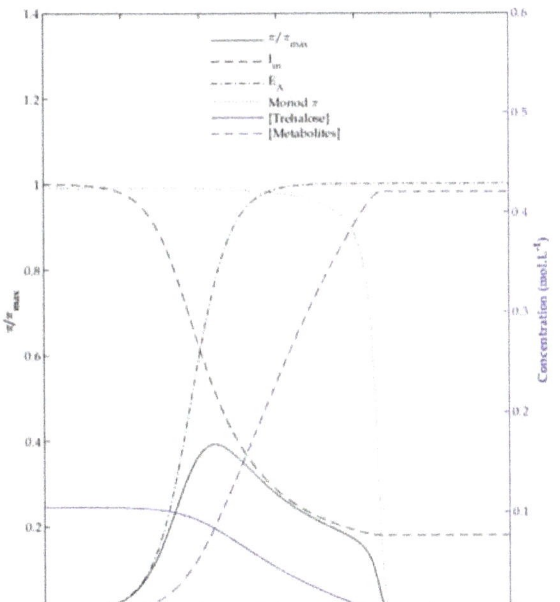

Figure 4. Typical evolution of the relative production rate over time.

3.2. Model Parameter Identification

The system of kinetic equations for a single fermentation experiment included 24 parameters: five yield coefficients, five inhibition parameters, five growth/production rates, five Monod-like saturation constants, three shape factors, and one lag time. Due to a limited number of experimental data and to facilitate the identification procedure, a single value was adopted for the inhibition parameter (K_{Im}) and the Monod saturation constant (K_{Sm}) of the four identified metabolites. Moreover, 10 parameters were fixed for all experiments: the shape factors, the yield coefficients, and the Monod saturation constants (K_{SX} and K_{Sm}). For each fermentation, lag time was determined by graphical readout. This simplification of fixing parameters independent of operating conditions is supported by two assumptions often used in the literature: (1) metabolite production yields are constant and therefore independent of culture conditions [41] and (2) the saturation constant of the Monod model depends only on the nature of the substrate [33,38], which was the same in all experiments of this study.

The remaining group of seven parameters ($\mu_{max,X}$, $\pi_{max,F}$, $\pi_{max,A}$, $\pi_{max,L}$, $\pi_{max,E}$, K_{IX}, K_{Im}) were identified for each fermentation of the experimental design by nonlinear regression. Here, the Levenberg–Marquardt algorithm [42] was used to minimize the sum of squares of the errors between experimental and predicted concentrations. However, since the ranges and the number of measurements were slightly different among the metabolites, the values compared in the least squares function were normalized by dividing by their maximum value and were weighted by the relevant number of experimental measurements.

The quality of the model representation was quantified with two error indicators, defined as follows:

Root mean square error:

$$\mathrm{RMSE} = \left[\frac{1}{N}\sum_{i=1}^{N}\left(C_{model,i} - C_{exp,i}\right)^2\right]^{1/2} \tag{15}$$

Relative mean error (as a percentage):

$$\mathrm{RME} = \frac{1}{N}\sum_{i=1}^{N}\frac{\left|C_{model,i} - C_{exp,i}\right|}{C_{exp,max} - C_{exp,min}}\cdot 100\% \tag{16}$$

where N is the number of available measurements, C_{model} and C_{exp} are respectively the values of the concentration variables calculated with the model and measured experimentally.

3.3. Response Surface Model for Parameter Dependence on Fermentation Conditions

Nonlinear regression was performed to model the relationship between the seven parameters of the dynamic model specific to each experiment and the fermentation operating conditions—namely, temperature (T) and pH. The regression model had a similar form for all parameters, the logarithm of the parameter being expressed as a second-order polynomial with interaction:

$$\log_{10}\mathrm{Par}_i = \beta_{0i} + \beta_{1i}T + \beta_{2i}\mathrm{pH} + \beta_{3i}T^2 + \beta_{4i}\mathrm{pH}^2 + \beta_{5i}T\mathrm{pH} \tag{17}$$

The regression coefficients (β) for all parameters depending on operating conditions ($\mu_{max,X}$, $\pi_{max,F}$, $\pi_{max,A}$, $\pi_{max,L}$, $\pi_{max,E}$, K_{IX}, K_{Im}) were simultaneously computed by least-squares optimization based on all available concentration measurements. In this way, the accuracy and standard errors of the coefficients were statistically acceptable, due to a large number of degrees of freedom: several hundreds of concentration data were used to estimate 42 coefficients. Initial guesses for these coefficients were obtained using Equation (17), and parameter values were determined separately for each experiment.

In this procedure, two sets of data from the experimental design were defined as indicated in Figure 2: eight calibration experiments, located in extreme positions of the experimental domain, used simultaneously for coefficients (β) estimation, and eight validation experiments, only used a posteriori to verify the accuracy of the complete dynamic model.

4. Results and Discussion

4.1. Model Parameter Identification

The values of the parameters that are independent of operating conditions, summarized in Table 1, were determined from the experimental data of experiment F10. This run was placed in a central position in the composite experimental design (T = 30 °C, pH = 8) (Figure 2). Monod saturation constants are usually difficult to determine from batch experiments because the number of measurements is typically very low in the substrate limitation zone. Saturation constants were thus fixed to a common value with a typical order of magnitude [43]. As for yields, they were found to differ from the theoretical ones defined through standard stoichiometric reactions of anabolism and catabolism. These differences can be due to other reactions involving the carbon substrate, whose products were not analytically measured and were not considered in the model.

Table 1. Model parameters independent of operating conditions, determined from the experimental data of experiment F10 (T = 30 °C, pH = 8) with t_{lag} = 10 h.

Parameter	Constant Value
$Y_{X/S}$ ($mol_C.mol^{-1}$)	6.9
$Y_{F/S}$ ($mol.mol^{-1}$)	5.6
$Y_{A/S}$ ($mol.mol^{-1}$)	3.8
$Y_{L/S}$ ($mol.mol^{-1}$)	7.0
$Y_{E/S}$ ($mol.mol^{-1}$)	4.7
K_{SX} (mol L^{-1})	0.001
K_{Sm} (mol L^{-1})	0.001
n	3
p	1
r (h^{-1})	0.8

After fixing the parameters in Table 1 for the whole set of experiments, the group of seven adjustable parameters of the model ($\mu_{max,X}$, $\pi_{max,F}$, $\pi_{max,A}$, $\pi_{max,L}$, $\pi_{max,E}$, K_{IX}, K_{Im}) were identified for each run by nonlinear regression.

The parameters obtained by this procedure are summarized in Table 2. Standard errors were computed from the variance–covariance matrix of the nonlinear optimization algorithm. These errors represented between 5% and 13% of the value of the identified parameters, a reasonable uncertainty level for a biological model.

Table 2. Model parameters determined for each experiment by nonlinear regression.

	Fermentation	μ_{maxX} (h^{-1})	π_{maxF} (h^{-1})	π_{maxA} (h^{-1})	π_{maxL} (h^{-1})	π_{maxE} (h^{-1})	KI_X ($Mol.L^{-1}$)	KI_m ($Mol.L^{-1}$)
F01	Value	0.224	0.152	0.064	0.215	0.078	0.117	0.069
	Standard error	0.003	0.003	0.003	0.005	0.001	0.004	0.003
F02	Value	0.096	0.122	0.064	0.053	0.064	0.143	0.142
	Standard error	0.021	0.012	0.007	0.006	0.006	0.083	0.036
F03	Value	0.164	0.137	0.064	0.131	0.072	0.099	0.102
	Standard error	0.012	0.009	0.004	0.009	0.005	0.006	0.013
F04	Value	0.078	0.063	0.031	0.072	0.034	0.144	0.092
	Standard error	0.012	0.004	0.002	0.004	0.002	0.047	0.016
F05	Value	0.130	0.130	0.060	0.196	0.070	0.160	0.100
	Standard error	0.005	0.002	0.002	0.004	0.001	0.004	0.004
F06	Value	0.094	0.129	0.063	0.159	0.063	0.163	0.092
	Standard error	0.001	0.002	0.001	0.003	0.001	0.003	0.002
F07	Value	0.074	0.028	0.010	0.089	0.022	0.122	0.139
	Standard error	0.002	0.001	0.001	0.004	0.002	0.005	0.004
F08	Value	0.060	0.055	0.024	0.126	0.048	0.149	0.060
	Standard error	0.002	0.002	0.000	0.003	0.002	0.001	0.002
F09	Value	0.071	0.097	0.048	0.114	0.071	0.108	0.044
	Standard error	0.009	0.007	0.004	0.010	0.006	0.021	0.007

Table 2. *Cont.*

	Fermentation	μ_{maxX} (h^{-1})	π_{maxF} (h^{-1})	π_{maxA} (h^{-1})	π_{maxL} (h^{-1})	π_{maxE} (h^{-1})	KI_X (Mol.L^{-1})	KI_m (Mol.L^{-1})
F10	Value	0.220	0.230	0.100	0.300	0.120	0.193	0.091
	Standard error	0.051	0.013	0.007	0.022	0.007	0.044	0.015
F11	Value	0.121	0.127	0.060	0.116	0.066	0.179	0.133
	Standard error	0.004	0.007	0.004	0.005	0.003	0.006	0.003
F12	Value	0.132	0.155	0.082	0.076	0.086	0.043	0.059
	Standard error	0.010	0.026	0.003	0.002	0.006	0.002	0.007
F13	Value	0.147	0.162	0.077	0.219	0.092	0.164	0.106
	Standard error	0.006	0.006	0.002	0.008	0.004	0.009	0.003
F14	Value	0.047	0.045	0.020	0.112	0.032	0.140	0.130
	Standard error	0.005	0.004	0.001	0.006	0.003	0.016	0.007
F15	Value	0.160	0.180	0.080	0.330	0.110	0.260	0.170
	Standard error	0.056	0.013	0.008	0.030	0.010	0.022	0.010
F16	Value	0.110	0.110	0.050	0.200	0.070	0.320	0.280
	Standard error	0.003	0.004	0.002	0.007	0.003	0.051	0.104

For the whole set of experiments, the prediction errors are reported in Appendix A Table A1. Except for some runs for variables S, F, and A, all RME were lower than 15%. Additionally, the average RMSE and RME values for each concentration were of the same order of magnitude as the experimental variability, here defined as the biological repeatability for run F01, for which three independent replicates were performed. These results validate the formulation and accuracy of the proposed model under the operating conditions included in the experimental design.

In the specific case of reference run F10, a comparison between the model simulation (using the corresponding parameters from Table 2) and experimental data is illustrated in Figure 5.

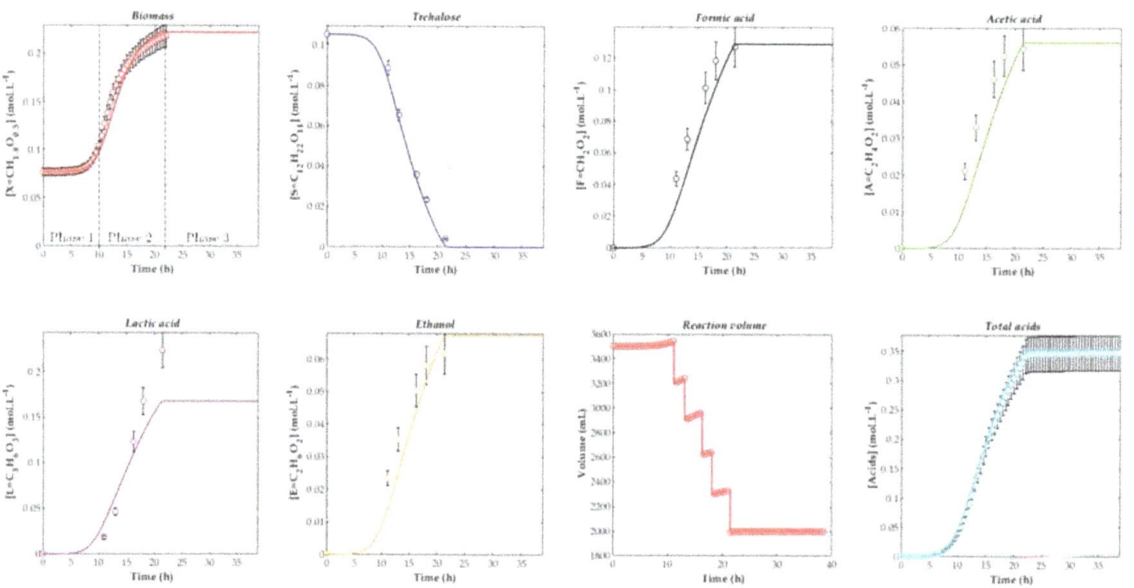

Figure 5. Evolution of concentrations over time for experiment F10 (T = 30 °C, pH = 8). Comparison between model (continuous line, using parameters from Tables 1 and 2) and experimental data (symbols). The error bars for data represent the biological standard deviation, calculated from three independent repetitions of the run F01.

Three growth phases are apparent in Figure 5: a lag phase (phase 1, between 0 and 10 h); a phase of active growth, substrate consumption, and metabolite production (phase 2, between 10 h and 21 h); and a final phase where concentrations do not change over time, owing to the depletion of the carbon source or growth inhibition by metabolites (phase 3, after 21 h). Regarding culture volume evolution, as already mentioned, the discrete variations at regular intervals were due to sampling for analysis of the culture medium and the gradual increase was due to NaOH addition for pH control. One can also observe that the four metabolites were produced simultaneously, with no gap for the growth dynamics. The metabolites were thus primary end products generated during a single trophophase [44]. This justifies the choice of a global inhibitor concentration (Equation (13)), which included four correlated concentrations.

In consideration of the visual fit from Figure 5, the model representation is reasonably satisfactory. The most pronounced discrepancy between the model and experimental data appears for lactic acid, for which the model predicted a lower concentration before substrate depletion. This is related to a slightly underestimated yield factor $Y_{L/S}$.

4.2. *Response Surface Model for Parameter Dependence on Fermentation Conditions*

Model parameters were expressed as a function of temperature and pH, according to the surface model (Equation (17)). The values of the β regression coefficients were adjusted globally using the whole set of calibration data.

The resulting response surfaces for the seven model parameters are plotted in Figure 6. For the five kinetic parameters, (i.e., the maximum specific growth and production rates), the response surfaces have the same convex shape, with a well-defined maximum value at intermediate T and pH conditions. These maxima likely indicate the optimal temperatures and pH for cellular growth, as well as the enzymatic activity catalyzing each of the reactions, leading to the production of the different metabolites (Figure 1).

Concerning the inhibition concentrations, the response surface for K_{Im} has a concave shape with a local minimum, whereas that of K_{IX} resembles a saddle surface. For this latter case, the surface shape indicates that for every pH there is a T where K_{IX} is minimal, and for every T there is a pH where K_{IX} is maximal. Both K_{Im} and K_{IX} represent the combined effect of several inhibiting metabolites (Equations (11)–(13)) with potentially different inhibition mechanisms.

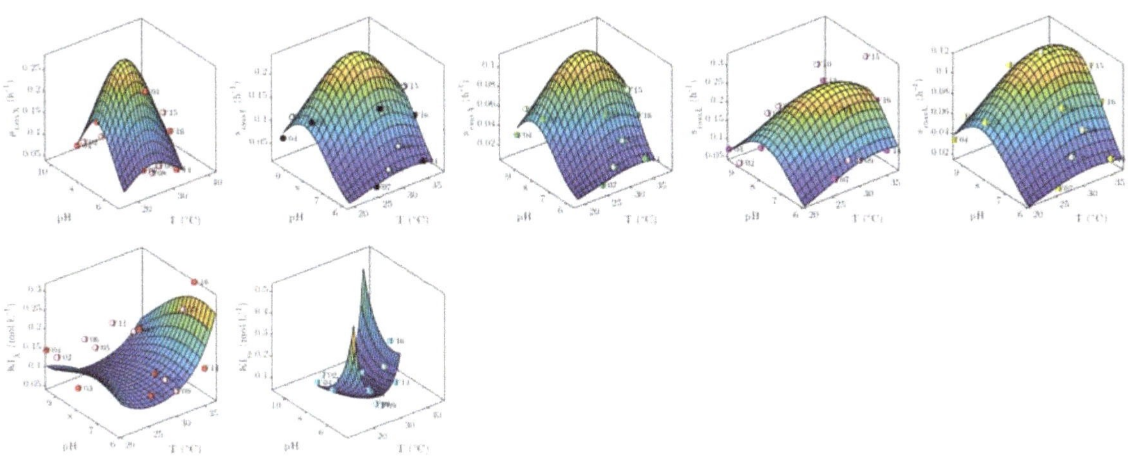

Figure 6. Response surfaces for model parameters, calculated with globally adjusted β coefficients in Equation (17).

For completeness, the final values of the regression coefficients of Equation (17) for the seven adjustable parameters of the dynamic model are reported in Appendix A Table A2. All coefficients in Equation (17) for each model parameter were significantly different from zero at a 0.05 level. A comparison between the parameter values determined for each experiment (Section 4.1) and the parameter values computed with Equation (17) (from globally adjusted β coefficients) is depicted in Appendix A Figure A1. The goodness of the fit was assessed through the coefficient of determination, R^2. This coefficient is higher than 0.89 for six out of seven model parameters, which is a high threshold for biological data. In the case of K_{Im}, only 66% of the variance of this parameter was explained by variables T and pH. The remaining 34% could be associated with inherent experimental variability and factors not included in the model, for instance transient variability of the inhibition and kinetics parameters and actual dependence of the fixed parameters (Table 1) with T and pH [45]. From a more general point of view, differences from experimental data could be due to features that were not represented by the mathematical model, such as population segregation, internal pH variability, and concentration gradients in the culture medium [46,47].

4.3. Model Validation

The ability of the dynamic model including the parameters calculated from operating conditions (Equation (17)) to predict data of independent experiments was assessed with a set of validation experiments.

A comparison between the average RMSE values obtained in Section 4.1 (determined for each experiment) and Section 4.2 (calculated from operating conditions) for calibration and validation sets is depicted in Figure 7. In most cases, RMSE values were higher than the corresponding experimental variabilities, indicating that more complex models could capture additional phenomena not included in the present model, such as dependence of yields, saturation constants, or lag time (Table 1) on operating conditions. As one might expect, RMSE was generally lower for the calibration experiments than for the validation experiments, not used for parameter determination. However, the relative difference remained small (less than 30%), indicating a satisfactory ability of the developed model to predict time evolution of the considered biomass, substrate, and metabolites under new conditions within the explored experimental range.

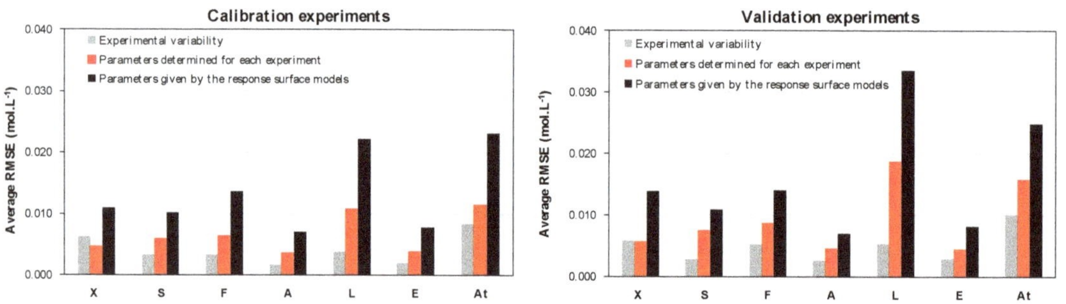

Figure 7. Comparison between experimental variability and average RMSE values for concentrations computed using parameters determined for each experiment (Table 2) and the response surface models (Table A2 and Equation (17)).

It also appears in Figure 7 that average RMSE values with parameters given by the response surface model (Table A3) are about 50% higher than with parameters determined separately for each experiment (Table A1), for both calibration and validation sets. This result could be expected since in the global calibration step, data from eight independent experiments were combined as a whole for the least squares estimation, with a detrimental effect on the individual representation of each experiment. However, results with the parameters calculated from operating conditions are the most useful in engineering purposes since they enable a quick prediction of growth and metabolites production dynamics, based on the selected combination of temperature and pH.

In light of this quantitative analysis, the prediction accuracy of the empirical dynamic model coupled to the regression model may be considered satisfactory within the operating domain covered in this study.

4.4. Model-Based Optimization of Fermentation Operating Conditions for Industrial Use

Optimal conditions for growth and metabolite production of *C. maltaromaticum* calculated using the developed model are summarized in Table 3. Two optimization criteria were considered: final concentrations and final productivities calculated for a 99.9% substrate consumption.

For a detailed representation of the evolution of final concentrations and productivities for biomass and metabolites with temperature and pH, the reader is referred to Appendix A Figures A2 and A3. As a general trend, the highest productivities were obtained around 35 °C and pH 7.5, although the exact optimal conditions depended on the considered metabolite (Table 3). No general trend was readily apparent for the maximization of the final concentrations.

Table 3. Optimal conditions for growth and production of metabolites according to the developed model. **In bold**: targeted metabolite for each set of operating conditions. Final concentrations and productivities calculated with initial conditions: $[X]_0 = 80$ mmol L^{-1}, $[S]_0 = 100$ mmol L^{-1}, $[F, A, L, E]_0 = 0$.

A. Target		T (°C)	pH	Final Concentrations (mmol L^{-1})						Final Productivities (mmol $L^{-1}.h^{-1}$)					
				X	F	A	L	E	A_T	X	F	A	L	E	A_T
Biomass	$B_{conc.}\uparrow$	20.0	7.8	**227**	129	56	177	73	363	6.56	3.73	1.61	5.11	2.09	10.46
Formic acid	$F_{conc.}\uparrow$	28.0	9.5	133	**176**	89	123	94	387	1.68	2.20	1.12	1.54	1.18	4.86
Acetic acid	$A_{conc.}\uparrow$	28.0	9.5	133	176	**89**	123	94	387	1.68	2.20	1.12	1.54	1.18	4.86
Lactic acid	$L_{conc.}\uparrow$	37.0	6.0	180	95	42	**296**	72	433	1.68	2.20	1.12	1.54	1.18	4.86
Ethanol	$E_{conc.}\uparrow$	37.0	9.5	118	147	77	128	**139**	352	2.35	2.90	1.54	2.54	2.76	6.98
Ethanol	$E_{conc.}\downarrow$	27.0	7.6	217	143	62	166	**68**	372	6.93	4.57	1.98	5.31	2.16	11.86
Total acids	$A_{Tconc.}\uparrow$	37.0	6.0	180	95	42	296	72	**433**	3.33	1.76	0.77	5.48	1.33	8.01
Total acids	$A_{Tconc.}\downarrow$	37.0	9.5	118	147	77	128	139	**352**	2.35	2.90	1.54	2.54	2.76	6.98
Biomass	$B_{prod.}\uparrow$	33.5	7.5	199	139	64	178	73	382	**7.49**	5.25	2.41	6.71	2.77	14.38
Formic acid	$F_{prod.}\uparrow$	34.5	8.0	178	148	71	163	82	381	6.77	**5.61**	2.70	6.18	3.12	14.48
Acetic acid	$A_{prod.}\uparrow$	35.0	8.1	172	148	72	161	85	381	6.49	5.58	**2.71**	6.06	3.21	14.35
Lactic acid	$L_{prod.}\uparrow$	37.0	7.1	188	123	58	215	78	395	6.60	4.31	2.04	**7.54**	2.73	13.89
Ethanol	$E_{prod.}\uparrow$	37.0	8.3	158	144	73	162	97	378	5.57	5.08	2.56	5.70	**3.43**	13.34
Ethanol	$E_{prod.}\downarrow$	28.0	6.0	175	108	41	278	78	427	2.14	1.32	0.50	3.38	**0.95**	5.20
Total acids	$A_{Tprod.}\uparrow$	35.5	7.7	184	139	67	177	81	384	7.16	5.41	2.59	6.90	3.13	**14.90**
Total acids	$A_{Tprod.}\downarrow$	25.0	9.5	146	173	88	126	87	386	1.71	2.03	1.03	1.48	1.02	**4.53**

↑ maximization, ↓ minimization, conc.: final concentration, prod.: batch-average productivity.

These data can be useful in optimizing industrial processes involving the growth of *C. maltaromaticum* cells in a trehalose-based substrate. A first application consists of producing *C. maltaromaticum* concentrates, regardless of metabolite production. In this case two conditions of cultivation appear advisable: 20 °C and pH 7.8 to maximize concentration (227 $mmol_C \cdot L^{-1}$) or 33.5 °C and pH 7.5 in order to maximize productivity (7.49 $mmol_C \cdot L^{-1} \cdot h^{-1}$) and thus the biomass production per unit of time, at the expense of a 12% reduction of the final biomass concentration (199 $mmol_C \cdot L^{-1}$).

A second application deals with the development and parametrization of time–temperature integrators (TTI), labels in which a pH decline, associated with acids synthesis, entails an irreversible color change from green to red. Modulating the acidifying activity of *C. maltaromaticum* thus allows a reliable shelf-life estimation of different food products. Long shelf-lives can be tracked using TTI composed of concentrates exhibiting low acidifying activities (minimal production of total acids), while short shelf-lives can be tracked using concentrates exhibiting high acidifying activities. In the scenario of maximizing acidifying activity, the production of total acids must be favored, and thus fermentation should be carried out under two possible conditions: 37.0 °C and pH 6.0 to maximize their final concentration (433 mmol·L^{-1}) or 35.5 °C and pH 7.7 to maximize their productivity (14.90 mmol·L^{-1}·h^{-1}). Under these conditions, the biomass production decreases respectively by 20% and 4% with respect to its optimal values. If the objective is, on the contrary, to minimize acidifying activity, two conditions can be envisaged to favor the lowest production of total acids: 37.0 °C and pH 9.5 for a final concentration of 352 mmol L^{-1} or 25.0 °C and pH 9.5 for a final productivity of 4.53 mmol·L^{-1}·h^{-1}. Under these conditions, the mean biomass production would decrease respectively by 48% and 77% with respect to the maximal values.

Data from Table 3 show that the conditions to minimize the total acids concentration (37.0 °C and pH 9.5) coincide with those to maximize the ethanol concentration (the non-acidifying metabolite, 139 mmol L^{-1}) and lead to a lactic acid concentration close to its minimal value (128 mmol L^{-1} versus the minimum around 120 mmol L^{-1}). Conversely, when the production of total acids is maximized, the lactic acid concentration is also maximal (296 mmol L^{-1}) and that of ethanol is close to its minimum (72 mmol L^{-1} versus 68 mmol L^{-1}).

Furthermore, it should be noted that the condition 27 °C and pH 7.6 leads both to a good biomass productivity (6.93 mmol·L^{-1}·h^{-1} versus the maximum 7.49 mmol·L^{-1}·h^{-1}) and a low total acids concentration (372 mmol L^{-1} versus the minimum 352 mmol L^{-1}). Cultivation under this condition turns out be advantageous to ally a high biomass production and a relatively low total acidification.

5. Conclusions

The dynamic model developed in this study is able to predict with satisfactory accuracy the growth of *C. maltaromaticum* CNCM I-3298 (average error of 10%) as well as the conversion of trehalose into four primary metabolites (average error of 14%) under a wide range of conditions of temperature and pH. The interpolation capability of the model was verified with a set of eight independent validation experiments, for which the average relative error was 13%.

This model constitutes a useful tool for optimizing *C. maltaromaticum* cultures. Based on two easily controllable parameters, pH and temperature, it could be implemented in industrial applications of food technology to define optimal growth and metabolite production conditions with various objectives, such as the maximization of biomass for production of bacterium concentrates or the maximization or minimization of the acidifying activity. A typical operating condition for this bacterium could be, for instance, 30.0 °C and pH 7.0. If the goal is to produce bacterium concentrates, to maximize final biomass concentration, our results suggest that a quite different condition should be selected (20.0 °C and pH 7.8), while for maximum biomass productivity, 33.5 °C and pH 7.5 is most appropriate. Such results are quite difficult to anticipate from the qualitative knowledge of the bacterium alone, and a large number of time-consuming experiments would be required to locate these optimal conditions experimentally without constructing a dynamic model of the process.

The effort of developing the model is especially cost effective when a variety of scenarios are explored. If the goal is to develop time–temperature integrators (TTI) to track the cold-chain of food products, a set of labels with specific shelf-lives has to be designed for various target products. The range of desired shelf-lives can be as large as 1 to 30 days, requiring very different TTI designs. In a traditional approach, for each desired shelf-life duration, a range of factors such as the initial bacterium concentration and the buffer properties of the medium have to be explored in a series of relatively time-consuming experiments. In such an environment, temperature varies in an arbitrary but known way, and pH depends on the produced acids. The presented dynamic model can be extended to predict the moment when a specific amount of acids is produced, corresponding to the pH-induced color change of the TTI label and hence to the desired shelf-life. Model-based design of the TTI labels is expected to be faster and more accurate than a trial and error procedure.

On a more fundamental level, further work is required to incorporate the effect of other culture parameters, such as aeration, nutrient concentrations, or the use of a different carbon source, which may modify growth kinetics and metabolite production. Additionally, it would be relevant to deepen the understanding of inhibition mechanisms of the metabolites to give more biological significance to the associated parameters in the model.

Author Contributions: Conceptualization, C.P., A.G., F.F. and I.-C.T.; methodology, C.P., A.G., F.F., and I.-C.T.; software, C.P. and I.-C.T.; validation, C.P., A.G., S.P., F.F. and I.-C.T.; formal analysis, C.P. and I.-C.T.; investigation, C.P. and A.G.; resources, F.F. and I.-C.T.; data curation, C.P. and A.G.; writing—original draft preparation, C.P. and A.G.; writing—review and editing, C.P., A.G., S.P., F.F. and I.-C.T.; visualization, C.P., A.G. and I.-C.T.; supervision, S.P., F.F. and I.-C.T.; project administration, S.P., F.F. and I.-C.T.; funding acquisition, S.P. and F.F. All authors have read and agreed to the published version of the manuscript.

Funding: This study received funding from the European Union's Horizon 2020 Marie Skłodowska-Curie research and innovation program under grant agreement no. 777657.

Data Availability Statement: The data presented in this study are available on request from the corresponding author. The data are not publicly available due to an ongoing research project.

Acknowledgments: The authors thank Marie-Nöelle Lecercq-Perlat for HPLC analysis.

Conflicts of Interest: The authors declare no conflict of interest. The funders had no role in the design of the study; in the collection, analyses, or interpretation of data; in the writing of the manuscript; or in the decision to publish the results.

Nomenclature

A	Acetic acid
E	Ethanol
F	Formic acid
L	Lactic acid
S	Carbon substrate (trehalose)
X	Biomass
E_A	Enzymatic activation factor
C ($mol \cdot L^{-1}$)	Concentration (in the calculation of errors and the definition of the inhibition factors)
[i] ($mol \cdot L^{-1}$)	Concentration of species i (substrate, metabolite, biomass) in the culture medium (in the system of differential equations)
I_m	Production inhibition factor of metabolite m
I_X	Biomass growth inhibition factor
K_{Im} ($mol \cdot L^{-1}$)	Concentration for 50% production rate inhibition of metabolite m
K_{IX} ($mol_C \cdot L^{-1}$)	Concentration for 50% growth rate inhibition of biomass

K_{Sm} (mol·L^{-1})	Concentration of production rate saturation of metabolite m
K_{SX} (mol$_C$·L^{-1})	Concentration of biomass growth rate saturation
mol$_C$	Carbon-mol of biomass
n	Shape factor of the growth inhibition function
p	Shape factor of the production inhibition function
pH	Potential of hydrogen
Q (L·h^{-1})	Rate of base addition for pH control
R	Shape factor of the enzymatic activation function
T (K)	Temperature
TTI	Time–temperature indicator
RMSE	Root-mean square error
RME	Relative mean error
SE	Standard error
t (h)	Time
t_{lag} (h)	Lag time
V (L)	Culture medium volume
$Y_{i/S}$ (mol·mol^{-1})	Yield of product i on substrate S
$Y_{X/S}$ (mol$_C$·mol^{-1})	Biomass yield on substrate S
μ_X (h^{-1})	Specific growth rate
$\mu_{max,X}$ (h^{-1})	Maximal specific growth rate
π_m (h^{-1})	Specific production rate of metabolite m
$\pi_{max,m}$ (h^{-1})	Maximum specific production rate of metabolite m

Appendix A.

Appendix A.1. Model Parameter Identification

Table A1. Residual modelling error with model parameters determined for each experiment and summarized in Table 2.

	RMSE (Mol L^{-1})							RME (%)						
	X	S	F	A	L	E	A Total	X	S	F	A	L	E	A Total
F01	0.008	0.006	0.010	0.007	0.022	0.007	0.033	4	4	7	13	6	10	7
F02	0.005	0.007	0.006	0.003	0.004	0.003	0.005	4	16	8	8	8	7	3
F03	0.005	0.006	0.003	0.001	0.003	0.001	0.006	4	9	3	2	3	2	1
F04	0.006	0.006	0.003	0.001	0.005	0.002	0.005	7	13	5	5	6	6	3
F05	0.008	0.005	0.014	0.007	0.027	0.007	0.034	4	4	9	11	10	9	8
F06	0.006	0.004	0.012	0.005	0.024	0.006	0.012	5	2	7	7	8	8	3
F07	0.006	0.007	0.011	0.006	0.009	0.006	0.007	4	5	14	33	4	11	2
F08	0.006	0.007	0.011	0.005	0.030	0.004	0.003	4	5	17	23	8	7	1
F09	0.004	0.005	0.002	0.001	0.002	0.002	0.003	7	17	6	7	5	6	2
F10	0.005	0.009	0.006	0.004	0.028	0.004	0.026	3	7	4	7	9	5	6
F11	0.003	0.018	0.009	0.005	0.011	0.006	0.032	3	27	9	10	9	11	6
F12	0.005	0.013	0.002	0.001	0.003	0.001	0.004	13	61	3	4	8	2	2
F13	0.002	0.006	0.012	0.005	0.024	0.005	0.007	2	4	7	6	6	5	2
F14	0.003	0.003	0.009	0.005	0.012	0.006	0.015	4	3	16	20	5	12	4
F15	0.009	0.005	0.009	0.006	0.023	0.005	0.011	5	4	7	13	8	8	2
F16	0.003	0.001	0.001	0.001	0.008	0.001	0.014	2	1	2	4	6	2	3
Mean	0.005	0.007	0.008	0.004	0.015	0.004	0.014	5	11	8	11	7	7	4

Response surface model for parameter dependence on fermentation conditions.

Table A2. Response surface coefficients fitted to experimental data by multiple regression.

Variable	Coefficient	μ_{maxX} (h^{-1})	π_{maxF} (h^{-1})	π_{maxA} (h^{-1})	π_{maxL} (h^{-1})	π_{maxE} (h^{-1})	KI_X (Mol L^{-1})	KI_m (Mol L^{-1})
Constant	Value (β_0)	-1.38×10^{1}	-1.64×10^{1}	-1.74×10^{1}	-1.21×10^{1}	-1.12×10^{1}	-8.61×10^{-1}	1.19×10^{1}
	Standard error	0.05×10^{1}	0.07×10^{1}	0.09×10^{1}	0.09×10^{1}	0.07×10^{1}	0.05×10^{-1}	0.03×10^{1}
T	Value (β_1)	2.63×10^{-1}	2.66×10^{-1}	2.71×10^{-1}	1.82×10^{-1}	1.27×10^{-1}	-1.87×10^{-1}	-5.80×10^{-1}
	Standard error	0.13×10^{-1}	0.26×10^{-1}	0.28×10^{-1}	0.23×10^{-1}	0.14×10^{-1}	0.05×10^{-1}	0.17×10^{-1}
pH	Value (β_2)	2.42×10^{0}	2.89×10^{0}	3.01×10^{0}	2.27×10^{0}	2.02×10^{0}	6.65×10^{-1}	-1.27×10^{0}
	Standard error	0.11×10^{0}	0.14×10^{0}	0.19×10^{0}	0.16×10^{0}	0.13×10^{0}	0.45×10^{-1}	0.16×10^{0}
T^2	Value (β_3)	-4.26×10^{-3}	-3.45×10^{-3}	-3.26×10^{-3}	-2.05×10^{-3}	-1.92×10^{-3}	3.97×10^{-3}	7.95×10^{-3}
	Standard error	0.13×10^{-3}	0.32×10^{-3}	0.33×10^{-3}	0.81×10^{-3}	0.22×10^{-3}	0.09×10^{-3}	0.32×10^{-3}
pH^2	Value (β_4)	-1.46×10^{-1}	-1.63×10^{-1}	-1.66×10^{-1}	-1.34×10^{-1}	-1.22×10^{-1}	-4.03×10^{-2}	4.35×10^{-2}
	Standard error	0.06×10^{-1}	0.10×10^{-1}	0.12×10^{-1}	0.10×10^{-1}	0.09×10^{-1}	0.31×10^{-2}	0.55×10^{-2}
T·pH	Value (β_5)	-3.53×10^{-3}	-7.58×10^{-3}	-9.17×10^{-3}	-7.22×10^{-3}	-4.68×10^{-3}	-3.54×10^{-3}	1.86×10^{-2}
	Standard error	0.11×10^{-3}	0.97×10^{-3}	1.17×10^{-3}	0.91×10^{-3}	0.05×10^{-3}	0.06×10^{-3}	0.01×10^{-2}

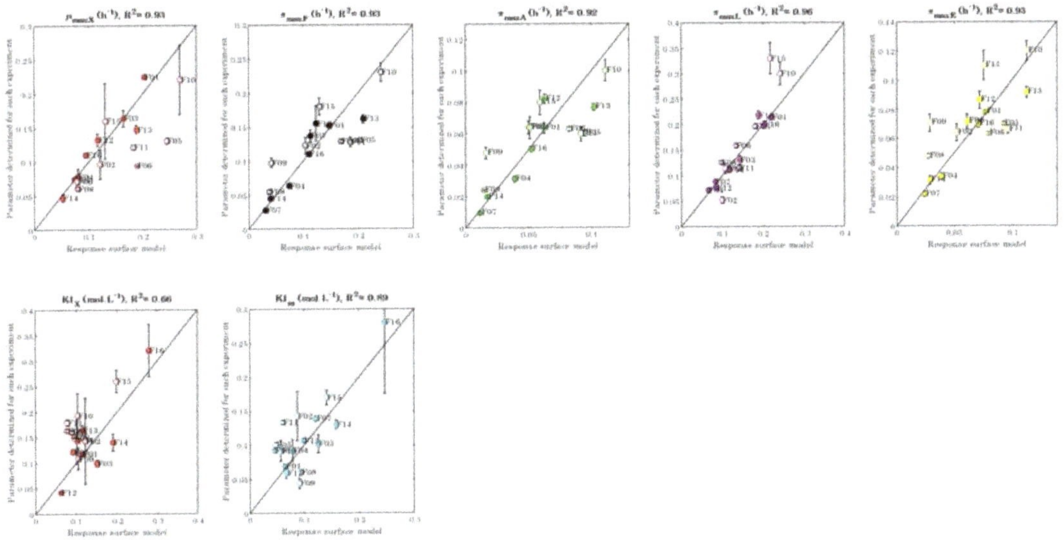

Figure A1. Comparison between model parameters determined for each experiment (Table 2) and parameters computed with the response surface models (Equation (17) using coefficients in Table A2).

Appendix A.2. Model Validation

Error indicators RMSE and RME for calibration and validation sets are summarized in Appendix A Table A3. The RME values varied from 1% to 53%, with an average of 15% for calibration set and 13% for validation set. Likewise, the average RME was lower than 15% for most of experiments, except for runs F12 (17%), F15 (18%), F09 (20%), and F16 (34%). Considering both calibration and validation sets, the average RME was 10% for biomass and 14% for substrate and metabolites.

Table A3. Quality of fit of the model with parameters computed with the response surface models.

Fermentation		RMSE (Mol L^{-1})							RME (%)						
		X	S	F	A	L	E	At	X	S	F	A	L	E	At
Calibration	**F01**	0.010	0.010	0.021	0.010	0.041	0.014	0.033	5	6	14	20	10	18	7
	F03	0.021	0.009	0.008	0.005	0.008	0.003	0.012	15	14	7	9	7	5	4
	F04	0.005	0.010	0.009	0.006	0.007	0.005	0.015	5	26	18	23	9	17	9
	F07	0.012	0.009	0.010	0.006	0.015	0.006	0.009	9	8	13	31	5	10	3
	F12	0.003	0.011	0.008	0.005	0.003	0.004	0.009	8	53	13	16	8	11	6
	F13	0.010	0.007	0.006	0.003	0.036	0.003	0.011	7	5	4	4	8	4	2
	F14	0.015	0.002	0.013	0.007	0.004	0.009	0.012	20	2	23	28	2	16	3
	F16	0.011	0.023	0.033	0.014	0.064	0.018	0.085	9	33	50	45	48	38	18
	Mean	0.011	0.010	0.014	0.007	0.022	0.008	0.023	10	18	18	22	12	15	7
Validation	**F02**	0.005	0.003	0.017	0.009	0.008	0.009	0.007	5	8	20	22	15	21	4
	F05	0.011	0.008	0.007	0.006	0.045	0.005	0.016	7	7	4	8	13	8	4
	F06	0.008	0.009	0.013	0.005	0.044	0.007	0.032	6	5	7	6	11	9	6
	F08	0.009	0.010	0.012	0.006	0.026	0.008	0.003	7	8	19	25	8	14	1
	F09	0.011	0.006	0.005	0.004	0.018	0.004	0.009	19	22	15	24	35	15	7
	F10	0.021	0.014	0.025	0.011	0.063	0.014	0.044	11	10	17	18	20	20	10
	F11	0.018	0.013	0.007	0.003	0.017	0.004	0.050	16	20	7	6	12	8	9
	F15	0.028	0.023	0.026	0.010	0.047	0.014	0.036	15	20	22	21	17	21	7
	Mean	0.014	0.011	0.014	0.007	0.033	0.008	0.025	11	13	14	16	16	15	6

Model-based optimization of fermentation conditions.

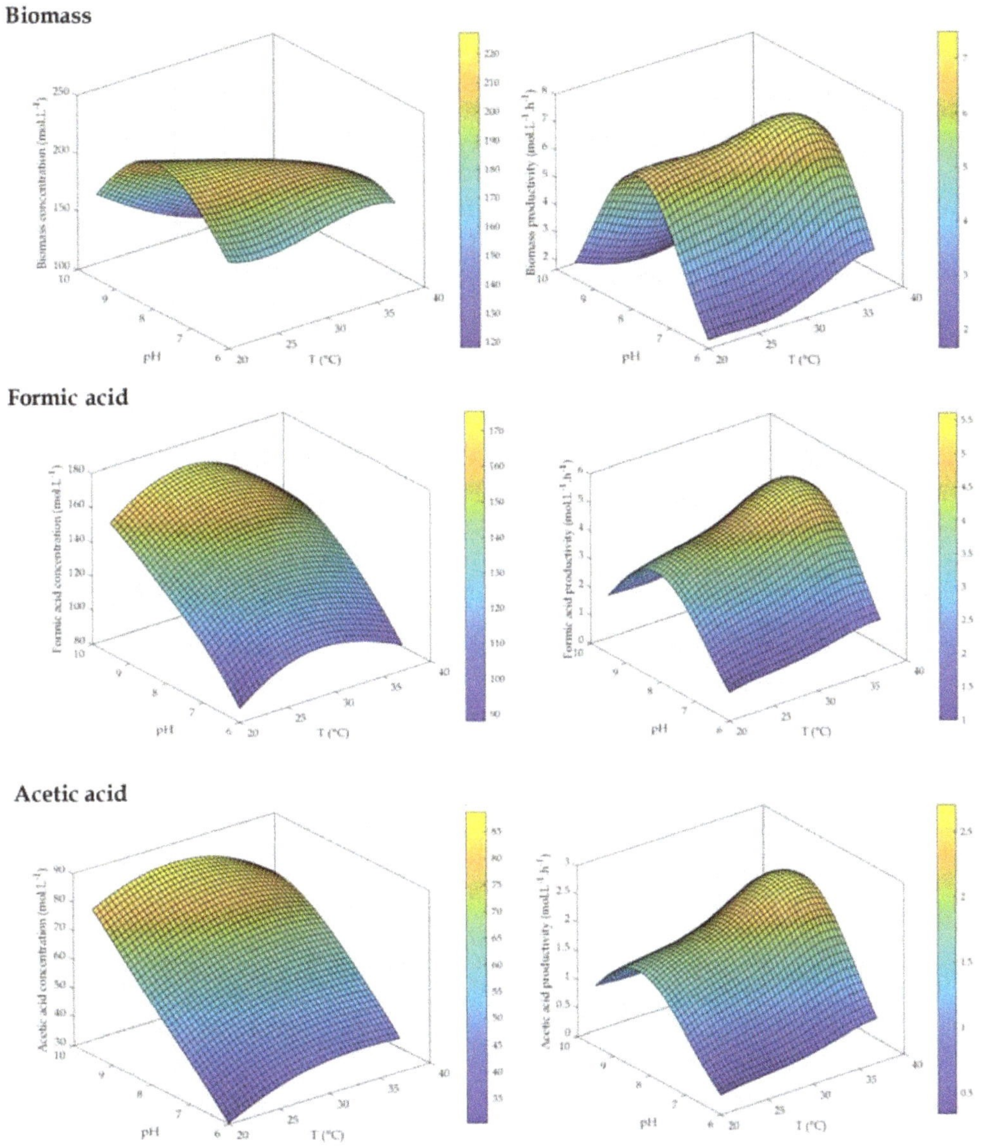

Figure A2. Evolution of final concentrations (left) and batch-average productivities (right) with temperature and pH for biomass, formic acid, and acetic acid.

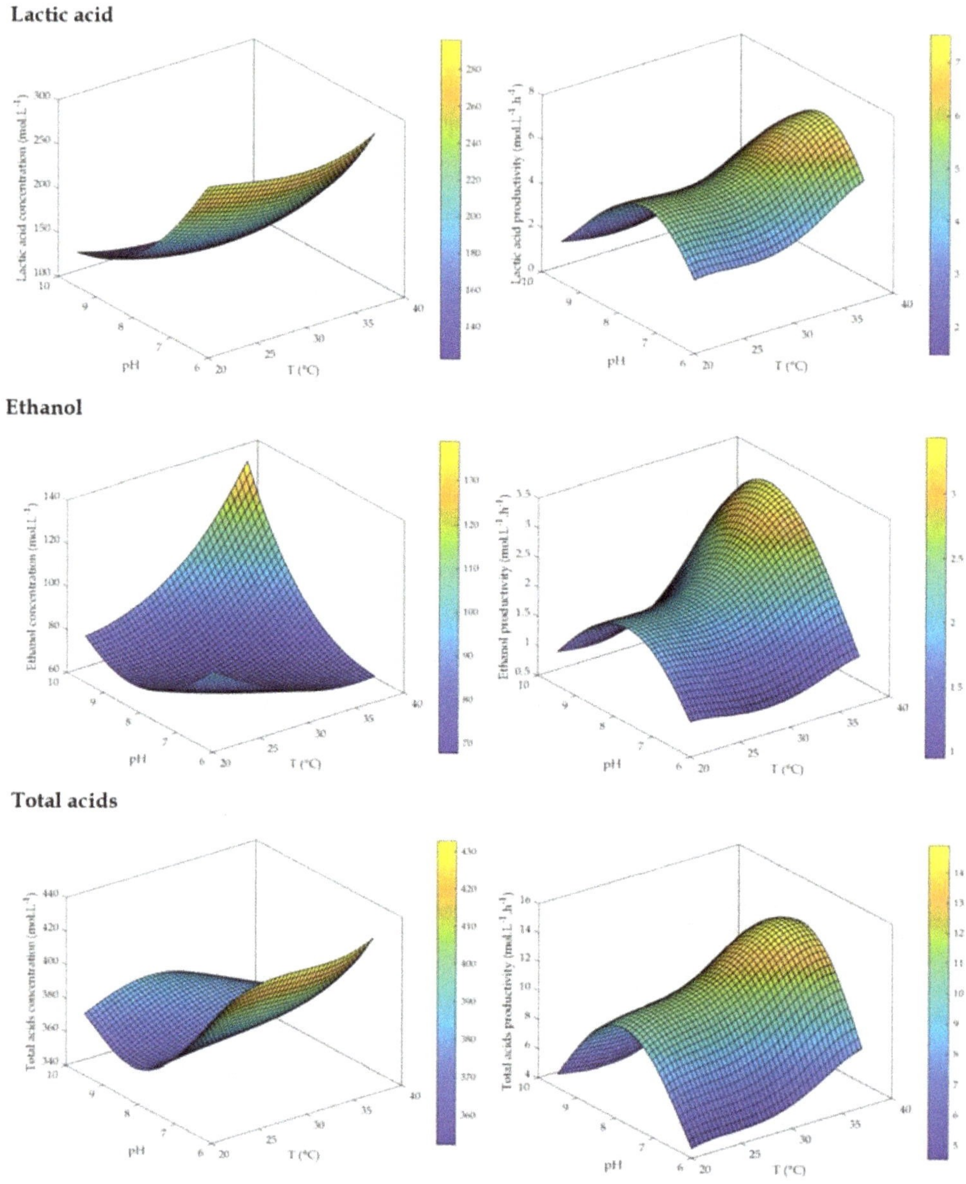

Figure A3. Evolution of final concentrations (left) and batch-average productivities (right) with temperature and pH for lactic acid, ethanol, and total acids.

References

1. Edima, H.; Cailliez-Grimal, C.; Revol-Junelles, A.-M.; Tonti, L.; Linder, M.; Millière, J.B. A selective enumeration medium for *Carnobacterium maltaromaticum*. *J. Microbiol. Methods* **2007**, *68*, 516–521. [CrossRef]
2. Afzal, M.; Jacquet, T.; Delaunay, S.; Borges, F.; Millière, J.B.; Revol-Junelles, A.M.; Cailliez-Grimal, C. *Carnobacterium maltaromaticum*: Identification, isolation tools, ecology and technological aspects in dairy products. *Food Microbiol.* **2010**, *27*, 573–579. [CrossRef]

3. Cailliez-Grimal, C.; Edima, H.; Revol-Junelles, A.-M.; Millière, J.-B. *Carnobacterium maltaromaticum*: The Only *Carnobacterium* Species in French Ripened Soft Cheeses as Revealed by Polymerase Chain Reaction Detection. *J. Dairy Sci.* **2007**, *90*, 1133–1138. [CrossRef]
4. Edima, H.; Cailliez-Grimal, C.; Revol-Junelles, A.M.; Rondags, E.; Millière, J.B. Impact of pH and Temperature on the Acidifying Activity of *Carnobacterium Maltaromaticum*. *J. Dairy Sci.* **2008**, *91*, 3806–3813. [CrossRef]
5. Taoukis, P.; Labuza, T. Applicability of Time-Temperature Indicators as Shelf Life Monitors of Food Products. *J. Food Sci.* **1989**, *54*, 783–788. [CrossRef]
6. Ellouze, M.; Pichaud, M.; Bonaiti, C.; Coroller, L.; Couvert, O.; Thuault, D.; Vaillant, R. Modelling pH evolution and lactic acid production in the growth medium of a lactic acid bacterium: Application to set a biological TTI. *Int. J. Food Microbiol.* **2008**, *128*, 101–107. [CrossRef]
7. Girardeau, A.; Puentes, C.; Keravec, S.; Peteuil, P.; Trelea, I.C.; Fonseca, F. Influence of culture conditions on the technological properties of *Carnobacterium maltaromaticum* CNCM I-3298 starters. *J. Appl. Microbiol.* **2019**, *126*, 1468–1479. [PubMed]
8. Borch, E.; Molin, G. The aerobic growth and product formation of Lactobacillus, Leuconostoc, Brochothrix, and Carnobacterium in batch cultures. *Appl. Microbiol. Biotechnol.* **1989**, *30*, 81–88. [CrossRef]
9. dos Reis, F.B.; de Souza, V.M.; Thomaz, M.R.S.; Fernandes, L.P.; de Oliveira, W.P.; de Martinis, E.C.P. Use of *Carnobacterium maltaromaticum* cultures and hydroalcoholic extract of *Lippia sidoides* Cham. Against *Listeria monocytogenes* in fish model systems. *Int. J. Food Microbiol.* **2011**, *146*, 228–234. [CrossRef] [PubMed]
10. Afzal, M.I.; Delaunay, S.; Paris, C.; Borges, F.; Revol-Junelles, A.M.; Cailliez-Grimal, C. Identification of metabolic pathways involved in the biosynthesis of flavor compound 3-methylbutanal from leucine catabolism by *Carnobacterium maltaromaticum* LMA 28. *Int. J. Food Microbiol.* **2012**, *157*, 332–339. [CrossRef]
11. Afzal, M.I.; Boulahya, K.A.; Paris, C.; Delaunay, S.; Cailliez-Grimal, C. Effect of oxygen on the biosynthesis of flavor compound 3-methylbutanal from leucine catabolism during batch culture in *Carnobacterium maltaromaticum* LMA 28. *J. Dairy Sci.* **2013**, *96*, 352–359. [CrossRef]
12. Rahman, A.; Gleinser, M.; Lanhers, M.-C.; Riedel, C.U.; Foligne, B.; Hanse, M.; Yen, F.T.; Klouj, A.; Afzal, M.I.; Back, A.; et al. Adaptation of the lactic acid bacterium *Carnobacterium maltaromaticum* LMA 28 to the mammalian gastrointestinal tract: From survival in mice to interactions with human cells. *Int. Dairy J.* **2014**, *34*, 93–99.
13. Rahman, A.; El Kheir, S.M.; Back, A.; Mangavel, C.; Revoljunelles, A.-M.; Borges, F. Repeat-based bequence typing of *Carnobacterium maltaromaticum*. *Int. J. Food Microbiol.* **2016**, *226*, 1–4. [CrossRef] [PubMed]
14. Zhang, P.; Badoni, M.; Gänzle, M.; Yang, X. Growth of *Carnobacterium* spp. isolated from chilled vacuum-packaged meat under relevant acidic conditions. *Int. J. Food Microbiol.* **2018**, *286*, 120–127.
15. De Bruyn, I.N.; Louw, A.I.; Visser, L.; Holzapfel, W.H. *Lactobacillus divergens* is a homofermentative organism. *Syst. Appl. Microbiol.* **1987**, *9*, 173–175. [CrossRef]
16. De Bruyn, I.N.; Holzapfel, W.H.; Visser, L.; Louw, A.I. Glucose metabolism by *Lactobacillus divergens*. *J. Gen. Microbiol.* **1988**, *134*, 2103–2109. [CrossRef]
17. Leisner, J.J.; Laursen, B.G.; Prévost, H.; Drider, D.; Dalgaard, P. *Carnobacterium*: Positive and negative effects in the environment and in foods. *FEMS Microbiol. Rev.* **2007**, *31*, 592–613. [CrossRef]
18. Schillinger, U.; Holzapfel, W.H. The genus *Carnobacterium*. In *The Genera of Lactic Acid Bacteria. The Lactic Acid Bacteria*; Wood, B.J.B., Holzapfel, W.H., Eds.; Springer: Boston, MA, USA, 1995; Volume 2, pp. 307–326.
19. Gänzle, M.G. Lactic metabolism revisited: Metabolism of lactic acid bacteria in food fermentations and food spoilage. *Curr. Opin. Food Sci.* **2015**, *2*, 106–117. [CrossRef]
20. Ward, B. Bacterial energy metabolism. In *Molecular Medical Microbiology*, 2nd ed.; Tang, Y.W., Liu, D., Schwartzman, J., Sussman, M., Poxton, I., Eds.; Elsevier: Waltham, MA, USA, 2015; pp. 201–233.
21. Laursen, B.G.; Leisner, J.J.; Dalgaard, P. *Carnobacterium* Species: Effect of Metabolic Activity and Interaction with *Brochothrix thermosphacta* on Sensory Characteristics of Modified Atmosphere Packed Shrimp. *J. Agric. Food Chem.* **2006**, *54*, 3604–3611. [CrossRef]
22. Zhang, P.; Gänzle, M.; Yang, X. Complementary Antibacterial Effects of Bacteriocins and Organic Acids as Revealed by Comparative Analysis of *Carnobacterium* spp. from Meat. *Appl. Environ. Microbiol.* **2019**, *85*, e01227-19. [CrossRef] [PubMed]
23. Andersson, U.; Rådström, P. Beta-Glucose 1-phosphate-interconverting enzymes in maltose- and trehalose-fermenting lactic acid bacteria. *Environ. Microbiol.* **2002**, *4*, 81–88. [CrossRef]
24. Levisauskas, D.; Galvanauskas, V.; Henrich, S.; Wilhelm, K.; Volk, N.; Lübbert, A. Model-based optimization of viral capsid protein production in fed-batch culture of recombinant *Escherichia coli*. *Bioprocess. Biosyst. Eng.* **2003**, *25*, 255–262. [CrossRef]
25. Antelo, L.; Passot, S.; Fonseca, F.; Trelea, I.C.; Alonso, A.A. Towards optimal operation conditions of freeze-drying processes via a multi-level approach. *Dry. Technol.* **2012**, *30*, 1432–1448. [CrossRef]
26. Kiparissides, A.; Pistikopoulos, E.N.; Mantalaris, A. On the model-based optimization of secreting mammalian cell (GS-NS0) cultures. *Biotechnol. Bioeng.* **2015**, *112*, 536–548. [CrossRef] [PubMed]
27. Balsa-Canto, E.; Alonso, A.A.; Arias-Méndez, A.; Garcia, M.; López-Núñez, A.; Mosquera-Fernández, M.; Vázquez, C.; Vilas, C. Modeling and optimization techniques with applications in food processes, bio-processes and bio-systems. In *Numerical Simulation in Physics and Engineering*; Higueras, I., Roldán, T., Torrens, J., Eds.; Springer: Cham, Switzerland, 2016; Volume 9, pp. 187–216.

28. Sanchez-Castaneda, A.; Athès, V.; Moussa, M.; Lopez-Miranda, J.; Paez-Lerma, J.; Soto-Cruz, N.; Trelea, I.-C. Modeling of isoamyl acetate production by fermentation with *Pichia fermentans* in an aerated system coupled to in situ extraction. *Process. Biochem.* **2018**, *65*, 11–20. [CrossRef]
29. Sinner, P.; Kager, J.; Daume, S.; Herwig, C. Model-based analysis and optimisation of a continuous *Corynebacterium glutamicum* Bioprocess Utilizing Lignocellulosic Waste. *IFAC-Pap.* **2019**, *52*, 181–186. [CrossRef]
30. Battley, E.H. Growth-Reaction Equations for *Saccharomyces cerevisiae*. *Physiol. Plant.* **1960**, *13*, 192–203. [CrossRef]
31. Bastin, G.; Dochain, D. *On-Line Estimation and Adaptive Control of Bioreactors*; Elsevier: Amsterdam, The Netherlands, 1990; pp. 1–82.
32. Bailey, J.E. Mathematical modeling and analysis in biochemical engineering: Past accomplishments and future opportunities. *Biotechnol. Prog.* **1998**, *14*, 8–20. [CrossRef]
33. Monod, J. *Recherches sur la Croissance des Cultures Bactériennes*; Hermann & Cie: Paris, France, 1942.
34. Levenspiel, O. The monod equation: A revisit and a generalization to product inhibition situations. *Biotechnol. Bioeng.* **1980**, *22*, 1671–1687. [CrossRef]
35. Baranyi, J.; Roberts, T.A. A dynamic approach to predicting bacterial growth in food. *Int. J. Food Microbiol.* **1994**, *23*, 277–294. [CrossRef]
36. Aiba, S.; Shoda, M. Reassessment of the product inhibition in alcohol fermentation. *J. Ferment. Technol.* **1969**, *47*, 790–803.
37. Claret, C.; Bories, A.; Soucaille, P. Inhibitory effect of dihydroxyacetone on *Gluconobacter oxydans*: Kinetic aspects and expression by mathematical equations. *J. Ind. Microbiol.* **1993**, *11*, 105–112. [CrossRef]
38. Monod, J. The growth of bacterial cultures. *Annu. Rev. Microbiol.* **1949**, *3*, 371–394. [CrossRef]
39. Baranyi, J. Comparison of Stochastic and Deterministic Concepts of Bacterial Lag. *J. Theor. Biol.* **1998**, *192*, 403–408. [CrossRef] [PubMed]
40. Rolfe, M.; Rice, C.; Lucchini, S.; Pin, C.; Thompson, A.; Cameron, A.; Alston, M.; Michael, F.; Stringer, M.; Betts, R.; et al. Lag Phase Is a Distinct Growth Phase that Prepares Bacteria for Exponential Growth and Involves Transient Metal Accumulation. *J. Bacteriol.* **2012**, *194*, 686–701. [CrossRef]
41. Bouville, M. Fermentation kinetics including product and substrate inhibitions plus biomass death: A mathematical analysis. *Biotechnol. Lett.* **2007**, *29*, 737–741. [CrossRef]
42. Seber, G.; Wild, C. *Nonlinear Regression*, 2nd ed.; Wiley: Hoboken, NJ, USA, 2003.
43. Owens, J.D.; Legan, J.D. Determination of the Monod substrate saturation constant for microbial growth. *FEMS Microbiol. Rev.* **1987**, *3*, 419–432. [CrossRef]
44. Brock, T.D.; Madigan, M.T. *Biology of Microorganisms*, 6th ed.; Prentice Hall: London, UK, 1991; p. 353.
45. Rodríguez, J.; Lema, J.M.; Kleerebezem, R. Energy-based models for environmental biotechnology. *Trends Biotechnol.* **2008**, *26*, 366–374. [CrossRef]
46. Bellgardt, K.H. Bioprocess Models. In *Bioreaction Engineering: Modeling and Control*; Schügerl, K., Bellgardt, K.H., Eds.; Springer: Berlin/Heidelberg, Germany, 2000; pp. 44–105.
47. González-Cabaleiro, R.; Lema, J.M.; Rodríguez, J. Metabolic Energy-Based Modelling Explains Product Yielding in Anaerobic Mixed Culture Fermentations. *PLoS ONE* **2015**, *10*, e0126739. [CrossRef]

foods

MDPI

Article

A Population Balance Model to Describe the Evolution of Sublethal Injury

Simen Akkermans [1,2,3], Davy Verheyen [1,2,3], Cindy Smet [1,2,3] and Jan F. M. Van Impe [1,2,3,*]

1 BioTeC, Chemical and Biochemical Process Technology and Control, Department of Chemical Engineering, KU Leuven, 9000 Ghent, Belgium; simen.akkermans@kuleuven.be (S.A.); davy.verheyen@kuleuven.be (D.V.); cindy.smet@kuleuven.be (C.S.)
2 OPTEC, Optimization in Engineering Center-of-Excellence, KU Leuven, 3000 Leuven, Belgium
3 CPMF2, Flemish Cluster Predictive Microbiology in Foods-www.cpmf2.be, 9000 Ghent, Belgium
* Correspondence: jan.vanimpe@kuleuven.be

Abstract: The detection and quantification of sublethal injury (SI) of pathogenic microorganisms has become a common procedure when assessing the efficiency of microbial inactivation treatments. However, while a plethora of studies investigates SI in function of time, no suitable modelling procedure for SI data has been proposed thus far. In this study, a new SI model structure was developed that relies on existing microbial inactivation models. This model is based on the description of inactivation kinetics between the subpopulations of healthy, sublethally injured and dead cells. The model was validated by means of case studies on previously published results, modelled by different inactivation models, i.e., (i) log-linear inactivation; (ii) biphasic inactivation; and (iii) log-linear inactivation with tailing. Results were compared to those obtained by the traditional method that relies on calculating SI from independent inactivation models on non-selective and selective media. The log-linear inactivation case study demonstrated that the SI model is equivalent to the use of independent models when there can be no mistake in calculating SI. The biphasic inactivation case study illustrated how the SI model avoids unrealistic calculations of SI that would otherwise occur. The final case study on log-linear inactivation with tailing clarified that the SI model provides a more mechanistic description than the independent models, in this case allowing the reduction of the number of model parameters. As such, this paper provides a comprehensive overview of the potential and applications for the newly presented SI model.

Keywords: food safety; predictive microbiology; mathematical models; microbial inactivation; sublethal injury

Citation: Akkermans, S.; Verheyen, D.; Smet, C.; Van Impe, J.F.M. A Population Balance Model to Describe the Evolution of Sublethal Injury. *Foods* **2021**, *10*, 1674. https://doi.org/10.3390/foods10071674

Academic Editors: Carlos Vilas, Míriam R. García and Jose A. Egea

Received: 17 June 2021
Accepted: 14 July 2021
Published: 20 July 2021

1. Introduction

In predictive microbiology, the effects of intrinsic and/or extrinsic factors on the microbial behaviour (e.g., growth, inactivation) in foods are studied and quantified to predict the effect of varying environmental conditions on the microbial response. Accumulated knowledge on microbial behaviour is distilled into mathematical models, which can be integrated into user-friendly software tools. These tools can be used by food producers, governments and scientists to determine food safety and quality aspects [1,2].

Due to the health consequences related to the ingestion of pathogenic foodborne microorganisms, food industries design processing treatments to inactivate microorganisms that may be present in food products [3]. Exposing food products containing bacterial populations to an inactivation treatment (e.g., heating, irradiation, antimicrobial agents) leads to the formation of three bacterial subpopulations, i.e., (i) healthy cells that are uninjured; (ii) sublethally injured cells that can potentially still recover from their injuries; and (iii) dead cells that are completely inactivated/killed [4]. Sublethal injury (SI) is generally defined as "a consequence of exposure to a chemical or physical process that damages but does not kill a microorganism" [5–7]. The injury can involve a loss of the permeability barrier in the cell wall and/or membrane (i.e., structural damage), damage to functional cell components, such as ribosomes and structural DNA (i.e., metabolic damage), or a combination of both types [8]. Sublethally injured cells are sensitive to selective components to which uninjured cells show resistance, making them unable to grow on selective media commonly used for the detection of foodborne pathogens in the food industry [9]. Consequently, the challenges related to the occurrence of sublethally injured cells in foods are twofold. The first challenge relates to food diagnostics, as the number of microorganisms is underestimated when solely enumerated by plating on selective media. As a second challenge, sublethally injured microorganisms might recover from their damage if exposed to optimal conditions for a sufficient amount of time, for example during food storage [10–13].

The degree of SI of a microbial population is traditionally assessed by the difference in plate counts on non-selective and selective media [7,8,14]. Non-selective media allow the growth of the total population of all viable culturable cells (i.e., uninjured and sublethally injured), while selective media only allow the growth of uninjured cells [15]. Hence, SI can be quantified as a percentage of the entire population by means of formulas similar to Equation (1), introduced by the likes of Busch et al. and Dykes [16,17].

$$\mathrm{SI} = \left(1 - \frac{\text{Counts on selective medium}}{\text{Counts on nonselective medium}}\right) \cdot 100\% \tag{1}$$

Following the increased attention to SI in the last decades, the scientific literature experienced a rise in the number of predictive microbiology studies that include a representation of the SI evolution of microorganisms during inactivation. In most studies, SI is calculated at different time points, regardless of the inactivation behaviour of the cells [18–32]. To investigate the SI of cells during inactivation treatments in more depth, some researchers have calculated the SI evolution over time based on the modelled inactivation of the total culturable population and the healthy subpopulation. Apart from visually illustrating the SI evolution in a continuous manner, calculating the SI evolution as a function of the inactivation treatment time allows a comparison of total amounts of SI among different treatment conditions, e.g., using the time-averaged injured cells coefficient (TICC) [33].

SI cells should not be confused with so-called viable but non-culturable (VBNC) cells. VBNC cells are not regarded as dead cells because their cell membrane and genetic material is intact and they are metabolically active [34]. SI cells are not dead either. They have sustained injuries that cause them not to be viable on stress-containing selective media, but they are still viable on non-selective media. As such, the difference between VBNC and SI cells is that the former cannot be cultured on any media, while the latter can still be cultured on non-selective media [35]. The use of a combination of optimal rich media and selective media therefore allows differentiating between SI cells, on the one hand, and VBNC and dead cells on the other hand.

Thus far, different approaches for calculating the SI have been used with no consensus on a standard method having been reached. In the most common method, the SI evolution is calculated by using the non-log transformed model outputs corresponding to non-selective and selective media as input data for the SI equation [36–43]. The major disadvantage related to this methodology is the frequent occurrence of SI trends that are (in part) artefacts of the methodology, rather than accurate representations of physical phenomena, e.g., (i) extended periods with constant values of SI caused by consecutive differences in model fits on the non-selective and selective medium, e.g., during tailing; (ii) strange SI trends caused by differences in the model output behaviour on the non-selective and selective medium; and (iii) negative values of SI caused by intersecting model fits on the non-selective and selective media. Examples of these artefacts are illustrated in Figure 1. Given these artefacts, the methodology requires the assumption of zero SI when the model output on the selective medium is higher than on the non-selective medium. To avoid some of the aforementioned artefacts, Noriega et al. used a slightly different methodology; they directly used the log transformed model fit values on the non-selective and selective medium as input for the SI equation [12]. On the one hand, the adapted approach resulted in more smooth SI evolutions characterised by less abruptly changing SI evolutions. On the other hand, the subpopulation of sublethally injured cells was not represented as a percentage of the total cell population, essentially not corresponding to the SI equation. Moreover, SI evolutions not explained by physical phenomena still occurred due to the SI evolution solely being based on the difference between model fits on non-selective and selective media. Therefore, Verheyen et al. used the raw cell count data directly as an input for the SI equation and fitted a third degree polynomial to the resulting SI data points [44,45]. Log transformed cell count data were used because fitting a model to the absolute values of an exponential process would lead to a large variance in SI values over the course of inactivation treatments. While this methodology was able to describe the SI-behaviour of microbial inactivation fairly well, some artefacts of using the third-degree polynomial as an SI model were observed, e.g., a sudden increase in SI at the end of inactivation treatments. In addition, the subpopulation of sublethally injured cells was not represented as a percentage of the total cell population due to the use of log transformed cell counts, similar to what was the case for the methodology of Noriega et al. [12]. Moreover, this method is a black box approach to describing the evolution of SI, ignoring any mechanistic knowledge that is available on the process. As such, there is also no guarantee that the selected polynomial is appropriate to represent the true underlying phenomena.

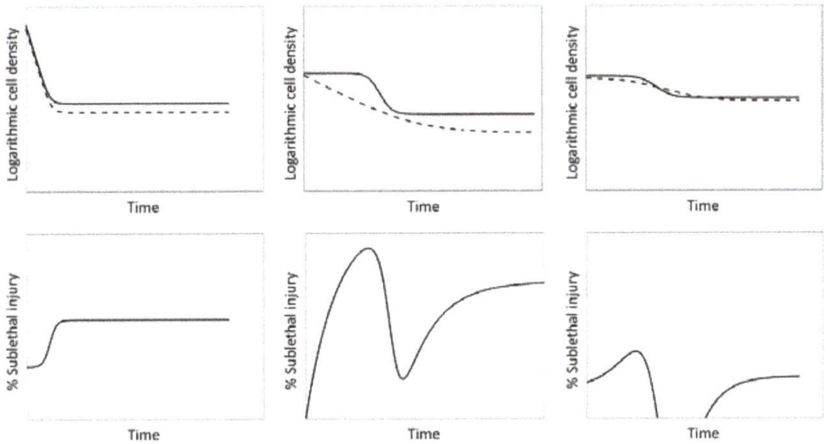

Figure 1. Examples of artefacts occurring due to calculating SI directly from non-log transformed model fit values. The modelled evolution of the population density on non-selective (—) and selective (- - -) medium is presented in the top row and the resulting SI in the bottom row. Both are presented as a function of unitless time. Left: long periods of constant SI caused by consecutive differences in model fits on the non-selective and selective medium. Middle: strange SI evolution caused by differences in model output shapes on non-selective and selective medium. Right: negative values in the SI evolution caused by intersecting model fits on the non-selective and selective media. For each case, top figures represent the model fits to non-selective and selective media, while bottom figures represent the SI calculated according to Equation (1).

The objective of this study was to develop a SI modelling method specifically tailored to use in combination with existing microbial inactivation models, to avoid disadvantages related to the approaches discussed above. In essence, all these disadvantages can be avoided if a more mechanistic modelling approach is used that can guarantee an appropriate relationship between the model fits to the non-selective and selective media. Therefore, a method was proposed to model the evolution of experimental data both on non-selective and selective media by a single model that describes the relationship between the corresponding populations of all culturable and healthy culturable cells. This methodology assumes that cells are first sublethally injured prior to inactivation; an assumption that can be incorporated directly into primary inactivation models. The newly developed modelling concept was validated by means of three parameter estimation case studies from literature, comparing integrated SI modelling results to the results obtained with the traditional methodology that relies on independent models, as used in the respective papers.

2. Materials and Methods

2.1. Datasets

Three datasets from previous research by the authors' research group were used to evaluate the implementation of the novel sublethal injury (SI) model (presented in Section 3.1). For each dataset, the SI model was compared with the classical method of using independent models. The three datasets were selected specifically because they display different inactivation kinetics and therefore require different inactivation models as well. Dataset 1 was published by Smet et al. and displays simple log-linear inactivation kinetics [42]. Specifically, this dataset was obtained by applying cold atmospheric plasma treatments to *Salmonella* typhimurium that was grown as planktonic cells in an environment at pH 5.5 with 60 g/L NaCl. The reported inactivation curves were the result of the storage of samples at 8 °C after they had been treated with this plasma in a liquid carrier. Samples were collected over a period of 770 h and there were 22 samples for both the non-selective medium (TSA; Tryptic Soy Agar; Oxoid, Basingstoke, UK) and selective medium (XLD; Xylose Lysine Deoxycholate agar; Merck & Co, Rahway, NJ, USA). Dataset 2 and Dataset 3 have been published by Noriega et al. in the context of studying the effect of cell immobilization on SI during mild heat treatments (54 °C) [12]. Dataset 2 was from the inactivation of low-density colonies of *Listeria innocua* and displays a biphasic log-linear inactivation. There were 104 datapoints on non-selective medium and 87 on selective medium. The non-selective medium was TSA supplemented with 6 g/L yeast extract (Merck, Darmstadt, Germany) and the selective medium was the same supplemented with an additional 65 g/L NaCl (VWR, Leuven, Belgium). It should be noted that *L. innocua* was used as an innocuous surrogate organism for the food pathogen *L. monocytogenes*. Although this is a commonly used surrogate, it is not as tolerant to stress factors as *L. monocytogenes*. Dataset 3 was obtained from the inactivation of high-density colonies of *Salmonella* Typhimurium and displays log-linear inactivation with a tail. This dataset contained 130 datapoints on non-selective medium (TSA) and 89 datapoints on selective medium (TSA supplemented with 50 g/L NaCl). In this work, all data were converted to be expressed in the natural logarithm instead of the common logarithm of the cell density.

2.2. Parameter Estimation and Uncertainty

Model parameters were estimated using the function *lsqnonlin* of MATLAB R2019b (The MathWorks). The parameter estimation was based on the minimization of the sum of squared errors SSE:

$$\mathrm{SSE} = \sum_{i=1}^{\nu_m} \left(n_{m,i}(t_i) - n_{p,i}(t_i, p)\right)^2 \quad (2)$$

with ν_m the number of measurements in the dataset, $n_{m,i}$ the logarithm of the measured cell concentration, $n_{p,i}$ the logarithm of the predicted cell concentration, t_i the time point corresponding to a specific measurement and p the vector of model parameters. The different models will be described in the results and discussion section when used. The 95% confidence bounds of the estimated parameters were estimated using Equation (3) [46]:

$$\left[\, p_i \pm t_{0.975,\ \nu_m - \nu_p} \cdot \sigma_{p_i} \right] \quad (3)$$

with t as the inverse Student's t-distribution, ν_p the number of model parameters, and σ_{p_i} the standard deviation of a model parameter p_i. This standard deviation was calculated through Equations (4)–(7) [47]:

$$\sigma_{p_i} = \sqrt{V(i,i)} \quad (4)$$

$$V = F^{-1} \tag{5}$$

$$F = \frac{1}{MSE} J' \cdot J \tag{6}$$

$$MSE = \frac{SSE}{\nu_m - \nu_p} \tag{7}$$

where V is the variance-covariance matrix of the model parameters, F is the Fisher Information Matrix, J is the Jacobian matrix and MSE is the mean sum of squared errors.

2.3. *Sublethal Injury*

Sublethal injury (SI) is defined here as the percentage of cells that cannot grow on selective media, but are able to grow on non-selective media:

$$SI = \frac{N_{non-selective} - N_{selective}}{N_{non-selective}} \cdot 100\% \tag{8}$$

with $N_{non-selective}$ and $N_{selective}$ the quantity of cells as detected on non-selective and selective media. It should be noted that the actual quantity of cells is used and not a logarithmic transformation of this quantity. As such, SI can be expressed as a percentage.

3. Results and Discussion

The current study presents a new modelling concept to describe sublethal injury (SI) as a function of time through sets of differential equations. A variety of inactivation models is described in literature. Together, these models have the ability to produce a wide variety of inactivation kinetics and they are selected based on the requirements of the specific scenario that is under study. Consequently, the goal of this work is not to present a specific model, but rather to present a method that allows the user to develop a new SI model based on any existing inactivation model that is available as a (set of) differential equation(s). The global concept of this modelling method is unveiled in Section 3.1. Then, this concept is applied to three different scenarios based on different datasets that required different inactivation models (Sections 3.2–3.4). For each scenario, the SI model is composed, the model parameters to be estimated are identified, the resulting model is tested on an available dataset and the results are compared with what would have been obtained if the data would be treated independently as done in the conventional method.

3.1. *The Sublethal Injury Model*

The main lack of the conventional approach to describe the evolution of SI is that the evolution of all culturable cells and of the healthy culturable cells as a function of time are treated as independent phenomena. Therefore, the main requirement for the new modelling concept is to link these inactivation kinetics. Composing this model concept is done by defining different subpopulations within the model. The distinct subpopulations are the healthy culturable cells N_H, the sublethally injured (SI) culturable cells N_{SI} and the dead (and non-culturable) cells N_D. Additionally, the healthy cells and injured cells can be grouped together in the class of all culturable cells N_C. This classification is presented in the Venn diagram of Figure 2. Based on these subpopulations, the inactivation of the microbial population can be described as the transition of cells from one subpopulation to another. A healthy cell can become a SI cell due to the inactivation method or survival conditions that it undergoes. It can be assumed that, under the studied conditions that cause inactivation of the cells, injured cells will be unable to become healthy cells again. Therefore, the rate of cells becoming injured is equal to the change in the quantity of healthy cells. This is described mathematically by a differential equation for dN_H/dt. Experimentally, the evolution of the healthy population of cells is monitored through viable plate counting on selective media. Consequently, if injured cells are harmed further, their injuries become irreparable and they are thus converted to dead cells. If one assumes that the healthy cells are always first injured before they can be killed, the rate of injured cells dying off is equal to the change in the population of all culturable cells. This is mathematically expressed by a differential equation for dN_L/dt and the process can be monitored experimentally by using viable plate counts on non-selective medium. This assumption has been confirmed, e.g., in the research of Perni et al. on the inactivation of *E. coli* by pulsed electric fields [48]. If cells first have to become injured before being killed, this also means that the size of dN_C/dt is proportional to the quantity of sublethally injured cells, not to the total quantity of culturable cells. The change of the population of injured cells is then a combination of the transition of healthy cells to injured cells and of injured cells to dead cells ($dN_{SI}/dt = -dN_H/dt + dN_C/dt$). Within this modelling framework, two differential equations are needed to describe the evolution of the three subpopulations: (i) the change in the number of healthy cells dN_H/dt and (ii) the change in the number of culturable cells dN_C/dt. These two populations are monitored experimentally by plating respectively on selective and non-selective medium. As such, it is possible to identify the model equations on the data that are available out of the same types of experiments that are commonly being done to study SI. The data that are already being collected is sufficient, but the model itself needs to link the subpopulations as described above. By linking these subpopulations in a semi-empirical population balance model, there is a guarantee that the model will deliver reasonable predictions of the phenomena where, e.g., the number of healthy cells can never be higher than the total number of culturable cells. To make the proposed modelling concept more tangible, the next three subsections will present how to compose the model equations for the SI model based on three different inactivation models that are taken from literature.

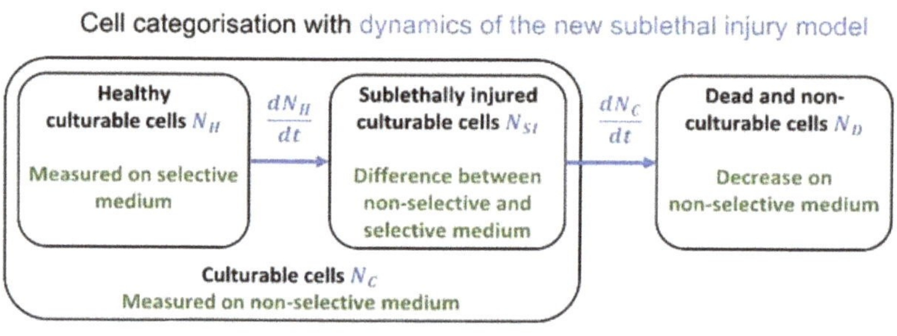

Figure 2. Overview of the microbial subpopulations that are considered in the sublethal injury model. All cells can be divided in culturable and dead cells. The culturable cells can be subdivided in healthy and injured cells. The mathematical notations and conversions are marked in blue and the experimental quantifications are written in green.

3.2. Case Study 1: Log-Linear Inactivation

The first SI model that is composed here is based on simple log-linear inactivation for both the healthy and sublethally injured populations. As such, the model of Bigelow and Esty can be used to describe the evolution of the healthy population as a function of time [49]:

$$\frac{dN_H(t)}{dt} = -k_{H,max} \cdot N_H(t) \tag{9}$$

with $k_{H,max}$ the maximum specific inactivation rate of the healthy population, which is expressed in an inverse time unit (e.g., 1/min, 1/s). For computational purposes, it is better to work with the logarithmic transformation of the population size:

$$\frac{dn_H(t)}{dt} = -k_{H,max} \tag{10}$$

with n_H the logarithm of the size of the population of healthy cells. The next equation that can be composed is that for the total population of culturable cells:

$$\frac{dN_C(t)}{dt} = -k_{SI,max} \cdot N_{SI}(t) \tag{11}$$

with $k_{SI,max}$ the maximum specific inactivation rate of the population of injured cells. This equation demonstrates that the decrease of the total quantity of culturable cells is only a function of the quantity of injured cells. This is a consequence of the assumption that cells first have to become injured before they can be killed. The quantity of injured cells can be calculated as the difference between the total population of culturable cells and the population of healthy cells ($N_{SI}(t) = N_C(t) - N_H(t)$). As such, on a logarithmic scale, Equation (11) becomes:

$$\frac{dn_C(t)}{dt} = -k_{SI,max} \cdot (1 - \exp(n_H(t) - n_C(t))) \tag{12}$$

with n_C the logarithm of the size of the population of culturable cells. The differential equation for the evolution of the population of injured cells itself is a combination of the increase due to the healthy cells becoming injured (Equation (9)) and a decrease due to the injured cells dying off (Equation (11)). The Equation (14) for the evolution of the injured population can therefore be composed as:

$$\frac{dN_{SI}(t)}{dt} = -\frac{dN_H(t)}{dt} + \frac{dN_C(t)}{dt} \tag{13}$$

$$\frac{dN_{SI}(t)}{dt} = k_{H1,max} \cdot N_H(t) - k_{SI1,max} \cdot N_{SI}(t) \tag{14}$$

Converting Equation (14) to the logarithmic scale results in the following equation:

$$\frac{dn_{SI}(t)}{dt} = k_{H1,max} \cdot \frac{\exp(n_H(t))}{\exp(n_{SI}(t))} - k_{SI,max} \tag{15}$$

Given (i) that the quantity of injured cells at any point in time can be calculated from the difference between the amount of culturable and healthy cells and (ii) that the experimental data of viable plate counts on non-selective and selective media provides direct information about the quantity of culturable and healthy cells, it is not necessary to include Equation (15) on the evolution of the injured population when solving the differential equations that describe this SI model. Instead, the quantity of injured cells can be calculated from the results of the culturable and healthy populations. As such, the SI model for log-linear inactivation can be described by Equations (10) and (12). To solve this system of differential equations, the initial conditions are required. These initial conditions are:

$$n_H(t=0) = n_{H,0} \tag{16}$$

$$n_C(t=0) = n_{C,0} \tag{17}$$

with $n_{H,0}$ and $n_{C,0}$ the logarithm of the population sizes at the initial time point. These initial conditions cannot take any value as the initial quantity of all culturable cells should of course be larger than its subpopulation of initial healthy cells. This constraint can be added into the model by converting the first initial condition to:

$$n_H(t=0) = f_{H,0} \cdot n_{C,0} \quad \text{with} \quad f_{H,0} \in [0;1] \tag{18}$$

with $f_{H,0}$ the fraction of the total culturable population of cells that are healthy cells. The full model can thus be summarised as:

$$\frac{dn_H(t)}{dt} = -k_{H,max} \quad \text{with} \quad n_H(t=0) = f_{H,0} \cdot n_{C,0} \quad \text{and} \quad f_{H,0} \in [0;1] \tag{19}$$

$$\frac{dn_C(t)}{dt} = -k_{SI,max} \cdot (1 - \exp(n_H(t) - n_C(t))) \quad \text{with} \quad n_C(t=0) = n_{L,0} \tag{20}$$

with the following model parameters to be estimated: $f_{H,0}$, $n_{C,0}$, $k_{H,max}$ and $k_{SI,max}$. In contrast, when considering the healthy and culturable populations as independent populations, they would be described by the following independent models:

$$\frac{dn_H(t)}{dt} = -k_{H,max} \quad \text{with} \quad n_H(t=0) = n_{H,0} \tag{21}$$

$$\frac{dn_C(t)}{dt} = -k_{C,max} \quad \text{with} \quad n_C(t=0) = n_{C,0} \tag{22}$$

where $k_{C,max}$ is the maximum specific inactivation rate of the population of culturable cells. The model parameters in these independent models are $n_{H,0}$, $n_{C,0}$, $k_{H,max}$ and $k_{C,max}$.

The SI model of Equations (19) and (20) is compared with the independent model of Equations (21) and (22) based on Dataset 1 (see Section 2.1). For this comparison, the model parameters of each set of equations were estimated and the results are presented in Table 1. The parameters $n_{C,0}$ and $k_{H,max}$, which appear in both models, have identical values and confidence bounds. The parameters $k_{SI,max}$ and $k_{C,max}$ of respectively the SI model and the independent models also have identical values and confidence bounds. These parameters are essentially each other's equivalent for the dataset that was used in this example. In contrast, the initial number of cells in the independent models is described by the parameter $n_{H,0}$, whereas for the SI model the parameter $f_{H,0}$ determines the fraction of the $n_{C,0}$ that are healthy cells. Relative to their sizes, the uncertainty on the parameter $f_{H,0}$ appears to be high compared to that of $n_{H,0}$. However, the parameter $f_{H,0}$ expresses the fraction of the population size on a linear scale whereas the parameter $n_{H,0}$ follows a logarithmic scale. On the other hand, there is indeed additional uncertainty in the calculation of the initial population of healthy cells in the SI model because its value depends on both $f_{H,0}$ and $n_{C,0}$, which are both marked by uncertainty. The model predictions of the SI model and the independent models are visually indistinguishable and are therefore both represented by the graphs in Figure 3a. The sublethal injury (SI) is calculated as a function of time based on the following equation:

$$SI(t) = \frac{\exp(n_C(t)) - \exp(n_H(t))}{\exp(n_C(t))} \tag{23}$$

Table 1. Parameter estimation results from fitting the independent models (Equations (21) and (22)) and sublethal injury model (Equations (19) and (20)) based on log-linear inactivation on Dataset 1.

Model Parameter	Parameter Estimate	95% Confidence Bounds
	Independent models	
$n_{H,0}$	11.0	[10.6; 11.4]
$n_{C,0}$	11.9	[11.4; 12.3]
$k_{H,max}$	7.15×10^{-3}	$[6.19 \times 10^{-3}; 8.11 \times 10^{-3}]$
$k_{C,max}$	5.95×10^{-3}	$[4.74 \times 10^{-3}; 7.15 \times 10^{-3}]$
MSE	0.344	
	Sublethal injury model	
$f_{H,0}$	0.402	[0.143; 0.662]
$n_{C,0}$	11.9	[11.4; 12.3]
$k_{H,max}$	7.15×10^{-3}	$[6.19 \times 10^{-3}; 8.11 \times 10^{-3}]$
$k_{SI,max}$	5.95×10^{-3}	$[4.74 \times 10^{-3}; 7.15 \times 10^{-3}]$
MSE	0.344	

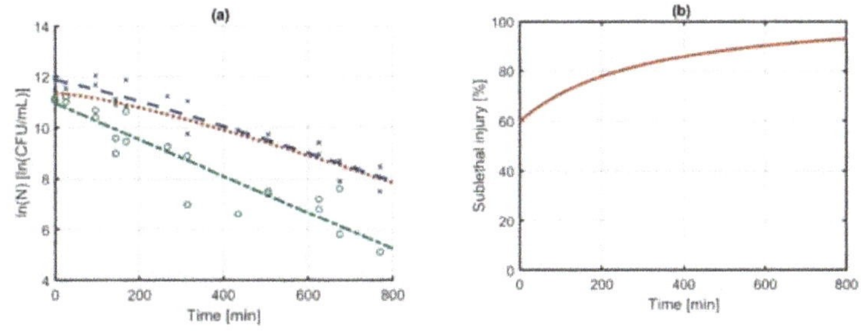

Figure 3. (**a**) Log-linear inactivation of all culturable (—) and healthy culturable (- - -) cells as described by the independent models or sublethal injury model that is constructed with the model of Bigelow and Esty [49]. Based on these modelling results, the evolution of the data are sublethally injured population (···) and (**b**) the percentage sublethal injury (—) is calculated. The based on counts of culturable (x) and healthy (o) cells on non-selective and selective media as published in Smet et al. [42]. The dataset was constructed by monitoring the survival of *Salmonella* Typhimurium, which was grown as planktonic cells in an environment at pH 5.5 with 60 g/L NaCl, after being treated in a liquid carrier with cold atmospheric plasma and stored at 8 °C (Dataset 1).

The evolution of SI with time as calculated by both models is illustrated in Figure 3b. This comparison of the model parameters and predictions of the SI model and independent models based on Dataset 1 demonstrates that the SI model is in a way equivalent to the independent models in cases where there can be no mistake in the calculation of the SI. Even though the equations and the model parameter constraints of both models are different, the model predictions are quasi-identical and the model parameters are equivalent. As will be demonstrated in the next section, these models start to differ when the independent model would yield wrongful predictions of the SI.

3.3. Case Study 2: Biphasic Inactivation

Dataset 2 portrays inactivation data that follows biphasic inactivation kinetics, as can be seen from Figure 4. Biphasic microbial inactivation is typically modelled with the model of Cerf [50]. Independent models for the healthy and culturable cells can be defined based on the model of Cerf as follows:

$$\frac{dn_{H1}(t)}{dt} = -k_{H1,max} \quad \text{with} \quad n_{H1}(t=0) = n_{H1,0} \tag{24}$$

$$\frac{dn_{H2}(t)}{dt} = -k_{H2,max} \quad \text{with} \quad n_{H2}(t=0) = n_{H2,0} \tag{25}$$

$$\frac{dn_{C1}(t)}{dt} = -k_{C1,max} \quad \text{with} \quad n_{C1}(t=0) = n_{C1,0} \tag{26}$$

$$\frac{dn_{C2}(t)}{dt} = -k_{C2,max} \quad \text{with} \quad n_{C2}(t=0) = n_{C2,0} \tag{27}$$

with n_{H1} and n_{H2} the healthy cells and n_{C1} and n_{C2} the culturable cells of, respectively, the first and second subpopulation. If one subpopulation has both a lower initial size and a lower inactivation rate than the other, the typical biphasic inactivation is observed [51]. The parameters of the independent models that have to be estimated are $n_{H1,0}$, $n_{H2,0}$, $n_{C1,0}$, $n_{C2,0}$, $k_{H1,max}$, $k_{H2,max}$, $k_{C1,max}$, and $k_{C2,max}$. When constructing the SI model based on the model of Cerf, the only change in the equations for the evolution of healthy cells in each subpopulation compared to Equations (24) and (25) is that the initial number of cells is defined as a fraction of the initial number of total culturable cells in the respective subpopulation ($f_{H1,0}$ and $f_{H2,0}$):

$$\frac{dn_{H1}(t)}{dt} = -k_{H1,max} \quad \text{with} \quad n_{H1}(t=0) = f_{H1,0} \cdot n_{C1,0} \text{ and } f_{H1,0} \in [0;1] \tag{28}$$

$$\frac{dn_{H2}(t)}{dt} = -k_{H2,max} \quad \text{with} \quad n_{H2}(t=0) = f_{H2,0} \cdot n_{C2,0} \text{ and } f_{H2,0} \in [0;1] \tag{29}$$

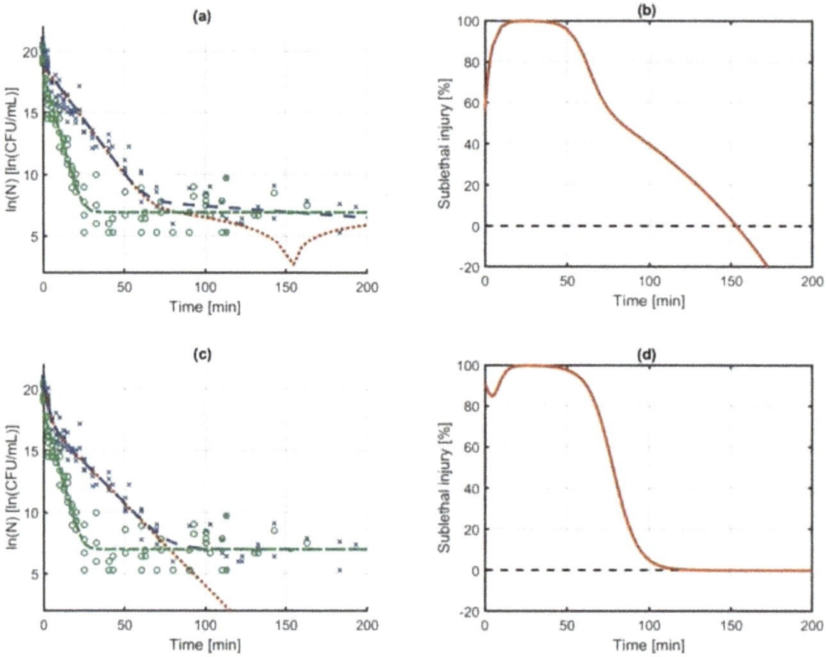

Figure 4. Biphasic inactivation of all culturable (—) and healthy culturable (- - -) cells as described by (**a**) the independent models and (**c**) the sublethal injury model that are constructed with the model of Cerf [50]. Based on these modelling results, the evolution of the sublethally injured population (···) and the percentage sublethal injury (—, (**b**,**d**) are calculated. The data are based on counts of culturable (x) and healthy (o) cells on non-selective and selective media as published by Noriega et al. [12]. The data were obtained by heat-treating low-density colonies of *Listeria innocua* at 54 °C (Dataset 2).

The differential equations for the two subpopulations of culturable cells are analogous to the differential equation for the culturable cells in the SI model for log-linear inactivation (Equation (20)):

$$\frac{dn_{C1}(t)}{dt} = -k_{SI1,max} \cdot (1 - \exp(n_{H1}(t) - n_{C1}(t))) \quad \text{with} \quad n_{C1}(t=0) = n_{C1,0} \tag{30}$$

$$\frac{dn_{C2}(t)}{dt} = -k_{SI2,max} \cdot (1 - \exp(n_{H2}(t) - n_{C2}(t))) \quad \text{with} \quad n_{C2}(t=0) = n_{LC,0} \tag{31}$$

The model parameters of the SI model in Equations (28)–(31) are $f_{H1,0}$, $f_{H2,0}$, $n_{C1,0}$, $n_{C2,0}$, $k_{H1,max}$, $k_{H2,max}$, $k_{C1,max}$ and $k_{C2,max}$. When comparing the model structure and parameterization between the two models, the differences in the SI model (Equations (28)–(31)) compared to the independent model (Equations (23)–(27)) are (i) that the initial condition of the healthy subpopulation is defined as a fraction of the initial culturable subpopulation and (ii) that the inactivation rate of the culturable cells depends on the number of injured cells instead of on the number of culturable cells.

Both models were fitted on the Dataset 2. The resulting model outputs and the calculated SI evolution are illustrated in Figure 4. In the results for the independent models, the model output for the culturable cells decreases below the model output for the healthy cells. In practice, it is of course not possible that the number of healthy cells would ever be higher than the number of culturable cells since the former is a subpopulation of the latter. However, due to this error, the number of injured cells that is calculated becomes negative and its logarithm a complex number. This is all because the independent models do not consider that the data are meant to describe the phenomenon of SI. The model output of the SI model on the other hand provides a good approximation of the experimental data and a reasonable calculation of the evolution of SI. The parameter estimation results for the independent models in Table 2 do not reveal the problems that occur when calculating the SI. Since the independent models fit the respective datasets for culturable and healthy cells well, the model parameter uncertainty is relatively low and the model fit is decent (as seen from the MSE). The problems in this approach only occurs when using the model outputs for calculating the SI. When comparing the parameters of the independent models and the SI model, the large difference in the initial populations catches the eye. These values of the initial populations are part of predicting the biphasic behaviour. In the SI model, the decrease of the culturable population is bound by the number of healthy cells. As such, there is no need for the biphasic behaviour of the inactivation model to predict the levelling off of the culturable cells towards the healthy cells. Consequently, when fitting the model parameters of the SI, the flexibility of the biphasic model is used to predict faster inactivation in the first 10 h compared to the time interval between 10 and 70 h (see Figure 4c). This phenomenon can also be observed from comparing the inactivation rate of the culturable population at time zero in the SI model (as determined by $k_{SI2,max}$) with that of the independent models (as determined by $k_{C2,max}$). This comparison shows that the initial inactivation rate in the SI model is much higher than in the independent models. These differences in the model structure also resulted in a better model fit of the SI model, as indicated from the MSE. Due to its constraints, the SI model will not always provide a better approximation of experimental data than the independent models. On the other hand, these constraints cause the SI model to be able to describe some phenomena with less model complexity than the independent models. As an example, in the current modelling results for the SI model, the tail in the culturable population of cells is a result from the tail in the healthy population of cells. As such, this tailing is described by the model without the need for it to be included separately in the differential equations of the culturable population of cells. In this case, the additional model complexity of the two phases is used to obtain a better approximation of the microbial inactivation in the initial stage of the experiment. The next section will further demonstrate the possibility to work with a decreased model complexity in the SI model.

Table 2. Parameter estimation results from fitting the independent models (Equations (24) to (27)) and sublethal injury model (Equations (28) to (31)) based on biphasic inactivation on Dataset 2.

Model Parameter	Parameter Estimate	95% Confidence Bounds
	Independent models	
$n_{H1,0}$	18.2	[17.7; 18.7]
$n_{H2,0}$	6.96	[5.96; 7.96]
$n_{C1,0}$	8.40	[6.64; 10.2]
$n_{C2,0}$	19.0	[18.6; 19.4]
$k_{H1,max}$	4.52×10^{-1}	$[4.03 \times 10^{-1}; 5.02 \times 10^{-1}]$
$k_{H2,max}$	2.34×10^{-14}	$[-1.15 \times 10^{-2}; 1.15 \times 10^{-2}]$
$k_{C1,max}$	9.41×10^{-3}	$[-4.50 \times 10^{-3}; 2.33 \times 10^{-2}]$
$k_{C2,max}$	1.80×10^{-1}	$[1.62 \times 10^{-1}; 1.99 \times 10^{-1}]$
MSE	1.41	

Table 2. *Cont.*

Model Parameter	Parameter Estimate	95% Confidence Bounds
	Sublethal injury model	
$f_{H1,0}$	2.91×10^{-5}	$[1.17 \times 10^{-6}; 5.71 \times 10^{-5}]$
$f_{H2,0}$	9.62×10^{-2}	$[1.44 \times 10^{-2}; 1.78 \times 10^{-1}]$
$n_{C1,0}$	17.5	[16.8; 18.1]
$n_{C2,0}$	20.6	[19.8; 21.3]
$k_{H1,max}$	1.09×10^{-10}	$[-7.16 \times 10^{-3}; 7.16 \times 10^{-3}]$
$k_{H2,max}$	4.53×10^{-1}	$[4.09 \times 10^{-1}; 4.97 \times 10^{-1}]$
$k_{SI1,max}$	1.34×10^{-1}	$[1.15 \times 10^{-1}; 1.52 \times 10^{-1}]$
$k_{SI2,max}$	7.48×10^{-1}	$[3.59 \times 10^{-1}; 1.14]$
MSE	1.14	

This example demonstrates that the proposed SI model provides a guarantee for reasonable predictions on the evolution of SI by incorporating basic knowledge on the relationship of the different microbial subpopulation into the model structure.

3.4. Case Study 3: Log-Linear Inactivation with Tailing

Figure 5a illustrates the log-linear inactivation behaviour with tailing of both the population of culturable cells and the subpopulation of healthy cells in Dataset 3. Based on the inactivation model of Geeraerd et al., the following independent models are proposed to describe the populations of culturable and healthy cells [52]:

$$\frac{dn_{H(t)}}{dt} = k_{H,max} \cdot (1 - \exp(n_{H,res} - n_H(t))) \quad \text{with} \quad n_H(t=0) = n_{H,0} \tag{32}$$

$$\frac{dn_{C(t)}}{dt} = k_{C,max} \cdot (1 - \exp(n_{C,res} - n_C(t))) \quad \text{with} \quad n_C(t=0) = n_{C,0} \tag{33}$$

with $n_{H,res}$ and $n_{C,res}$ the resistant subpopulations of the healthy and culturable cells. Converting these models into a linked SI model, results in the following set of equations:

$$\frac{dn_{H(t)}}{dt} = k_{H,max} \cdot (1 - \exp(n_{H,res} - n_H(t))) \tag{34}$$

$$\frac{dn_{C(t)}}{dt} = k_{SI,max} \cdot \left(1 - \frac{\exp(n_{SI,res})}{\exp(n_{SI}(t))}\right) \cdot \frac{\exp(n_{SI}(t))}{\exp(n_C(t))} \tag{35}$$

with $n_{SI,res}$ the resistant population of injured cells. To obtain a set of differential equations that is only a function of the healthy and culturable cells, Equation (35) is rewritten to:

$$\frac{dn_{C(t)}}{dt} = k_{I,max} \cdot \left(1 - \frac{\exp(n_{C,res}) - \exp(n_{H,res})}{\exp(n_C(t)) - \exp(n_H(t))}\right) \cdot \frac{\exp(n_C(t)) - \exp(n_C(t))}{\exp(n_C(t))} \tag{36}$$

If the model in Equation (36) is used, the resistant population of healthy cells needs to be described as a fraction of the resistant population of culturable cells $f_{H,res}$:

$$n_{H,res} = f_{H,res} \cdot n_{C,res} \tag{37}$$

When assuming all cells that can become injured can also be killed, there would be no residual population of injured cells. Consequently, the SI model for log-linear inactivation with tailing could be written as:

$$\frac{dn_{H(t)}}{dt} = k_{H,max} \cdot (1 - \exp(n_{H,res} - n_H(t))) \quad \text{with} \quad n_H(t=0) = n_{H0} \tag{38}$$

$$\frac{dn_{C(t)}}{dt} = k_{SI,max} \cdot (1 - \exp(n_H(t) - n_C(t))) \quad \text{with} \quad n_C(t = 0) = n_{C0} \tag{39}$$

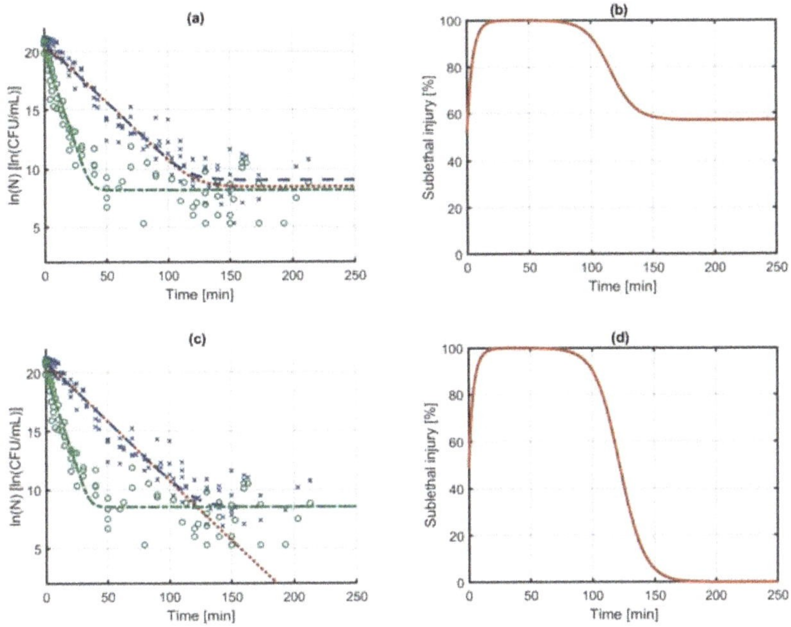

Figure 5. Log-linear inactivation with tailing of all culturable (—) and healthy culturable (-·-) cells as described by (**a**) the independent models and (**c**) the sublethal injury model that are constructed with the model of Geeraerd et al. [52]. Based on these modelling results, the evolution of the sublethally injured population (···) and the percentage sublethal injury (—), (**b**,**d**) are calculated. The data are based on counts of culturable (x) and healthy (o) cells on non-selective and selective media as published by Noriega et al. [12]. The data were obtained by heat-treating high-density colonies of *Salmonella* Typhimurium at 54 °C (Dataset 3).

These two equations basically describe log-linear inactivation with tailing for the subpopulation of healthy culturable cells, while the decrease of the total culturable population depends on the log-linear inactivation of the injured subpopulation. In this model, the decrease of the population of culturable cells will asymptotically approach the resistant population of healthy cells. The independent models of Equations (32) and (33) and the SI model of Equations (38) and (39) are fitted to Dataset 3. The model outputs of both models and the calculated SI are presented in Figure 5. For the independent models, the resistant population of the culturable cells is slightly higher than that of the healthy cells, suggesting that a fraction of the injured cells would be resistant as well. On the other hand, according to the SI model, the total population of culturable cells approaches the resistant population of healthy cells under the assumption that all injured cells are killed. Based on the current data and modelling results, there is of course no way of telling which model predicts the correct inactivation of each subpopulation. However, it would be perfectly possible to use the extended SI model of Equations (34) and (36), which would give practically the same model output as the simplified model of Equations (38) and (39). Because the estimation of the resistant population often comes with quite some uncertainty, it often happens that the resistant population of the culturable cells drops below that of the healthy cells, resulting in a negative SI percentage (similar to Case study 2). This problem will in any case be avoided when using the SI model that is presented here. The model parameters that are estimated are listed in Table 3. For the main part, the parameters are again equivalent between both models, with the main difference that the SI model now has one parameter less than the independent models. Even though the number of model parameters is reduced from six to five, the model accuracy only reduces slightly as shown by the MSE that changes from 1.59 to 1.63.

Table 3. Parameter estimation results from fitting the independent models (Equations (32) and (33)) and sublethal injury model (Equations (38) and (39)) based on log-linear inactivation with tailing on Dataset 3.

Model Parameter	Parameter Estimate	95% Confidence Bounds
	Independent models	
$n_{H,0}$	20.0	[19.5; 20.5]
$n_{C,0}$	20.7	[20.3; 21.1]
$n_{H,res}$	8.21	[7.81; 8.61]
$n_{C,res}$	9.06	[8.56; 9.56]
$k_{H,max}$	3.19×10^{-1}	$[2.79 \times 10^{-1}; 3.59 \times 10^{-1}]$
$k_{C,max}$	1.01×10^{-1}	$[9.28 \times 10^{-1}; 1.08 \times 10^{-1}]$
MSE	1.59	
	Sublethal injury model	
$f_{H,0}$	$9.67 \cdot 10^{-1}$	$[9.32 \times 10^{-1}; 1.00]$
$n_{C,0}$	20.7	[20.3; 21.2]
$n_{H,res}$	8.54	[8.22; 8.86]
$k_{H,max}$	3.29×10^{-1}	$[2.87 \times 10^{-1}; 3.72 \times 10^{-1}]$
$k_{SI,max}$	1.02×10^{-1}	$[9.46 \times 10^{-2}; 1.09 \times 10^{-1}]$
MSE	1.63	

4. Conclusions

A new method was proposed to model the evolution of sublethal injury (SI) during microbial inactivation. This method relies on defining the two subpopulations of healthy and injured cells within the culturable cells and describing inactivation as a two-step mechanism: (i) injuring healthy cells and (ii) killing injured cells. Based on three case studies, it was illustrated that this model concept can be applied to any existing dynamic inactivation model that is available in literature. Moreover, these case studies demonstrated, respectively, three important properties of the SI model: (i) the model is equivalent to the conventional method of using independent models when there is sufficient difference between the total quantity of culturable cells and the subpopulation of healthy cells, i.e., when there can be no mistake in the calculated SI; (ii) the new modelling method avoids unrealistic calculations of the SI by using its mechanistic definition of this process; and (iii) in some cases, it is possible to reduce the number of model parameters in the SI model because there is no need for additional model parameters to describe phenomena that are already hard coded into the more mechanistic description of the model. These properties make the SI model suitable for describing SI as a function of time during food processes. As such, this model can contribute to the optimization and control of the safety of food processes by considering SI dynamics.

Author Contributions: Conceptualization, S.A. and D.V.; methodology, S.A.; validation, S.A.; investigation, S.A.; data curation, D.V. and C.S.; writing—original draft preparation, S.A., D.V., C.S. and J.F.M.V.I.; writing—review and editing, S.A., D.V., C.S. and J.F.M.V.I.; visualization, S.A. and C.S.; supervision, J.F.M.V.I.; project administration, S.A. and J.F.M.V.I.; funding acquisition, S.A., D.V. and J.F.M.V.I. All authors have read and agreed to the published version of the manuscript.

Funding: Authors Simen Akkermans and Davy Verheyen were supported by the Research Foundation Flanders (FWO, grants 1224620N and 1254421N, respectively). This work was supported by the KU Leuven Research Fund through project C24/18/046, by the Research Foundation - Flanders through project G0B4121N and by the EU H2020 research and innovation program under the Marie Skłodowska-Curie grant agreement no. 956126.

Institutional Review Board Statement: Not applicable.

Informed Consent Statement: Not applicable.

Data Availability Statement: All data is available without restrictions upon request through contact with the corresponding author.

Conflicts of Interest: The authors declare no conflict of interest.

References

1. McMeekin, T.A.; Olley, J.; Ratkowsky, D.A.; Ross, T. Predictive microbiology: Towards the interface and beyond. *Int. J. Food Microbiol.* **2002**, *73*, 395–407. [CrossRef]
2. Pérez-Rodríguez, F.; Carrasco, E.; Pradhan, A.K.; Sant'Ana, A.S.; Valdramidis, V.P.; Valero, A. Special issue on 10th international conference of predictive modelling in foods: Towards a new paradigm in predictive microbiology. *Int. J. Food Microbiol.* **2019**, *291*, 65–66. [CrossRef] [PubMed]
3. Zwietering, M.H.; Garre, A.; Wiedmann, M.; Buchanan, R.L. All food processes have a residual risk, some are small, some very small and some are extremely small: Zero risk does not exist. *Curr. Opin. Food Sci.* **2021**, *39*, 83–92. [CrossRef]
4. Wu, V.C.H.; Fung, D.Y.C. Evaluation of Thin Agar Layer Method for Recovery of Heat-Injured Foodborne Pathogens. *J. Food Sci.* **2001**, *66*, 580–583. [CrossRef]
5. Hurst, A. Bacterial injury: A review. *Can. J. Microbiol.* **1977**, *23*, 935–944. [CrossRef]
6. Russell, A.D. Potential sites of damage in microorganisms exposed to chemical or physical agents. In *The Revival of Injured Microbes*; Andrew, M.H.E., Russell, A.D., Eds.; Academic Press: London, UK, 1984; pp. 1–18.
7. Wesche, A.M.; Gurtler, J.B.; Marks, B.P.; Ryser, E.T. Stress, Sublethal Injury, Resuscitation, and Virulence of Bacterial Foodborne Pathogens. *J. Food Prot.* **2009**, *72*, 1121–1138. [CrossRef]
8. Brashears, M.M.; Amezquita, A.; Stratton, J. Validation of Methods Used To Recover *Escherichia coli* O157:H7 and *Salmonella* spp. Subjected to Stress Conditions. *J. Food Prot.* **2001**, *64*, 1466–1471. [CrossRef] [PubMed]
9. Jasson, V.; Rajkovic, A.; Debevere, J.; Uyttendaele, M. Kinetics of resuscitation and growth of *L. monocytogenes* as a tool to select appropriate enrichment conditions as a prior step to rapid detection methods. *Food Microbiol.* **2009**, *26*, 88–93.
10. Deisingh, A.K.; Thompson, M. Strategies for the detection of *Escherichia coli* O157: H7 in foods. *J. Appl. Microbiol.* **2004**, *96*, 419–429. [CrossRef]
11. Jasson, V.; Uyttendaele, M.; Raykovic, A.; Debevere, J. Establishment of procedures provoking sub-lethal injury of *Listeria monocytogenes, Campylobacter jejuni* and *Escherichia coli* O157 to serve method performance testing. *Int. J. Food Microbiol.* **2007**, *118*, 241–249. [CrossRef]
12. Noriega, E.; Velliou, E.G.; Van Derlinden, E.; Mertens, L.; Van Impe, J.F. Effect of cell immobilization on heat-induced sublethal injury of *Escherichia coli, Salmonella Typhimurium* and *Listeria innocua*. *Food Microbiol.* **2013**, *36*, 355–364. [CrossRef]
13. Vermeiren, L.; Devlieghere, F.; Vandekinderen, I.; Rajtak, U.; Debevere, J. The sensory acceptability of cooked meat products treated with a protective culture depends on glucose content and buffering capacity: A case study with *Lactobacillu sakei* 10A. *Meat Sci.* **2006**, *74*, 532–545. [CrossRef]
14. Semanchek, J.J.; Golden, D.A. Influence of growth temperature on inactivation and injury of *Escherichia coli* O157: H7 by heat, acid, and freezing. *J. Food Prot.* **1998**, *61*, 395–401. [CrossRef] [PubMed]
15. Wu, V.C.H. A review of microbial injury and recovery methods in food. *Food Microbiol.* **2008**, *25*, 735–744. [CrossRef] [PubMed]
16. Busch, S.V.; Donnelly, C.W. Development of a Repair-Enrichment Broth for Resuscitation of Heat-Injured *Listeria monocytogenes* and *Listeria innocua*. *Appl. Environ. Microbiol.* **1992**, *58*, 14–20. [CrossRef] [PubMed]
17. Dykes, G.A. Physical and metabolic causes of sub-lethal damage in Listeria monocytogenes after long-term chilled storage at 4 °C. *J. Appl. Microbiol.* **1999**, *87*, 915–922. [CrossRef] [PubMed]
18. Bi, X.; Wang, Y.; Zhao, F.; Sun, Z.; Hu, X.; Liao, X. Sublethal injury and recovery of *Escherichia coli* O157:H7 by high pressure carbon dioxide. *Food Control* **2015**, *50*, 705–713. [CrossRef]
19. Carroll, L.M.; Bergholz, T.M.; Hildebrandt, I.M.; Marks, B.P. Application of a Nonlinear Model to Transcript Levels of Upregulated Stress Response Gene ipbA in Stationary-Phase *Salmonella enterica* Subjected to Sublethal Heat Stress. *J. Food Prot.* **2016**, *79*, 1089–1096. [CrossRef]
20. Ghate, V.; Leong, A.L.; Kumar, A.; Bang, W.S.; Zhou, W.; Yuk, H.-G. Enhancing the antibacterial effect of 461 and 521 nm light emitting diodes on selected foodborne pathogens in trypticase soy broth by acidic and alkaline pH conditions. *Food Microbiol.* **2015**, *48*, 49–57. [CrossRef] [PubMed]
21. Huang, M.; Zhuang, H.; Wang, J.; Yan, W.; Zhao, J.; Zhang, J. Inactivation Kinetics of *Salmonella* Typhimurium and *Staphylococcus aureus* in Different Media by Dielectric Barrier Discharge Non-Thermal Plasma. *Appl. Sci.* **2018**, *8*, 2087. [CrossRef]
22. Lv, R.; Wang, D.; Zou, M.; Wang, W.; Ma, X.; Chen, W.; Zhou, J.; Ding, T.; Ye, X.; Liu, D. Analysis of *Bacillus cereus* cell viability, sublethal injury, and death induced by mild thermal treatment. *J. Food Saf.* **2019**, *39*, e12581. [CrossRef]
23. Olszewska, M.A.; Zhao, T.; Doyle, M.P. Inactivation and induction of sublethal injury of *Listeria monocytogenes* in biofilm treated with various sanitizers. *Food Control* **2016**, *70*, 371–379. [CrossRef]
24. Pan, Y.; Cheng, J.-H.; Lv, X.; Sun, D.-W. Assessing the inactivation efficiency of Ar/O2 plasma treatment against *Listeria monocytogenes* cells: Sublethal injury and inactivation kinetics. *LWT Food Sci. Technol.* **2019**, *111*, 318–327. [CrossRef]

25. Sanz-Puig, M.; Moreno, P.; Pina-Pérez, M.C.; Rodrigo, D.; Martínez, A. Combined effect of high hydrostatic pressure (HHP) and antimicrobial from agro-industrial by-products against S. Typhimurium. *LWT Food Sci. Technol.* **2017**, *77*, 126–133. [CrossRef]
26. Shao, L.; Liu, Y.; Tian, X.; Wang, H.; Yu, Q.; Li, X.; Dai, R. Inactivation of *Staphylococcus aureus* in phosphate buffered saline and physiological saline ohmic heating with different voltage gradient and frequency. *J. Food Eng.* **2020**, *274*, 109834. [CrossRef]
27. Shi, H.; Chen, Z.; Chen, D.; Kan, J. Sublethal injury and recovery of *Escherichia coli* O157:H7 and K-12 after exposure to lactic acid. *Food Control* **2017**, *82*, 190–195. [CrossRef]
28. Silva, A.; Genovés, S.; Martorell, P.; Zanini, S.F.; Rodrigo, D.; Martinez, A. Sublethal injury and virulence changes in *Listeria monocytogenes* and *Listeria innocua* treated with antimicrobials carvacrol and citral. *Food Microbiol.* **2015**, *50*, 5–11. [CrossRef] [PubMed]
29. Thomas-Popo, E.; Mendonca, A.; Dickson, J.; Shaw, A.; Coleman, S.; Daraba, A.; Jackson-Davis, A.; Woods, F. Isoeugenol significantly inactivates *Escherichia coli* O157:H7, *Salmonella enterica*, and *Listeria monocytogenes* in refrigerated tyndallized pineapple juice with added *Yucca schidigera* extract. *Food Control* **2019**, *106*, 106727. [CrossRef]
30. Wang, X.; Devlieghere, F.; Geeraerd, A.; Uyttendaele, M. Thermal inactivation and sublethal injury kinetics of *Salmonella enterica* and *Listeria monocytogenes* in broth versus agar surface. *Int. J. Food Microbiol.* **2017**, *243*, 70–77. [CrossRef] [PubMed]
31. Zhang, H.; Zhao, Y.; Gong, C.; Jiao, S. Effect of radio frequency heating stress on sublethal injury of *Salmonella* Typhimurium in red pepper powder. *LWT Food Sci. Technol.* **2020**, *117*, 108700. [CrossRef]
32. Zhao, W.; Yang, R.; Shen, X.; Zhang, S.; Chen, X. Lethal and sublethal injury and kinetics of *Escherichia coli*, *Listeria monocytogenes* and *Staphylococcus aureus* in milk by pulsed electric fields. *Food Control* **2013**, *32*, 6–12. [CrossRef]
33. Miller, F.A.; Brandão, T.R.S.; Teixeira, P.; Silva, C.L.M. Recovery of heat-injured *Listeria innocua*. *Int. J. Food Microbiol.* **2006**, *112*, 261–265. [CrossRef] [PubMed]
34. Li, L.; Mendis, N.; Trigui, H.; Oliver, J.D.; Faucher, S.P. The importance of the viable but non-culturable state in human bacterial pathogens. *Front. Microbiol.* **2014**, *5*, 258. [CrossRef] [PubMed]
35. Pinto, D.; Santos, M.A.; Chabel, L. Thirty years of viable but nonculturable state research: Unsolved molecular mechanisms. *Crit. Rev. Microbiol.* **2013**, *41*, 61–76. [CrossRef] [PubMed]
36. Angarano, V.; Smet, C.; Akkermans, S.; Watt, C.; Chieffi, A.; Van Impe, J.F. Visible Light as an Antimicrobial Strategy for Inactivation of *Pseudomonas fluorescens* and *Staphylococcus epidermidis* Biofilms. *Antibiotics* **2020**, *9*, 171. [CrossRef] [PubMed]
37. Govaert, M.; Smet, C.; Baka, M.; Ećimović, B.; Walsh, J.L.; Van Impe, J. Resistance of L. monocytogenes and S. Typhimurium towards Cold Atmospheric Plasma as Function of Biofilm Age. *Appl. Sci.* **2018**, *8*, 2702. [CrossRef]
38. Govaert, M.; Smet, C.; Vergauwen, L.; Ećimović, B.; Walsh, J.L.; Baka, M.; Van Impe, J. Influence of plasma characteristics on the efficacy of Cold Atmospheric Plasma (CAP) for inactivation of *Listeria monocytogenes* and *Salmonella* Typhimurium biofilms. *Innov. Food Sci. Emerg. Technol.* **2019**, *52*, 376–386. [CrossRef]
39. Govaert, M.; Smet, C.; Graeffe, A.; Walsh, J.L.; Van Impe, J.F.M. Inactivation of *L. monocytogenes* and S. typhimurium Biofilms by Means of an Air-Based Cold Atmospheric Plasma (CAP) System. *Foods* **2020**, *9*, 157.
40. Smet, C.; Noriega, E.; Rosier, F.; Walsh, J.L.; Valdramidis, V.P.; Van Impe, J.F. Influence of food intrinsic factors on the inactivation efficacy of cold atmospheric plasma: Impact of osmotic stress, suboptimal pH and food structure. *Innov. Food Sci. Emerg. Technol.* **2016**, *38*, 393–406.
41. Smet, C.; Noriega, E.; Rosier, F.; Walsh, J.L.; Valdramidis, V.P.; Van Impe, J.F. Impact of food model (micro)structure on the microbial inactivation efficacy of cold atmospheric plasma. *Int. J. Food Microbiol.* **2017**, *240*, 47–56. [CrossRef]
42. Smet, C.; Baka, M.; Steen, L.; Fraeye, I.; Walsh, J.L.; Valdramidis, V.P.; Van Impe, J.F. Combined effect of cold atmospheric plasma, intrinsic and extrinsic factors on the microbial behavior in/on (food) model systems during storage. *Innov. Food Sci. Emerg. Technol.* **2019**, *53*, 3–17. [CrossRef]
43. Verheyen, D.; Baka, M.; Van Impe, J.F.M. Sublethal Injury Caused to *Listeria monocytogenes* by Natural Plant Extracts: Case Study on Grape Seed Extract and Garlic Extract. *Appl. Sci.* **2019**, *9*, 2731.
44. Verheyen, D.; Baka, M.; Akkermans, S.; Skåra, T.; Van Impe, J.F. Effect of microstructure and initial cell conditions on thermal inactivation kinetics and sublethal injury of Listeria monocytogenes in fish-based food model systems. *Food Microbiol.* **2019**, *84*, 103267. [PubMed]
45. Verheyen, D.; Govaert, M.; Seow, T.K.; Ruvina, J.; Mukherjee, V.; Baka, M.; Skåra, T.; Van Impe, J.F. The Complex Effect of Food Matrix Fat Contect on Thermal Inactivation of *Listeria monocytogenes*: Case Study in Emulsion and Gelled Emulsion Model Systems. *Front. Microbiol.* **2020**, *10*, 3149. [CrossRef] [PubMed]
46. Van Impe, J.F.; Bernaerts, K.; Geeraerd, A.H.; Poschet, F.; Versyck, K.J. Modelling and prediction in an uncertain environment. In *Food Process Modelling*; Tijskens, L.M.M., Hertog, M.L.A.T.M., Nicolaï, B.M., Eds.; Woodhead Publishing Limited: England, UK, 2001; pp. 156–179.
47. Walter, E.; Pronzato, L. *Identification of Parameteric Models from Experimental Data*; Springer: Paris, France, 1997.
48. Perni, S.; Chalise, P.R.; Shama, G.; Kong, M.G. Bacterial cells exposed to nanosecond pulsed electric fields show lethal and sublethal effects. *Int. J. Food Microbiol.* **2007**, *120*, 311–314. [PubMed]
49. Bigelow, W.D.; Esty, J.R. The thermal death point in relation to time of typical thermophilic organisms. *J. Infect. Dis.* **1920**, *27*, 602–617. [CrossRef]
50. Cerf, O. Tailing of survival curves of bacterial spores. *J. Appl. Bacteriol.* **1977**, *42*, 1–19.
51. Akkermans, S.; Smet, C.; Valdramidis, V.; Van Impe, J. Microbial Inactivation Models for Thermal Processing. In *Food Safety Engineering*; Demirci, A., Feng, H., Krishnamurthy, K., Eds.; Springer: Cham, Switzerland, 2020; pp. 399–420.

52. Geeraerd, A.H.; Herremans, C.H.; Van Impe, J.F. Structural model requirements to describe microbial inactivation during a mild heat treatment. *Int. J. Food Microbiol.* **2000**, *59*, 185–209.

foods

MDPI

Article

Prediction in the Dynamics and Spoilage of *Shewanella putrefaciens* in Bigeye Tuna (*Thunnus obesus*) by Gas Sensors Stored at Different Refrigeration Temperatures

Zhengkai Yi [1] and Jing Xie [1,2,3,4,*]

1 College of Food Science & Technology, Shanghai Ocean University, Shanghai 201306, China; zkyi121121@163.com
2 Shanghai Professional Technology Service Platform on Cold Chain Equipment Performance and Energy Saving Evaluation, Shanghai 201306, China
3 National Experimental Teaching Demonstration Center for Food Science and Engineering, Shanghai 201306, China
4 Shanghai Engineering Research Center of Aquatic Product Processing & Preservation, Shanghai 201306, China
* Correspondence: jxie@shou.edu.cn; Tel.: +86-021-6190-0391

Abstract: *Shewanella putrefaciens* have a faster growth rate and strong spoilage potential at low temperatures for aquatic products. This study developed a nondestructive method for predicting the kinetic growth and spoilage of *S. putrefaciens* in bigeye tuna during cold storage at 4, 7 and 10 °C by electronic nose. According to the responses of electronic nose sensor P30/2, the fitted primary kinetic models (Gompertz and logistic models) and secondary model (square root function model) were able to better simulate the dynamic growth of *S. putrefaciens*, with high R^2 and low RMSE values in the range of 0.96–0.99 and 0.021–0.061, respectively. A partial least squares (PLS) regression model based on both electronic nose sensor response values and electrical conductivity (EC) values predicted spoilage of *S. putrefaciens* in bigeye tuna more accurately than the PLS model based on sensor signal values only. In addition, SPME/GC-MS analysis suggested that 1-octen-3-ol, 2-nonanone, 2-heptanone, dimethyl disulfide and methylamine, N, N-dimethyl- are the key VOCs of tuna inoculated with *S. putrefaciens*.

Keywords: electronic nose; *Shewanella putrefaciens*; dynamic growth; spoilage prediction; GC-MS

Citation: Yi, Z.; Xie, J. Prediction in the Dynamics and Spoilage of *Shewanella putrefaciens* in Bigeye Tuna (*Thunnus obesus*) by Gas Sensors Stored at Different Refrigeration Temperatures. *Foods* **2021**, *10*, 2132. https://doi.org/10.3390/foods10092132

Academic Editors: Carlos Vilas and Barbara Cardazzo

Received: 27 July 2021
Accepted: 7 September 2021
Published: 9 September 2021

Publisher's Note: MDPI stays neutral with regard to jurisdictional claims in published maps and institutional affiliations.

1. Introduction

Bigeye tuna (*Thunnus obesus*) is a widely distributed and commercially important fish, favored by consumers because of its good taste and abundant nutrition [1]. However, bigeye tuna is an extremely perishable fish because of microbial spoilage and certain biochemical reactions during processing and storage. Its superior nutritional value and delicious meat make it important to preserve bigeye tuna [2]. Some methods have been used for the preservation of tuna, such as gas packaging, cool store, freezing processing, cryopreservation, etc. Refrigeration is currently an effective storage method used to slow down fish deterioration [3]. The main factor contributing to seafood spoilage during the refrigeration process is the activity of microorganisms. Many studies have reported that the specific spoilage organisms in refrigerated seafood were *Shewanella* spp., *Pseudomonas* spp., *Aeromonas* spp., and *Acinetobacter* spp. [3,4]. *S. putrefaciens* is the main spoilage microorganism of seafood in low-temperature storage, which has the potential for decomposing proteins and trimethyl-amine-N-oxide (TMAO) into ammonia, trimethylamine (TMA), and H_2S, producing a fishy odor [5]. *Shewanella* and *Pseudomonas* species isolated from spoiled tuna were considered as potential main contributors to spoilage in tuna during refrigerated storage [6]. Many studies have reported that the growth of *S. putrefaciens* may cause tuna to deteriorate during refrigeration [7,8]. Therefore, evaluating the spoilage potential of *S. putrefaciens* is crucial in the spoilage control of tuna at low temperatures.

Spoilage influences shelf life, marketing options, and safety of the product; no one buys spoiled foods nor should spoiled foods be on the market. Researchers usually use many conventional methods to identify fish spoilage, including sensory evaluation techniques, chemical methods [9], and microbiological methods [10]. Although conventional microbiological techniques are economical and simple to perform, these methods are time-consuming and cannot be continuously monitored in real time. The odor is an important indicator to evaluate the freshness of fish. Odors such as amines, ammonia, trimethylamine, and volatile sulfides are produced in marine fish during spoilage [11]. These volatile compounds can be potential indicators of spoilage in marine fish. Gas chromatography/mass spectrometry (GC-MS) has become a standard instrument for quantitative analysis of volatile substances in laboratories [12]. However, it is expensive, time-consuming, and unsuitable for large-scale detection. Moreover, the electronic nose (E-nose), a gas sensor array technology, has become an effective tool for predicting fish spoilage [13]. It can mimic the human olfactory system with sensitive sensors that interact with multiple odors to generate different electrical signals [14]. The E-nose has many advantages in predicting food spoilage, such as portability, non-destructive samples, low cost, short time consumption, and high sensitivity. For example, Semeano et al. [15] developed a system based on a gas sensing gel material coupled with an optical E-nose to detect tilapia deterioration, and the system predicted microbial growth well. In addition to the rapid detection of fish spoilage by the E-nose, fish spoilage can also be predicted using a simple physical index—electrical conductivity (EC). The decomposition of tissues and the outflow of electrolytes during the fish spoilage because of the catabolic activity of microorganisms and the oxidation of the fish body eventually lead to the rise of EC [16]. Heising et al. [17] found that EC values of aqueous solutions of volatile compounds produced by cod were positively correlated with freshness. Other researchers have also found a strong correlation between fish conductivity or electrical impedance and fish spoilage, and EC may be a good predictor of fish spoilage [18,19]. However, few studies have involved the use of electronic noses to predict the dynamic growth of spoilage bacteria in seafood and to predict the spoilage of seafood inoculated with spoilage organisms.

The dynamic growth of *S. putrefaciens* and the spoilage potential in aquatic products are of great importance. It is necessary to find a quick and easy method to study the spoilage of marine fish contaminated by *S. putrefaciens*. At present, there are few studies to predict the spoilage as well as the dynamic growth of specific spoilage bacteria in marine fish at different refrigeration temperatures. For this purpose, sterile tuna blocks were inoculated with *S. putrefaciens* and the changes in the total number of *S. putrefaciens* (TNS), total volatile basic nitrogen, EC, and volatiles at different refrigeration temperatures were investigated. The sensor P30/2 of the E-nose was selected to simulate the primary and secondary dynamic growth of *S. putrefaciens* in tuna. A partial least squares (PLS) regression model based on both E-nose sensor response values and EC values was used to predict the spoilage potential of *S. putrefaciens* in bigeye tuna. Finally, some key volatile organic compounds (VOCs) of tuna inoculated with *S. putrefaciens* were identified by GC/MS, and the correlation between VOCs and gas sensor signal values was analyzed.

2. Materials and Methods

2.1. Bacterial Strains and Cultural Conditions

S. putrefaciens was isolated and identified from spoiled bigeye tuna (Zhejiang Fenghui Ocean Fishing Company Ltd., Zhoushan, Zhejiang, China) and was identified based on 16S rRNA gene sequences, compared in GenBank using the BLAST function. Spoiled bigeye tuna was evaluated by trained panelists from the College of Food Science and Technology, Shanghai Ocean University, based on odor, color, and appearance. Bacteria were stored in tryptone soya broth (TSB) containing 25% glycerin at −80 °C. Before use, *S. putrefaciens* was precultured in brain–heart perfusion infusion (BHI) at 30 °C for 18 h and then cultured in TSB until the maximal concentration (10^8 CFU/mL).

2.2. Sample Preparation and Inoculation

The back muscle blocks of 20 kg tuna were purchased from Zhejiang Fenghui Ocean Fishing Company Ltd., Zhoushan, Zhejiang, China, and divided into rectangular blocks of about 30 g. Three replications were taken for TNS, pH, TVB-N, TMA, and VOC measurements, with ten replications for EC and e-nose measurement. Then, the blocks were sterilized by soaking in 0.5% (*v*/*v*) formalin solution for 40 s and washed in sterile water 2 times. Each sterile block was immersed in a bacterial suspension for 30 s of *S. putrefaciens* inoculation to achieve an inoculum level of 3.0–4.0 log CFU/g. Non-inoculated blocks immersed in sterile normal saline (0.85% NaCl) were used as the control check (CK) group. All samples were packed in a clean tray in a sterile environment and stored at 4, 7, and 10 °C.

2.3. Physicochemical Analysis

The physicochemical analysis included the measurement of pH, EC, total volatile basic nitrogen (TVB-N), and trimethylamine (TMA) values.

The EC of tuna blocks was measured using the method described by Yao et al. [19]. Briefly, tuna flesh (10 g) was homogenized and stirred for 30 min in 100 mL of distilled water. The mixture was filtered, and the EC of the filtrate was measured using an EC meter (Mettler Toledo FE20/EL20, Shanghai, China).

The pH value was determined by the method of [20]. The sample treatment was consistent with EC measurement and the pH of the filtrate was measured using a digital pH meter (Cyberscan Model 510; Eutech Instruments Pvt. Ltd., Singapore).

Total volatile basic nitrogen (TVB-N) was performed according to the method of [21]. Five grams of minced tuna flesh was accurately weighed. The TVB-N value was measured by an Automatic Kjeldahl Apparatus (KjeltecTM8400; FOSS Quality Assurance Co., Ltd., Copenhagen, Denmark).

TMA content was determined by Colorimetric Picric Acid Method [22]. Briefly, fish samples and trichloroacetic acid (TCA) were homogenized and mixed. After centrifugation, the supernatant was mixed with formaldehyde, saturated potassium carbonate solution, and toluene. The toluene layer solution and picric acid were mixed thoroughly, and absorbance readings were taken at 410 nm.

Measurements of pH, EC, TVB-N, and TMA were taken every 2 days for 12 days for samples stored at 4 °C, every 2 days for 10 days for samples stored at 7 °C, and every 1 day for 6 days for samples stored at 10 °C.

2.4. Microbiological Analysis and Growth Curve Fitting

The total number of *S. putrefaciens* (TNS) was determined by a basic method as described by Qian et al. [23] using iron agar. Briefly, 25 g of tuna flesh were put in 225 mL of sterilized saline water (NaCl, 0.85%, *w*/*v*) and homogenized for 2 min. Then, 0.1 mL of the dilution was spread on iron agar (IA) plates after serial dilution and incubated at 30 °C for 48 h. Black colonies were enumerated in IA plates. Plate counting agar was used to count the total viable count (TVC) in the CK group at 30 °C for 48 h. The measurement cycle of TNS and TVC was the same as the above physicochemical indexes.

2.4.1. Primary Models

The primary models, namely, Gompertz and logistic models, were used to simulate the growth curves of *S. putrefaciens* in tuna. They are represented by the following equations according to Gibson et al. [24]:

$$N = N_0 + (N_{max} - N_0) \times \exp\left(\quad \exp\left(\mu_{max} \times \frac{e}{N_{max} - N_0} \times (\lambda - t) + 1 \right) \right), \quad (1)$$

$$N = N_0 + (N_{max} - N_0)/(1 + \exp(\mu_{max} \times (\lambda - t))) \quad (2)$$

where N is the cell concentration (log CFU/g) at time t, N_0 and N_{max} represent the initial and maximum cell numbers (log CFU/g) of *S. putrefaciens*, respectively. λ is the lag time (h), t is real time, and μ_{max} represents the maximum growth rate (per h).

2.4.2. Secondary Models

To describe the temperature effect on μ_{max} and λ, a second model (square root model) was used as follows:

$$\sqrt{1/\lambda} = a_1 \times (T - T_{min1}), \quad (3)$$

$$\sqrt{\mu_{max}} = a_2 \times (T - T_{min2}) \quad (4)$$

where a_1 and a_2 are regression coefficients; T is the real temperature in °C; T_{min1} and T_{min2} are the theoretical minimum growth limits in °C.

2.5. E-Nose Analysis

Detection of the volatile compounds of the tuna flesh was performed by an electronic nose (E-nose, Fox 4000 Alpha-MOS, France). The 18 sensors are designed as follows: LY2/LG "Chlorine, fluoride, nitrogen oxide", LY2/G "Ammonia, amines, carbon oxides", LY2/AA "Alcohol", LY2/gCTL "Hydrogen sulphide", LY2/gCT "Propane and butane", T30/1 "Polar compounds, hydrogen chloride", P10/1 "Hydrocarbon, ammonia, chlorine", P10/2 "Methane and ethane", P40/1 "Fluoride and chlorine", T70/2 "Toluene and xylene", PA/2 "Ethanol, ammonia, amines", P30/1 "Hydrocarbon", P40/2 "Chlorine and fluoride", P30/2 "Hydrogen sulfide and ketones" T40/2 "chlorine", T40/1 "Fluoride", TA/2 "Alcohol". A total of 2.0 g of flesh was placed in a 10 mL vial for 10 min at 50 °C to generate balanced headspace samples. The gas flow rate was 2.5 mL/min, and the sensor cleaning time was 8 min. Then, the sensor response of the E-nose was determined as G/G0 (G0 and G represent the conductivity of the sensor exposed to the zero gas and sample gas). The measurement cycle of the E-nose of the sample was the same as above.

2.6. Headspace Solid Phase Microextraction Gas Chromatography/Mass Spectrometry (SPME-GC/MS) Analysis

According to the method of Li et al. [25] with minor modifications, 2 g of the minced sample was placed into a 20 mL glass vial and equilibrated at 40 °C for 20 min. The SPME extraction fiber was exposed to headspace for 30 min. Gas chromatography coupled with mass spectrometry (GC-MS) was used to measure the volatiles in bigeye tuna. The carrier gas was helium (high purity 99.999%), with a constant flow rate of 1 mL/min. The oven temperature program was as follows: initial temperature of 40 °C for 5 min, 5 °C/min to 120 °C, then 10 °C/min to 250 °C, and held for 5 min. Next, the volatiles were transferred to the MS system, MS source and quadrupole: 230 and 150 °C, respectively. Mass spectra were obtained within the mass range of 20–400 m/z at 70 eV. The VOCs of samples stored at 4 and 10 °C were measured. The samples at 4 °C were measured on days 4, 8 and 12, and the samples at 10 °C were measured on days 2, 4 and 6.

2.7. Statistical Analysis

The measurement experiments of pH, TVB-N, TMA, TNS, and SPME-GC/MS were repeated three times, and the EC and E-nose experiments were repeated ten times. The data of VOCs were expressed as the mean ± standard deviation. The growth kinetic model of *S. putrefaciens* was fitted using MATLAB 2017b (Math Works Inc., Natick, MA, USA). Pearson correlation analysis was used to evaluate the correlation between sensor response values and TNS to select appropriate sensors for predicting the dynamic growth of *S. putrefaciens*. Duncan's test and Pearson correlation coefficient were performed using SPSS 19.0 (SPSS Inc., Chicago, IL, USA). PLS regression was used to predict the spoilage potential of *S. putrefaciens* in tuna including TVB-N, TMA, and TNS. It is well known that the predictive performance of the calibration model cannot be determined merely by the internal validation but should also be externally validated based on predictions for samples not included in the calibration test. Data measured by the E-nose and electric conductivity meter were randomly divided into a training set (developing fitted models) and a testing set (validating models) in the ratio of 7:3. The training set of response values of the E-nose sensors was used as the data for building models, and the testing set was used as independent data to verify the accuracy of models. The data of pH, TVB-N, TMA, TNS, and SPME-GC/MS were not divided into training and validation sets. This was mainly because the data obtained from the E-nose and EC meter were used as source data of rapid detection to predict spoilage indicators (including TVB-N, TMA, TNS, and SPME-GC/MS) in the samples. Therefore, the spoilage indicators were used as a predicted object without being divided into training and validation sets (each indicator was considered as a whole). In addition, TNS was also one of the predicted indicators and was not divided into training and validation sets for the growth curve fitting. The model accuracy was evaluated by determination coefficient (R^2) and root-mean-square error (RMSE). RMSE was calculated as follows:

$$RMSE = \sqrt{\frac{1}{n}\sum_{i=1}^{n}(y_m - y_p)^2} \quad (5)$$

where y_m and y_p are measured and predicted values.

3. Results and Discussion

3.1. TVB-N and TMA

Changes in the TVB-N content of bigeye tuna blocks are shown in Figure 1A. The TVB-N values increased in inoculated tuna blocks throughout the storage period at different temperatures. TVB-N values of all groups showed a slow increase followed by a rapid increase. This result was consistent with Li et al. [26], who reported that when *S. putrefaciens* was inoculated into blunt snout bream flesh stored at 4 °C for 14 days, TVB-N values of samples were slow on the first 8 days, but increased rapidly on the last 4 days.

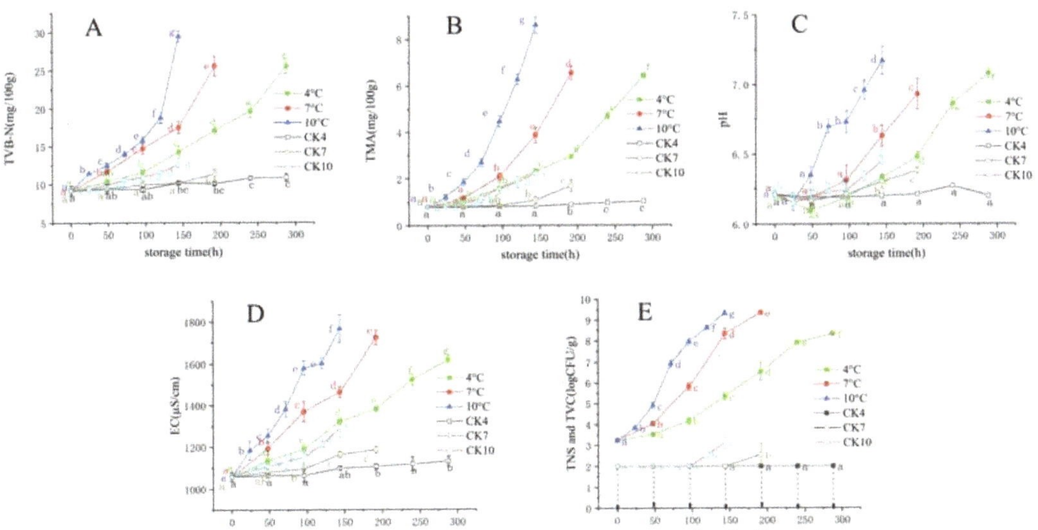

Figure 1. Changes in total volatile base nitrogen (TVB-N) (**A**), trimethylamine (TMA) (**B**), pH (**C**), Electrical conductivity (EC) (**D**), and total number of *S. putrefaciens* (TNS) of inoculation groups and total viable count (TVC) of control check (CK) groups (**E**) in bigeye tuna stored at different temperatures (each point is the mean value of three determinations). CK4, CK7, and CK10 are CK groups stored at 4, 7, and 10 °C. a–g in the same column with different superscripts are significantly different ($p < 0.05$). The arrow indicates that the viable count was below 2.0 log CFU/g.

The TMA value increased gradually during storage at 4, 7, and 10 °C (Figure 1B). Similar to the variation pattern of TVB-N values, the TMA values of inoculated tuna samples stored at 10 °C for 6 days were higher than those stored at 4 and 7 °C. This result indicated that temperature is a vital factor for microbial activity. The possible reason for the significant difference between the results for the CK and inoculated group may be that the production of TMA was promoted by *S. putrefaciens* in the inoculated samples, and the type of bacteria determined the ability to produce TMA. The TMA of samples increased exponentially with storage time, which was in agreement with [27], who reported that the change in TMA of yellowfin tuna fitted an exponential growth during the refrigeration.

3.2. pH and EC

The changes in the pH of aquatic products were closely associated with a series of chemical reactions caused by endogenous enzymes and microorganisms [28]. Changes in pH values in tuna samples at different temperatures are presented in Figure 1C. The initial pH value was 6.21 with a decreasing and then an increasing trend. A decrease in pH of the samples was due to the generation of lactic acid and the release of inorganic phosphate by decomposition of ATP [1], while the increase in the pH value was related to the release of alkaline decomposition products, such as histamine and TMA produced by spoilage microorganisms [29]. The pH of the sterile fish blocks (CK) stored at 4 °C fluctuated around 6.21, while those stored at 7 and 10 °C increased slowly. This may be due to the growth and metabolism of residual microorganisms in the CK group, and a similar phenomenon was observed in TVB-N and TMA.

EC can be used to detect meat quality from the efflux and excessive breakdown of body fluids from fish tissue during storage. Variations in EC value during cold storage are presented in Figure 1D. Initially, the EC in samples was 1057 μS/cm, and the EC of each group of samples increased significantly during storage. The change rate of EC values was higher for samples stored at higher temperatures. The significant increase in EC may be due to the autolysis of tuna cells after death and the decomposition of various nutrients in the cells into ions and small molecule metabolites with electrical conductivity under the action of enzymes and microorganisms, which enhances the electrical conductivity of the cell leachate [30]. Similar to the TVB-N and TMA curves, EC values increased slowly in the CK groups stored at 4, 7, and 10 °C.

3.3. Results of the E-Nose Analysis

As shown in Figure 2, odor maps were obtained with the E-nose from samples stored at 4, 7 and 10 °C. Response values from each sensor represent the average of 10 measurements. The E-nose responses of bigeye tuna samples stored at different storage temperatures showed a similar trend. Furthermore, sensors T70/2, PA/2, P30/1, and P30/2, which were sensitive to aromatic compounds, amines, hydrocarbons, and hydrogen sulfide, increased significantly at all storage temperatures. However, the signal values of sensors PA/2 and P30/2 increased more remarkably at 7 and 10 °C than those at 4 °C, indicating that storage temperature was an essential factor affecting the production of some compounds of tuna in storage. This finding was consistent with other studies that found temperature to be an important factor influencing the production of metamorphic substances in fish during refrigeration [31]. Changes in sensor signal values over time at different temperatures may be due to an increase in volatile gas concentrations and the production of new gas species [32], which may be related to the growth temperature and the number of *S. putrefaciens*.

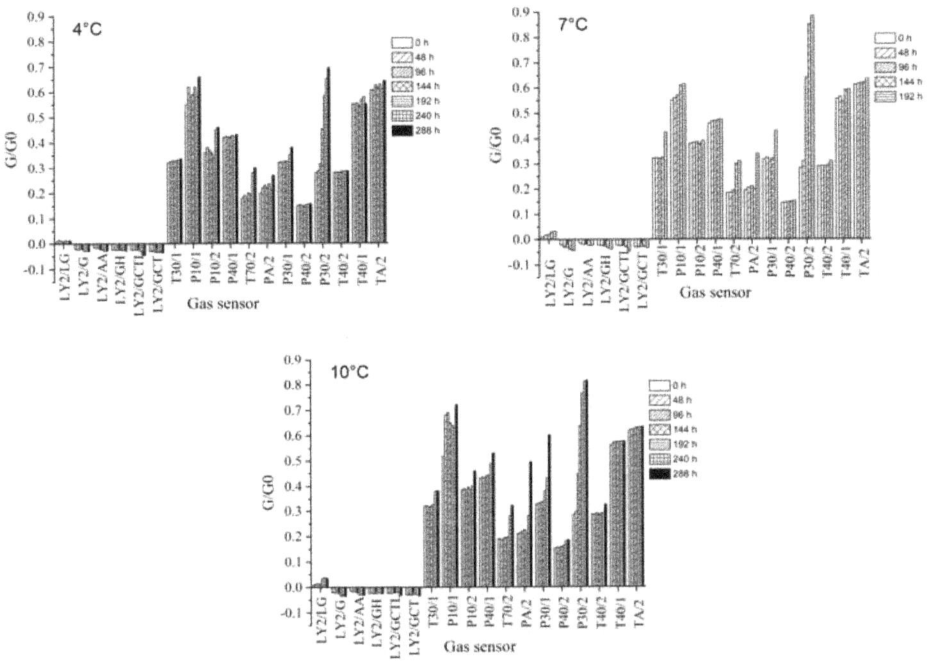

Figure 2. Average responses of 18 sensors in tuna samples inoculated with *S. Putrefaciens* at different temperatures during storage.

The corresponding values from the sensor arrays were explored to determine the differences in volatilization patterns of tuna during different cold storage periods using principal component analysis (PCA). To test whether the electronic nose could distinguish samples with different storage times, PCA was employed to investigate the feasibility of distinguishing tuna inoculated with *S. putrefaciens* sampled at different times and temperatures based on E-nose signals. As shown in Figure 3, the differences in tuna with different storage times can be represented using two main principal components (PCs), which accounted for 88.95% (4 °C), 93.78% (7 °C), and 80.8% (10 °C) of the total variance in PC1 and 5.22%, 4.71%, and 14.38% in PC2. Regarding the samples stored at 4 °C, the data points of groups 0, 2 and 4 d were placed in the first cluster, whereas the other groups were separated into another three clusters (6, 8 and 10–12 d). The data points of groups at 0, 2, and 4 d were similar, indicating that the odor profiles of the early contaminated samples were relatively similar. But the samples at 6, 8, 10, 12 d were clearly classified by PCA. For the samples stored at 7 °C, groups at 0 and 2 d were mixed in a cluster, and groups at 4 and 6 d had a clear right downshift along the ordinate (PC 1), located into the second cluster. Groups at 8 h were separated into another cluster along the abscissa (PC2), located away from other clusters. In 10 °C groups, the data of groups at 0–2 d located into the first cluster, and another two clusters contained 3–5 d and 6 d, respectively. For samples stored at 7 and 10 °C, PCA analysis indicated that the data points of the samples at the beginning of storage (0 and 2 d for 7 °C, 0 and 1 d for 10 °C) overlapped and were initially difficult to distinguish but could be distinguished at later time points. The samples stored at 4, 7, and 10 °C showed that the E-nose based on PCA analysis poorly identified very early contaminated samples but had good overall resolution for tuna inoculated with *S. putrefaciens*. Therefore, E-nose could be employed as a promising approach to realize the prediction of dynamics and spoilage of *Shewanella putrefaciens*.

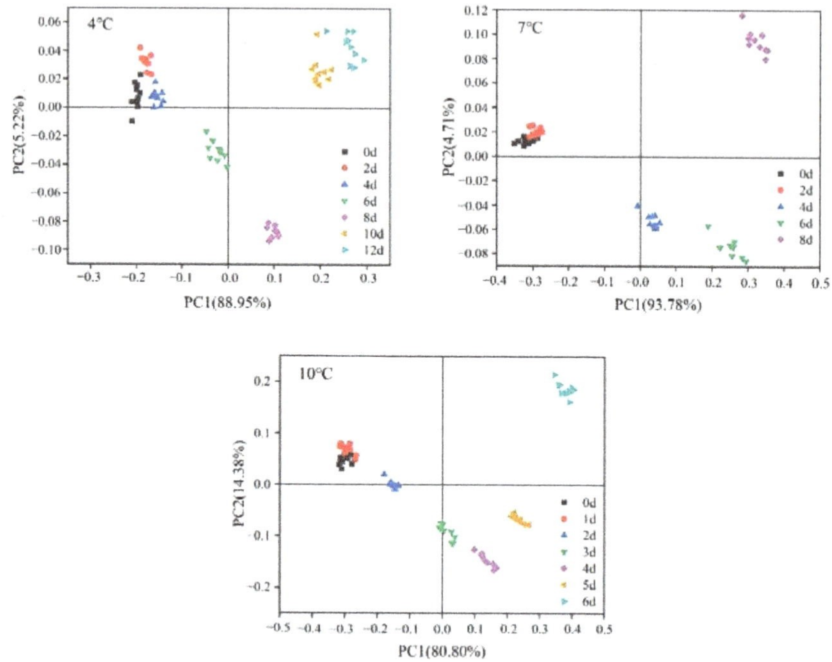

Figure 3. PCA score plots based on electronic nose measurements of tuna inoculated with *S. Putrefaciens* with different incubation times.

3.4. Dynamic Growth of *S. putrefaciens* in Tuna

The changes in the viable count of bigeye tuna are shown in Figure 1E. Gompertz and logistic models were fitted to the dynamic growth of *S. putrefaciens*. As shown in Table 1, the high R^2 and low RMSE values indicated a good fit of models, with ranges of 0.985–0.999 and 0.0654–0.283, respectively. The lag time of the lag phase (λ) and the maximum specific growth rate (μ_{max}) were two vital parameters for predicting the growth of microorganisms [33]. λ is especially vital to monitor food microorganisms and ensure food safety [34]. In our work, it was clear that the value of λ for *S. putrefaciens* decreased with increasing temperature, while μ_{maxe} was the opposite (Table 1). We observed that the storage temperature had a significant effect on the growth of *S. putrefaciens*, with slow growth rates at 4 and 7 °C, while a significant growth was observed at 10 °C based on λ_e and μ_{maxe} values. The growth of *S. putrefaciens* in tuna was in agreement with previous studies [35]. In addition, the initial colony count (N_0) of *S. putrefaciens* in tuna ranged from 2.3 to 3.3 log CFU/g; the value of the maximum colony count (N_{max}) peaked between 8.5 and 10.4 log CFU/g, according to two growth models.

Table 1. The primary growth models of *S. putrefaciens* in tuna at different temperatures based on modified Gompertz and logistic equation by CFU.

Fitted Models	T/°C	Equations	λ_e (h)	μ_{maxe} (h^{-1})	No	N_{max}	R^2	RMSE
Gompertz	4	$f(x) = 2.276 + 6.222 \times \exp(-\exp(0.07551/6.222 \times (68.8 - x) + 1))$	68.8	0.02778	2.276	8.498	0.997	0.101
	7	$f(x) = 3.292 + 7.149 \times \exp(-\exp(0.1375/7.149 \times (42.67 - x) + 1))$	42.67	0.05059	3.292	10.441	0.994	0.1387
	10	$f(x) = 3.252 + 6.331 \times \exp(-\exp(0.2/6.331 \times (24.36 - x) + 1))$	24.36	0.07358	3.252	9.583	0.986	0.283
Logistic	4	$f(x) = 3.053+5.783/(1 + \exp(0.02028 \times (166.6 - x)))$	166.6	0.02028	3.053	8.836	0.998	0.0763
	7	$f(x) = 3.092 + 6.657/(1 + \exp(0.03285 \times (105.9 - x)))$	105.9	0.03285	3.092	9.749	0.999	0.0654
	10	$f(x) = 2.83 + 6.48/(1 + \exp(0.04478 \times (63.6 - x)))$	63.6	0.04478	2.83	9.31	0.985	0.167

However, the primary models cannot estimate the effect of temperature on the growth of *S. putrefaciens* in tuna, but the secondary model can evaluate it. Therefore, the square root model was used to describe the relationship between the growth parameters ($\sqrt{\mu_{maxe}}$ and $\sqrt{\lambda e^{-1}}$) and the storage temperature for the microbe. As shown in Figure 4, the results showed a strong linear correlation between the kinetic parameters and the storage temperature, with R^2 values higher than 0.98. Our study also predicted the minimum growth temperature (T_{min}) of *S. putrefaciens* in tuna in the range of −8.5 to −4.6 °C based on a secondary model, which was slightly higher than −11.4 °C predicted by [36]. This may be due to the difference in fish samples and handling.

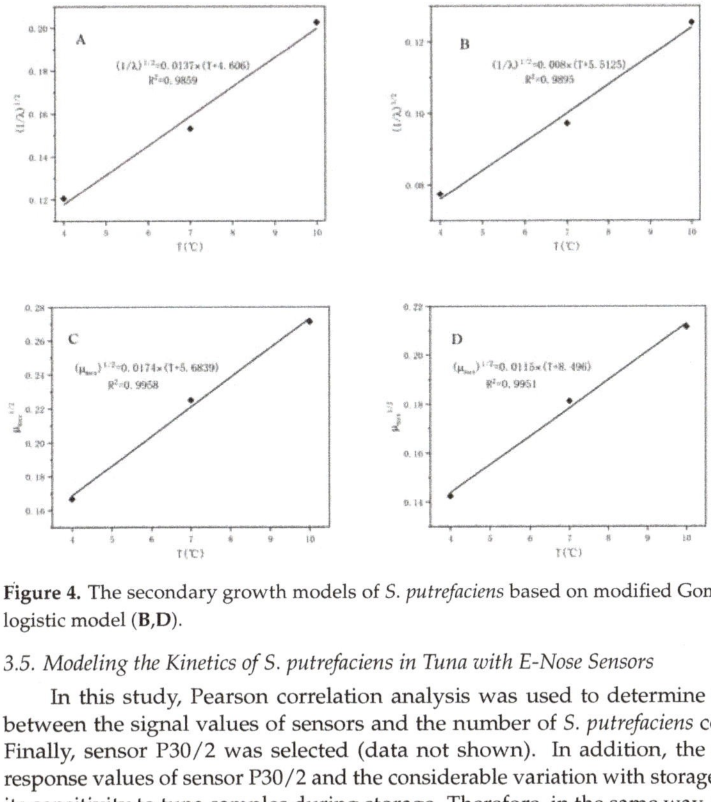

Figure 4. The secondary growth models of *S. putrefaciens* based on modified Gompertz (**A**,**C**) and logistic model (**B**,**D**).

3.5. Modeling the Kinetics of S. putrefaciens in Tuna with E-Nose Sensors

In this study, Pearson correlation analysis was used to determine the correlation between the signal values of sensors and the number of *S. putrefaciens* colonies in tuna. Finally, sensor P30/2 was selected (data not shown). In addition, the relatively high response values of sensor P30/2 and the considerable variation with storage time indicated its sensitivity to tuna samples during storage. Therefore, in the same way, an attempt was made to simulate the growth of *S. putrefaciens* by fitting the response of the sensor with Gompertz and logistic functions. Sensor P30/2 responses were fitted via Gompertz and logistic models to simulate *S. putrefaciens* growth according to the training set, and the validation set was used to verify the quality of the prediction models. The parameters of the generated mathematical equations are shown in Table 2, and the λ_e and μ_{maxe} of CFU were derived from Table 1. The sensor fitted both models well, with high R_c^2 and low $RMSE_c$ values, in a range of 0.971–0.994 and 0.0208–0.0472, respectively. Validated with the testing set, the fitting models were credible, with similar high R_p^2 and low $RMSE_p$ values of 0.963–0.987 and 0.0301–0.0613, respectively.

Table 2. Parameters of dynamic growth models of *S. putrefaciens* in tuna stored at different temperatures based on P30/2 by modified Gompertz and logistic equations.

Model	T/°C	Equations	Training Set		Testing Set		Sensor		CFU		r
			Rc^2	RMSEc	Rp^2	RMSEpc	λ_s (h)	μ_{maxs} (h^{-1})	λ_e (h)	μ_{maxe} (h^{-1})	
Gompertz	4	$f(x) = 0.280 + 0.401 \times \exp(-\exp(0.0102/0.4023 \times (90.05 - x) + 1))$	0.978	0.0261	0.968	0.0307	90.05	0.003753	68.8	0.02778	0.986
	7	$f(x) = 0.282 + 0.6125 \times \exp(-\exp(0.0229/0.6125 \times (52.27 - x) + 1))$	0.994	0.0208	0.985	0.0334	52.27	0.008425	42.67	0.05059	0.976
	10	$f(x) = 0.2872 + 0.509 \times \exp(-\exp(0.03126/0.509 \times (40.51 - x) + 1))$	0.983	0.0341	0.987	0.0324	40.51	0.01150	24.36	0.07358	0.986
Logistic	4	$f(x) = 0.2598 + 0.4307/(1 + \exp(0.03033 \times (156.50 - x)))$	0.971	0.0315	0.965	0.0329	156.50	0.03033	166.6	0.02028	0.996
	7	$f(x) = 0.2352 + 0.658/(1 + \exp(0.03812 \times (95.24 - x)))$	0.977	0.0472	0.963	0.0613	95.24	0.03812	105.9	0.03285	0.995
	10	$f(x) = 0.2629 + 0.5553/(1 + \exp(0.04864 \times (60.36 - x)))$	0.978	0.0391	0.987	0.0301	60.36	0.04864	63.6	0.04478	0.999

In addition, the P/30 λ_s and μ_{maxs} values obtained by the sensor based on the logistic model fit were closer to the results obtained from the actual growth model compared to the Gompertz model (Table 2). It indicated that the logistic model was more suitable for predicting of the growth of *S. putrefaciens* by the gas sensor in our study, which was different from the result reported by Gu et al. [34]. This difference may be due to differences in the gas sensors selected and strains. Kinetic parameters (λ_s and μ_{maxs}) are of special interest in predicting microbiology and are of high practical value in monitoring food quality and safety [37]. A high correlation coefficient (r) was obtained by comparing the growth fit models generated by sensor P/30 with the models in Table 1, which indicated that the response changes of sensor P/30 to the sample were similar to the growth of *S. putrefaction*. Microbial kinetic models according to microbial counting methods often have difficulty in obtaining λ_s and μ_{maxs} of microbial growth because of long training times [38]. Our study showed that a microbial odor response sensor may be used to simulate the dynamics of *S. putrefaciens* in tuna, but this requires more experimental verification.

A secondary model of the gas sensor P30/2 was also established and showed a good fit with R^2 in the range of 0.965–0.998, and the linear relationship based on logistic equation was slightly stronger than that of the Gompertz equation for λ_s and μ_{maxs} (Figure 5). Most studies predicted the number of microorganisms based on the chemometric method using multiple E-nose sensors [31,39,40], but rarely predicted the dynamic growth of spoilage microorganisms in aquatic products by a single sensor. In our study, primary and secondary kinetic models of *S. putrefaciens* in tuna were fitted by a single sensor, and the kinetic parameters were obtained.

3.6. Prediction of the Spoilage of S. putrefaciens in Tuna

To predict the spoilage level of inoculated bigeye tuna, the PLS algorithm was used to evaluate the correlation between the E-nose responses and spoilage indicators (TVB-N, TMA, and TNS) of the tuna. In addition, the changes in tissues caused by fish spoilage can be reflected by electrical conductivity [41,42], and the measurement method of EC is relatively simple. Therefore, the sensor responding values of E-nose and EC values were also combined to predict the spoilage of tuna. For PLS regression modeling, E-nose and EC measurement values from each group were randomly separated into two sets: seven samples used as the calibration set for and the remaining three samples as the prediction set. The leave-one-out cross-validation method was used to improve the accuracy of the PLS model.

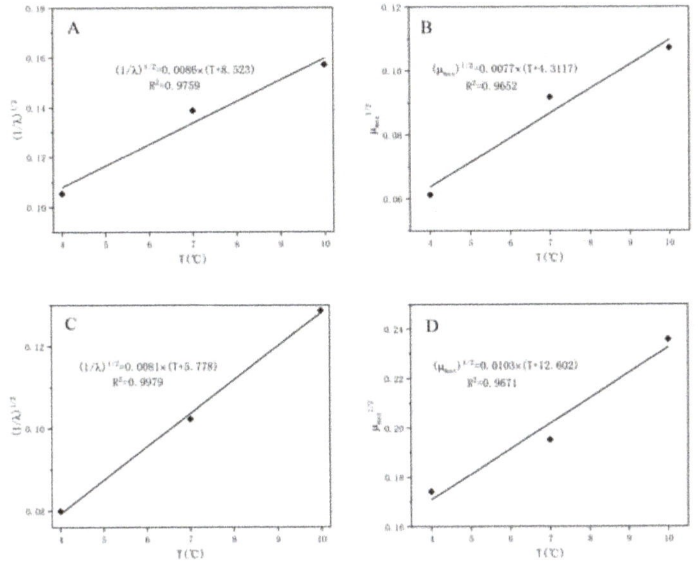

Figure 5. The secondary growth model of *S. putrefaciens* based on the response values of selected sensors P30/2. (**A**,**B**): Square root model based on Gompertz model; (**C**,**D**): Square root model based on logistic model.

Scatter graphs of the count of *S. putrefaciens* in tuna stored at 4 °C based on the PLS with and without EC values are given in Figure 6. The R^2 and RMSE values between the predicted and experimental values are shown in Table 3. It is well known that the key to evaluating the quality of a predictive model is not only internal validation (calibration) but also external validation of samples not included in the calibration set [43]. In the two sets, the predictions of the PLS models with and without EC values performed well, and the PLS model with EC values was better than those without EC values, except for the prediction of TMA values for tuna stored at 10 °C (but the difference was not significant). The reason for this result was that the accuracy of the PLS prediction model improved by adding the index (EC) that had a great correlation with the prediction object [44].

Table 3. Calibration and validation results in tuna stored at different temperatures based on the PLS model with and without EC values.

Indicators	Temperatures (°C)	PLS without EC Values: Calibration Set		PLS without EC Values: Validation Set		PLS with EC Values: Calibration Set		PLS with EC Values: Validation Set	
		R_c^2	$RMSE_c$	R_v^2	$RMSE_v$	R_c^2	$RMSE_c$	R_v^2	$RMSE_v$
TVB-N (mg/100 g)	4	0.9713	0.9327	0.9775	0.8673	0.9812	0.7519	0.9902	0.6448
TMA (mg/100 g)		0.99	0.1932	0.9862	0.2425	0.9905	0.189	0.988	0.2237
TNS (log_{10} CFU/mL)		0.9808	0.2679	0.9702	0.3589	0.9868	0.222	0.9843	0.2653
TVB-N (mg/100 g)	7	0.9874	0.6461	0.9863	0.6981	0.9925	0.4966	0.9919	0.5053
TMA (mg/100 g)		0.9956	0.1418	0.9851	0.322	0.9962	0.1326	0.9966	0.1292
TNS (log_{10} CFU/mL)		0.9956	0.1594	0.9923	0.225	0.9957	0.1565	0.995	0.1742
TVB-N (mg/100 g)	10	0.9932	0.5411	0.9896	0.6755	0.9958	0.4485	0.9963	0.4214
TMA (mg/100 g)		0.9897	0.276	0.9871	0.3478	0.9876	0.3020	0.9826	0.3837
TNS (log_{10} CFU/mL)		0.9857	0.2735	0.969	0.4185	0.9963	0.1396	0.9864	0.2705

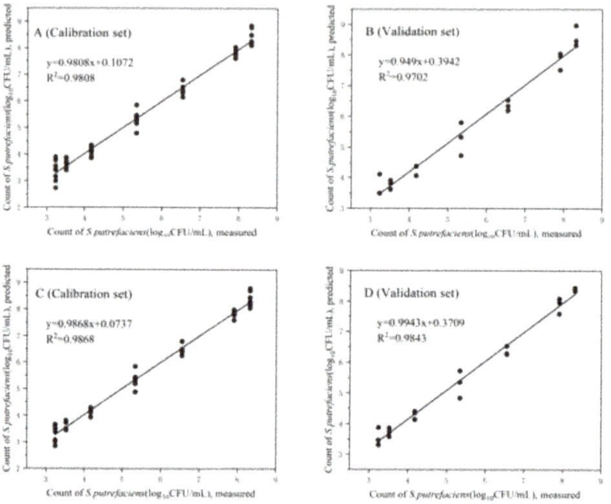

Figure 6. Reference measured data versus predicted data from the PLS models for prediction of the total number of *S. putrefaciens* (TNS) in tuna stored at 4 °C. (**A**,**B**): PLS models for calibration and validation set without EC values; (**C**,**D**): PLS models for calibration and validation set with EC values.

3.7. Volatile Compounds in Tuna Samples According to HS-SPME/GC-MS

In this work, a total of 30 VOCs were detected in the control and inoculated groups (Table 4). The changing trend of VOCs at 10 °C was greater than that at 4 °C, and various VOCs in this study were previously reported to be products of protein or lipid oxidation metabolism [45]. To exclude the effect of oxidation of the fish itself on VOCs, this study also measured the VOCs produced by non-inoculated tuna blocks during storage.

The increasing tendency of 1-Penten-3-ol, 1-Octen-3-ol, and 2-Hexen-1-ol, (Z)- was observed in tuna during storage, and their increase was associated with auto-oxidative derivatization of polyunsaturated fatty acids [46]. Ethanol was present only in the early stages of tuna storage (day 0, day 4–4, and day 2–10) and was not detected as the storage period increased, which was similar to the results of Liu et al. [10]. This may be related to the fact that the metabolic process of *Shewanella* does not produce ethanol but can use it [4]. In addition, hexanal, heptanal, octanal, nonanal, and propanal were detected at the early stage of storage, and the above substances were confirmed to be produced by fat oxidation and had a fishy odor. The content of hexanal increased with storage time and storage temperature, which may indicate that the fat in tuna was oxidized. The changes of aldehydes reflect the degree of oxidation of polyunsaturated fatty acids such as linoleic acid in bigeye tuna, which can be used as a basis for judging the freshness of tuna. Ketones including 2-nonanone, 2-undecanone, and 2-heptanone were significantly higher in inoculated tuna compared to the control group. These ketones may originate from lipolysis and dehydrogenation by spoilage bacteria [47], which explains the low ketone content of fresh fish samples (day 0). Some hydrocarbons were also detected in this study, which were mainly derived from the decomposition of alkoxy radicals of fatty acids. Various hydrocarbons were present in the volatiles of crustaceans and fish, but they had a high threshold and made little contribution to the flavor of bigeye tuna [48]. Methylamine, N, N-dimethyl-(so called trimethylamine) were detected only after 8–12 days at 4 °C and 4–6 days at 10 °C, which may be because of higher concentrations of TMA in the inoculated tuna during the late storage period. Dimethyl disulfide was detected in the inoculated tuna at the late storage period. This compound derived from the methionine catabolism that was produced by *S. putrefaciens* [49].

Table 4. Volatile organic compounds (VOCs) in tuna at 4 and 10 °C. Day 4–4, day 8–4, and day 12–4: inoculated tuna stored for 4, 8, and 12 days at 4 °C, respectively. Day 2–10, day 4–10, and day 6–10: inoculated tuna stored for 2, 4, and 6 days at 10 °C, respectively. CK 12–4 and CK 6–10: non-inoculated tuna stored for 12 and 6 days at 4 and 10 °C, respectively.

VOCs	Relative Concentration (Area 10^{-6})								
	Day 0	Day 4–4	Day 8–4	Day 12–4	CK 12–4	Day 2–10	Day 4–10	Day 6–10	CK 6–10
Alcohols									
1-Penten-3-ol	3.01 ± 1.13	3.56 ± 0.56	3.89 ± 1.36	4.65 ± 1.98	4.54 ± 0.64	6.89 ± 0.98	8.99 ± 2.36	9.29 ± 1.99	7.98 ± 2.36
1-Octen-3-ol	7.28 ± 1.86	21.69 ± 0.98	20.79 ± 3.25	46.13 ± 6.56	20.36 ± 2.56	18.61 ± 0.56	35.79 ± 2.11	66.98 ± 5.24	35.98 ± 4.65
Ethanol	9.65 ± 1.23	1.23 ± 0.21	ND	ND	ND	2.23 ± 0.56	ND	ND	ND
1-Hexanol	12.56 ± 0.33	6.22 ± 0.65	2.12 ± 0.32	ND	1.36 ± 0.05	3.56 ± 0.65	ND	ND	2.98 ± 0.06
(3-Methyl-oxiran-2-yl)-methanol	ND	0.66 ± 0.23	0.68 ± 0.24	1.65 ± 0.66	0.12 ± 0.04	ND	2.38 ± 0.56	3.01 ± 0.21	0.22 ± 0.08
2-Hexen-1-ol, (Z)-	ND	0.54 ± 0.04	1.89 ± 0.03	5.65 ± 0.69	ND	ND	2.32 ± 0.05	9.28 ± 0.32	ND
2-Nonen-1-ol	0.4 ± 0.02	1.25 ± 0.11	0.65 ± 0.01	0.33 ± 0.02	0.43 ± 0.03	0.54 ± 0.06	1.56 ± 0.02	0.51 ± 0.03	0.36 ± 0.02
Aldehydes									
Hexanal	7.86 ± 0.65	19.23 ± 1.89	33.21 ± 3.12	64.93 ± 5.36	13.54 ± 2.36	22.52 ± 1.35	35.26 ± 3.22	71.26 ± 5.62	21.22 ± 2.56
Heptanal	4.22 ± 0.12	4.12 ± 0.63	9.52 ± 1.32	5.24 ± 0.97	3.22 ± 0.86	8.79 ± 1.22	1.23 ± 0.06	3.54 ± 0.07	1.98 ± 0.21
Nonanal	2.21 ± 0.08	2.28 ± 0.21	4.60 ± 0.11	8.21 ± 0.99	3.21 ± 0.07	18.47 ± 2.36	12.11 ± 2.65	15.89 ± 3.06	11.21 ± 2.10
Propanal	2.35 ± 0.09	5.61 ± 0.46	7.22 ± 1.23	6.11 ± 0.12	2.12 ± 0.33	5.31 ± 0.22	4.33 ± 1.11	6.85 ± 0.64	5.11 ± 0.21
2-Decenal, (E)	ND	3.12 ± 0.18	2.47 ± 0.21	1.56 ± 0.06	ND	ND	0.43 ± 0.05	2.55 ± 0.65	0.35 ± 0.05
2-Nonanal, (E)-	ND	0.22 ± 0.03	3.56 ± 0.24	4.54 ± 0.68	3.72 ± 0.21	0.31 ± 0.05	5.33 ± 0.98	5.99 ± 1.33	3.56 ± 0.09
2-Dodecanal, (E)-	ND	ND	1.263 ± 0.05	2.112 ± 0.32	4.6 ± 0.23	5.68 ± 1.22	4.22 ± 0.09	3.68 ± 1.21	2.23 ± 0.21
Decanal	0.65 ± 0.02	1.36 ± 0.05	ND	ND	2.55 ± 0.65	1.55 ± 0.96	2.36 ± 0.12	1.32 ± 0.08	1.86 ± 0.09
Octanal	3.86 ± 0.33	1.69 ± 0.35	0.98 ± 0.04	9.98 ± 2.65	2.65 ± 0.22	1.98 ± 0.21	5.65 ± 0.69	3.23 ± 0.93	5.36 ± 0.97
4-Heptenal, (Z)-	ND	0.09 ± 0.01	0.98 ± 0.09	6.22 ± 1.23	0.28 ± 0.02	0.18 ± 0.04	2.32 ± 0.26	1.69 ± 0.66	0.56 ± 0.08
Ketones									
2,3-Octanedione	ND	ND	0.85 ± 0.15	1.7 ± 0.56	ND	0.08 ± 0.01	1.23 ± 1.23	2.56 ± 1.56	ND
2,3-Pentanedione	0.04 ± 0.01	0.06 ± 0.02	0.03 ± 0.01	1.56 ± 0.05	1.78 ± 0.58	2.53 ± 0.04	1.03 ± 0.05	4.22 ± 1.86	3.29 ± 0.12
2-Nonanone	ND	ND	0.06 ± 0.02	0.12 ± 0.04	ND	ND	0.08 ± 0.05	0.25 ± 0.08	ND
2-Undecanone	ND	ND	0.75 ± 0.02	2.67 ± 0.66	0.96 ± 0.06	ND	1.19 ± 0.28	4.55 ± 1.32	1.89 ± 0.64
2-Heptanone	ND	ND	0.12 ± 0.02	0.89 ± 0.13	0.06 ± 0.01	ND	1.35 ± 0.61	2.46 ± 0.05	0.09 ± 0.01
Hydrocarbons									
Heptacosane	ND	ND	1.23 ± 0.13	0.85 ± 0.04	2.35 ± 0.12	2.79 ± 0.13	0.46 ± 0.02	0.77 ± 0.07	1.23 ± 0.21
Pentadecane	2.23 ± 0.14	8.63 ± 1.36	4.62 ± 0.88	3.33 ± 0.21	3.26 ± 1.02	9.59 ± 1.33	4.35 ± 0.98	2.23 ± 0.29	1.9 ± 0.06
Tetradecane	4.23 ± 0.32	2.51 ± 0.11	0.56 ± 0.02	ND	1.03 ± 0.05	1.54 ± 0.21	ND	ND	0.81 ± 0.06
Others									
Ethyl acetate	5.03 ± 0.09	7.54 ± 1.32	3.21 ± 0.35	1.22 ± 0.35	2.13 ± 0.86	3.23 ± 0.78	0.77 ± 0.12	2.27 ± 0.65	1.22 ± 0.05
Methylamine, N, N-dimethyl-	ND	ND	3.31 ± 0.09	8.02 ± 1.04	ND	3.28 ± 0.39	10.65 ± 2.65	21.23 ± 4.65	ND
Methoxy-phenyl-oxime	ND	ND	0.64 ± 0.06	0.46 ± 0.03	0.11 ± 0.02	ND	0.83 ± 0.04	2.22 ± 0.25	0.23 ± 0.03
Naphthalene	0.07 ± 0.01	0.14 ± 0.02	0.08 ± 0.01	0.07 ± 0.02	0.12 ± 0.03	0.09 ± 0.01	0.16 ± 0.04	0.06 ± 0.01	0.16 ± 0.04
Dimethyl disulfide	ND	ND	ND	0.85 ± 0.06	ND	ND	ND	1.32 ± 0.31	ND

ND means not detected.

3.8. Relationship between E-Nose Results and VOCs

In this work, characteristic VOCs, including alcohols, aldehydes, ketones, amines, and sulfide compounds, played a significant role in distinguishing tuna samples infected with *S. putrefaciens* by GC-MS at different temperatures. Sensor P30/2 was sensitive to hydrogen sulfide and ketone. Comparing the VOCs with the sensor P30/2, the presence of alcohols, ketones, amines, and sulfide compounds had a significant impact on the sensor response. The significantly increased response values of the P30/2 were consistently correlated with the increased concentrations of ketones and sulfides. In addition, the relationship between sensor responses and VOCs was investigated by Pearson correlation coefficients (Figure 7). Correlation coefficient results indicated that the sensor responses were positively correlated with 1-octen-3-ol, 2-nonanone, 2-heptanone, dimethyl disulfide, and methylamine, N, N-dimethyl-, which revealed that the P30/2 was sensitive to representative VOCs of tuna inoculated with *S. putrefaciens*. In particular, dimethyl disulfide, and methylamine, N, N-dimethyl-, as the characteristic volatile compounds of *S. putrefaciens* [10], contained high content, which contributed significantly to the high signals of P30/2 to inoculated tuna. Therefore, the selected sensor could be used to discriminate tuna infected with *S. putrefaciens* through the specific response to characteristic VOCs.

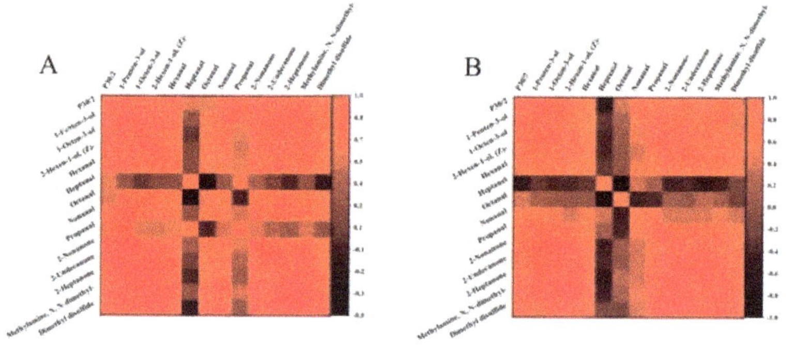

Figure 7. Correlations between sensor responses and GC-MS data (**A**): 4 °C; (**B**): 10 °C.

4. Conclusions

This work demonstrated that the growth of *S. putrefaciens* in tuna samples stored at 4, 7, and 10 °C was consistent with two primary kinetic models (Gompertz and logistic), with high R^2 and low RMSE values in the range of 0.985–0.999 and 0.0654–0.283, respectively, as well as a secondary kinetic model with high R^2 values in the range of 0.9859–0.9958. The selected sensor P30/2 accurately predicted the dynamic growth of *S. putrefaction*. In addition, the secondary model was used to characterize the relationship between the storage temperature of the samples and the growth kinetic parameters of *S. putrefaciens*. The secondary model fitted with the sensor P30/2 accurately estimated the influence of temperature on the kinetic parameters of *S. putrefaciens* and the minimum growth temperature range of *S. putrefaciens*. The PLS model based on the E-nose response values with the EC values was more accurate than the model without the EC values in predicting the spoilage of tuna inoculated with *S. putrefaciens*. Based on the GC-MS analysis, several alcohols, ketones, amines, and sulfide compounds, especially 1-octen-3-ol, 2 nonanone, 2-heptanone, dimethyl disulfide, and methylamine, N, N-dimethyl- were determined as characteristic VOCs in tuna infected with *S. putrefaciens* stored at 4 and 10 °C. These results revealed that the E-nose can have a wide range of applications for predicting the growth of spoilage microorganisms and performing a quantitative analysis of spoilage in tuna.

Author Contributions: Z.Y.: Conceptualization, Methodology, Software, Investigation, Writing. J.X.: Validation, Formal analysis, Writing—review and editing, Examination, Funding acquisition. All authors have read and agreed to the published version of the manuscript.

Funding: This research was financially supported by the National Key R&D Program of China (2019YFD0901604), the Key Project of the Science and Technology Commission of Shanghai Municipality (19DZ1207503), and the Shanghai Professional Technology Service Platform on Cold Chain Equipment Performance and Energy Saving Evaluation (19DZ2284000). All authors have read and agreed to the published version of the manuscript.

Institutional Review Board Statement: Not applicable.

Informed Consent Statement: Not applicable.

Data Availability Statement: Authors can confirm that all relevant data are included in the article.

Acknowledgments: We would like to thank the anonymous reviewers for their helpful comments and suggestions.

Conflicts of Interest: The authors declare no conflict of interest.

References

1. Kaewprachu, P.; Osako, K.; Benjakul, S.; Suthiluk, P.; Rawdkuen, S. Shelf life extension for Bluefin tuna slices (*Thunnus thynnus*) wrapped with myofibrillar protein film incorporated with catechin-Kradon extract. *Food Control* **2017**, *79*, 333–343. [CrossRef]
2. Rossi, S.; Lee, C.; Ellis, P.; Pivarnik, L. Biogenic amines formation in bigeye tuna steaks and whole skipjack tuna. *J. Food Sci.* **2002**, *67*, 2056–2060. [CrossRef]
3. Ruiz-Capillas, C.; Moral, A. Sensory and biochemical aspects of quality of whole bigeye tuna (*Thunnus obesus*) during bulk storage in controlled atmospheres. *Food Chem.* **2005**, *89*, 347–354. [CrossRef]
4. Parlapani, F.F.; Mallouchos, A.; Haroutounian, S.A.; Boziaris, I.S. Microbiological spoilage and investigation of volatile profile during storage of sea bream fillets under various conditions. *Int. J. Food Microbiol.* **2014**, *189*, 153–163. [CrossRef] [PubMed]
5. Wright, M.H.; Shalom, J.; Matthews, B.; Greene, A.C.; Cock, I.E. Terminalia ferdinandiana Exell: Extracts inhibit Shewanella spp. growth and prevent fish spoilage. *Food Microbiol.* **2019**, *78*, 114–122. [CrossRef] [PubMed]
6. Lee, B.-H.; Wu, S.-C.; Shen, T.-L.; Hsu, Y.-Y.; Chen, C.-H.; Hsu, W.-H. The applications of Lactobacillus plantarum-derived extracellular vesicles as a novel natural antibacterial agent for improving quality and safety in tuna fish. *Food Chem.* **2021**, *340*, 128104. [CrossRef]
7. Grande-Tovar, C.D.; Serio, A.; Delgado-Ospina, J.; Paparella, A.; Rossi, C.; Chaves-López, C. Chitosan films incorporated with Thymus capitatus essential oil: Mechanical properties and antimicrobial activity against degradative bacterial species isolated from tuna (*Thunnus* sp.) and swordfish (*Xiphias gladius*). *J. Food Sci. Technol.* **2018**, *55*, 4256–4265. [CrossRef] [PubMed]
8. Serio, A.; Fusella, G.C.; Lopez, C.C.; Sacchetti, G.; Paparella, A. A survey on bacteria isolated as hydrogen sulfide-producers from marine fish. *Food Control* **2014**, *39*, 111–118. [CrossRef]
9. Itoh, D.; Koyachi, E.; Yokokawa, M.; Murata, Y.; Murata, M.; Suzuki, H. Microdevice for on-site fish freshness checking based onk-value measurement. *Anal. Chem.* **2013**, *85*, 10962–10968. [CrossRef]
10. Liu, X.; Huang, Z.; Jia, S.; Zhang, J.; Li, K.; Luo, Y. The roles of bacteria in the biochemical changes of chill-stored bighead carp (*Aristichthys nobilis*): Proteins degradation, biogenic amines accumulation, volatiles production, and nucleotides catabolism. *Food Chem.* **2018**, *255*, 174–181. [CrossRef]
11. Odeyemi, O.A.; Burke, C.M.; Bolch, C.C.; Stanley, R. Seafood spoilage microbiota and associated volatile organic compounds at different storage temperatures and packaging conditions. *Int. J. Food Microbiol.* **2018**, *280*, 87–99. [CrossRef]
12. Wang, X.-Y.; Xie, J. Study on the volatile organic compounds and its correlation with water dynamics of bigeye tuna (*Thunnus obesus*) during cold storage. *Molecules* **2019**, *24*, 3119. [CrossRef] [PubMed]
13. Wu, L.; Pu, H.; Sun, D.-W. Novel techniques for evaluating freshness quality attributes of fish: A review of recent developments. *Trends Food Sci. Technol.* **2019**, *83*, 259–273. [CrossRef]
14. Rasekh, M.; Karami, H.; Wilson, A.; Gancarz, M. Classification and identification of essential oils from herbs and fruits based on a MOS electronic-nose technology. *Chemosensors* **2021**, *9*, 142. [CrossRef]
15. Semeano, A.T.; Maffei, D.F.; Palma, S.; Li, R.W.; Franco, B.D.; Roque, A.C.; Gruber, J. Tilapia fish microbial spoilage monitored by a single optical gas sensor. *Food Control* **2018**, *89*, 72–76. [CrossRef] [PubMed]
16. Luan, L.; Fu, S.; Yuan, C.; Ishimura, G.; Chen, S.; Chen, J.; Hu, Y. Combined effect of superchilling and tea polyphenols on the preservation quality of hairtail (*Trichiurus haumela*). *Int. J. Food Prop.* **2017**, *20*, S992–S1001. [CrossRef]
17. Heising, J.; Bartels, P.; van Boekel, M.; Dekker, M. Non-destructive sensing of the freshness of packed cod fish using conductivity and pH electrodes. *J. Food Eng.* **2014**, *124*, 80–85. [CrossRef]
18. Sun, J.; Zhang, R.; Zhang, Y.; Liang, Q.; Li, G.; Yang, N.; Xu, P.; Guo, J. Classifying fish freshness according to the relationship between EIS parameters and spoilage stages. *J. Food Eng.* **2018**, *219*, 101–110. [CrossRef]

19. Yao, L.; Luo, Y.; Sun, Y.; Shen, H. Establishment of kinetic models based on electrical conductivity and freshness indictors for the forecasting of crucian carp (*Carassius carassius*) freshness. *J. Food Eng.* **2011**, *107*, 147–151. [CrossRef]
20. Song, Y.; Liu, L.; Shen, H.; You, J.; Luo, Y. Effect of sodium alginate-based edible coating containing different anti-oxidants on quality and shelf life of refrigerated bream (*Megalobrama amblycephala*). *Food Control* **2011**, *22*, 608–615. [CrossRef]
21. Wang, X.-Y.; Xie, J. Growth kinetics and spoilage potential of co-culturing acinetobacter johnsonii and pseudomonas fluorescens from bigeye tuna (*Thunnus obesus*) during refrigerated storage. *Curr. Microbiol.* **2020**, *77*, 1637–1646. [CrossRef]
22. Dyer, W.J. Amines in Fish Muscle: I. Colorimetric determination of trimethylamine as the picrate salt. *J. Fish. Res. Board Can.* **1945**, *6d*, 351–358. [CrossRef]
23. Qian, Y.-F.; Xie, J.; Yang, S.-P.; Wu, W.-H.; Xiong, Q.; Gao, Z.-L. In vivo study of spoilage bacteria on polyphenoloxidase activity and melanosis of modified atmosphere packaged Pacific white shrimp. *Food Chem.* **2014**, *155*, 126–131. [CrossRef]
24. Gibson, A.M.; Bratchell, N.; Roberts, T.A. The effect of sodium chloride and temperature on the rate and extent of growth of Clostridium botulinum type A in pasteurized pork slurry. *J. Appl. Bacteriol.* **1987**, *62*, 479–490. [CrossRef]
25. Li, D.; Zhang, J.; Song, S.; Feng, L.; Luo, Y. Influence of heat processing on the volatile organic compounds and microbial diversity of salted and vacuum-packaged silver carp (*Hypophthalmichthys molitrix*) fillets during storage. *Food Microbiol.* **2018**, *72*, 73–81. [CrossRef] [PubMed]
26. Li, Y.; Jia, S.; Hong, H.; Zhang, L.; Zhuang, S.; Sun, X.; Liu, X.; Luo, Y. Assessment of bacterial contributions to the biochemical changes of chill-stored blunt snout bream (*Megalobrama amblycephala*) fillets: Protein degradation and volatile organic compounds accumulation. *Food Microbiol.* **2020**, *91*, 103495. [CrossRef]
27. Sofra, C.; Tsironi, T.; Taoukis, P.S. Modeling the effect of pre-treatment with nisin enriched osmotic solution on the shelf life of chilled vacuum packed tuna. *J. Food Eng.* **2018**, *216*, 125–131. [CrossRef]
28. Lyu, F.; Gao, F.; Ding, Y. Effects of gamma radiation combined with cinnamon oil on qualities of smoked salmon slices inoculated with *Shewanella putrefaciens*. *Food Sci. Nutr.* **2018**, *6*, 806–813. [CrossRef] [PubMed]
29. Feng, X.; Ng, V.K.; Mikš-Krajnik, M.; Yang, H. Effects of fish gelatin and tea polyphenol coating on the spoilage and degradation of myofibril in fish fillet during cold storage. *Food Bioprocess Technol.* **2017**, *10*, 89–102. [CrossRef]
30. Shen, S.; Jiang, Y.; Liu, X.; Luo, Y.; Gao, L. Quality assessment of rainbow trout (*Oncorhynchus mykiss*) fillets during super chilling and chilled storage. *J. Food Sci. Technol.* **2015**, *52*, 5204–5211. [CrossRef] [PubMed]
31. Shi, C.; Yang, X.; Han, S.; Fan, B.; Zhao, Z.; Wu, X.; Qian, J. Nondestructive prediction of tilapia fillet freshness during storage at different temperatures by integrating an electronic nose and tongue with radial basis function neural networks. *Food Bioprocess Technol.* **2018**, *11*, 1840–1852. [CrossRef]
32. Sun, J.; Wang, Q.J.; Huang, J.; Hou, Y.D.; Chen, Y.F.; Su, X.R. Influence of heating temperature on the development of volatile compounds in bigeye tuna meat (*Thunnus obesus*) as assessed by E-nose and SPME-GC/MS. *Int. Food Res. J.* **2013**, *20*, 3077–3083.
33. Medved'ová, A.; Šipošová, P.; Mančušková, T.; Valík, L. The effect of salt and temperature on the growth of fresco culture. *Fermentation* **2018**, *5*, 2. [CrossRef]
34. Gu, X.; Feng, L.; Zhu, J.; Li, Y.; Tu, K.; Dong, Q.; Pan, L. Application of gas sensors for modelling the dynamic growth of Pseudomonas in pork stored at different temperatures. *Meat Sci.* **2021**, *171*, 108282. [CrossRef] [PubMed]
35. Yang, Z.-Q.; Tao, X.-Y.; Zhang, H.; Rao, S.-Q.; Gao, L.; Pan, Z.-M.; Jiao, X.-A. Isolation and characterization of virulent phages infecting *Shewanella baltica* and *Shewanella putrefaciens*, and their application for biopreservation of chilled channel catfish (*Ictalurus punctatus*). *Int. J. Food Microbiol.* **2019**, *292*, 107–117. [CrossRef] [PubMed]
36. Taoukis, P.S.; Koutsoumanis, K.; Nychas, G.J.E. Use of time-temperature integrators and predictive modelling for shelf life control of chilled fish under dynamic storage conditions. *Int. J. Food Microbiol.* **1999**, *53*, 21–31. [CrossRef]
37. Tango, C.N.; Khan, I.; Park, Y.S.; Oh, D.H. Growth of Staphylococcus aureus in cooked ready-to-eat ground fish as affected by inoculum size and potassium sorbate as food preservative. *LWT Food Sci. Technol.* **2016**, *71*, 400–408. [CrossRef]
38. Ding, T.; Liao, X.-Y.; Dong, Q.-L.; Xuan, X.-T.; Chen, S.-G.; Ye, X.-Q.; Liu, D.-H. Predictive modeling of microbial single cells: A review. *Crit. Rev. Food Sci. Nutr.* **2018**, *58*, 711–725. [CrossRef] [PubMed]
39. Wang, H.; Hu, Z.; Long, F.; Guo, C.; Yuan, Y.; Yue, T. Early detection of Zygosaccharomyces rouxii—Spawned spoilage in apple juice by electronic nose combined with chemometrics. *Int. J. Food Microbiol.* **2016**, *217*, 68–78. [CrossRef] [PubMed]
40. Feng, H.; Zhang, M.; Liu, P.; Liu, Y.; Zhang, X. Evaluation of IoT-enabled monitoring and electronic nose spoilage detection for salmon freshness during cold storage. *Foods* **2020**, *9*, 1579. [CrossRef] [PubMed]
41. Zhu, S.; Luo, Y.; Hong, H.; Feng, L.; Shen, H. Correlation between electrical conductivity of the gutted fish body and the quality of bighead carp (*Aristichthys nobilis*) heads stored at 0 and 3 °C. *Food Bioprocess Technol.* **2013**, *6*, 3068–3075. [CrossRef]
42. Shi, C.; Cui, J.; Luo, Y.; Zhu, S.; Zhou, Z. Post-mortem changes of silver carp (*Hypophthalmichthys Molitrix*) stored at 0 °C assessed by electrical conductivity. *Int. J. Food Prop.* **2015**, *18*, 415–425. [CrossRef]
43. Huang, X.-C.; Yuan, Y.-H.; Wang, X.-Y.; Jiang, F.-H.; Yue, T.-L. Application of electronic nose in tandem with chemometric analysis for detection of Alicyclobacillus acidoterrestris-spawned spoilage in apple juice beverage. *Food Bioprocess Technol.* **2015**, *8*, 1295–1304. [CrossRef]
44. Prabhakar, P.K.; Vatsa, S.; Srivastav, P.P.; Pathak, S.S. A comprehensive review on freshness of fish and assessment: Analytical methods and recent innovations. *Food Res. Int.* **2020**, *133*, 109157. [CrossRef] [PubMed]

45. Edirisinghe, R.K.; Graffham, A.J.; Taylor, S.J. Characterisation of the volatiles of yellowfin tuna (*Thunnus albacares*) during storage by solid phase microextraction and GC–MS and their relationship to fish quality parameters. *Int. J. Food Sci. Technol.* **2007**, *42*, 1139–1147. [CrossRef]
46. Zhang, Q.; Ding, Y.; Gu, S.; Zhu, S.; Zhou, X.; Ding, Y. Identification of changes in volatile compounds in dry-cured fish during storage using HS-GC-IMS. *Food Res. Int.* **2020**, *137*, 109339. [CrossRef] [PubMed]
47. Ercolini, D.; Russo, F.; Nasi, A.; Ferranti, P.; Villani, F. Mesophilic and psychrotrophic bacteria from meat and their spoilage potential in vitro and in beef. *Appl. Environ. Microbiol.* **2009**, *75*, 1990–2001. [CrossRef]
48. Zhang, Y.; Ma, X.; Dai, Z. Comparison of nonvolatile and volatile compounds in raw, cooked, and canned yellowfin tuna (*Thunnus albacores*). *J. Food Process. Preserv.* **2019**, *43*, e14111. [CrossRef]
49. Wang, H.; Liu, X.; Zhang, Y.; Lu, H.; Xu, Q.; Shi, C.; Luo, Y. Spoilage potential of three different bacteria isolated from spoiled grass carp (*Ctenopharyngodon idellus*) fillets during storage at 4 °C. *LWT Food Sci. Technol.* **2017**, *81*, 10–17. [CrossRef]

foods

MDPI

Review

Assessment and Prediction of Fish Freshness Using Mathematical Modelling: A Review

Míriam R. García [1], Jose Antonio Ferez-Rubio [1,2] and Carlos Vilas [1,*]

[1] Research Group on Biosystems and Bioprocess Engineering (Bio^{2}eng), IIM-CSIC, 36208 Vigo, Spain; miriamr@iim.csic.es (M.R.G.); jaferez@cebas.csic.es (J.A.F.-R.)
[2] Research Group on Microbiology and Quality of Fruit and Vegetables, CEBAS-CSIC, 30100 Murcia, Spain
* Correspondence: carlosvf@iim.csic.es

Abstract: Fish freshness can be considered as the combination of different nutritional and organoleptic attributes that rapidly deteriorate after fish capture, i.e., during processing (cutting, gutting, packaging), storage, transport, distribution, and retail. The rate at which this degradation occurs is affected by several stress variables such as temperature, water activity, or pH, among others. The food industry is aware that fish freshness is a key feature influencing consumers' willingness to pay for the product. Therefore, tools that allow rapid and reliable assessment and prediction of the attributes related to freshness are gaining relevance. The main objective of this work is to provide a comprehensive review of the mathematical models used to describe and predict the changes in the key quality indicators in fresh fish and shellfish during storage. The work also briefly describes such indicators, discusses the most relevant stress factors affecting the quality of fresh fish, and presents a bibliometric analysis of the results obtained from a systematic literature search on the subject.

Keywords: mathematical modelling; fish quality; fish freshness; bibliometric analysis; predictive microbiology; stress variables; quality degradation

Citation: García, M.R.; Férez-Rubio, J.A.; Vilas, C. Assessment and Prediction of Fish Freshness Using Mathematical Modelling: A Review. *Foods* **2022**, *11*, 2312. https://doi.org/10.3390/foods11152312

Academic Editor: Robert L. Buchanan

Received: 16 June 2022
Accepted: 25 July 2022
Published: 2 August 2022

1. Introduction

The main causes of food discarding among consumers and retailers are the food aspect, outdating, and safety uncertainty [1]. Damage and spoilage of foods lead to around 15% of waste, which increases to 35% if food is subject to inadequate storage and transport conditions [2,3]. Mathematical modelling describing the evolution of food quality indicators, under given storage and transport conditions, is central to minimising food waste [2]. Therefore, the prediction of fresh fish quality is a major challenge for the food industry, distributors and retailers to adjust prices and minimise food waste.

Fresh fish and shellfish are highly perishable products due to their biological composition. Under normal handling chilled or refrigerated storage conditions, their shelf life is limited by enzymatic, chemical and microbiological spoilage. Fresh fish is stored, transported and distributed in boxes of high-density poly-ethylene filled with ice. Other common conservation methods for fresh fish are the transport and storage in tubs with water and ice or at superchilling temperatures [4,5]. From fish capture to consumer consumption, there are several factors affecting fish quality, being temperature the most relevant. The prediction of fish quality in all these cases is critical to determine the price of the product and sell it before it is not of sufficient quality, or even safety, for the consumer.

Quality in fresh fish, and generally in food, is a broad concept that involves different attributes (chemical, physical, microbiological, sensory) which can be measured either directly or indirectly. In the last decades, several analytical techniques have arisen [6] including biosensors to measure microbial pathogens or biogenic amines [7]; electronic noses or electronic tongues for volatile compounds, K-value or TVB-N [8,9]; or hiperespectral imaging to determine moisture content or texture [10,11]; among other. The selection of the analytical method depends on the selected indicator, but also on properties such as accuracy, reliability, portability, rapidity, easiness to use and analyse the results, time consumption and price. Ideally, the methods should be also non-destructive and non-invasive [12]. Recent works [6,13,14] present exhaustive reviews regarding the different analytical methods considered in the literature to measure the most commonly used quality indicators for fish freshness assessment. Typically, assessment methods focus on fish quality at the moment of measurement but are unable to predict quality changes in the following days. Prediction requires the use of appropriate mathematical models. It is important to mention that some authors, for instance, ref. [15–17], use the term *predictive model* to denote models that correlate freshness indicators with experimental measurements (pH, TVB-N, hyperspectral imaging, Electronic nose data, etc.). Typically, some type of regression is used to obtain these models. Although these works are common and necessary, in this review, we will use the term *predictive* for those models able to forecast the future evolution of the freshness indicators. Otherwise, we will use the term *assessment* or *estimation*. Mathematical models for the prediction of fresh fish quality are diverse and difficult to classify.

In this review, we propose the use of general features of the mathematical structure to organise the different modelling alternatives, as illustrated in Figure 1. The final objective of the model is either to estimate (using indirect online measurements) or to predict one or more chemical, physical, microbiological or sensory attributes that are indicators of the consumer's perception. Such attributes (odour, texture, TVB-N, spoilage bacteria, etc.) are the output of the model (blue part of the figure). The input of the model (red part of the figure) can be any external factor (temperature, pH, fish feed, etc.) affecting the quality indicators. The model (orange part of the figure) provides a quantitative relationship between inputs and outputs either using empirical relationships or inspired by known mechanisms of fish quality degradation.

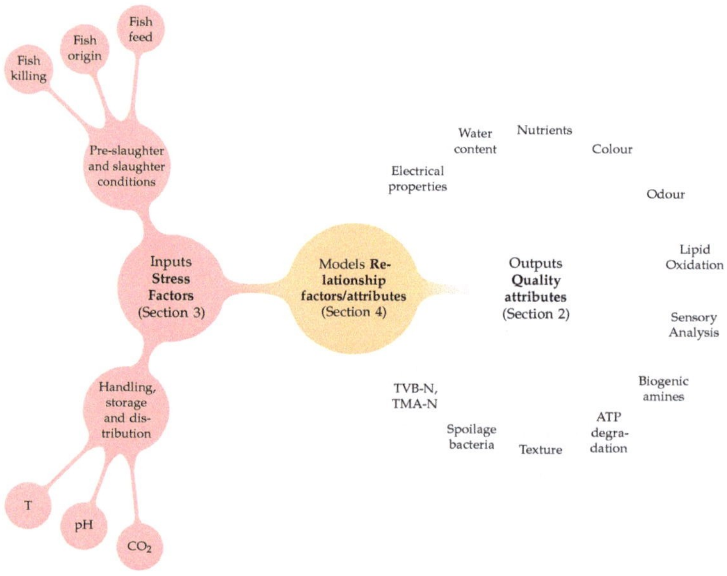

Figure 1. Fish quality models are described attending to their Quality attributes (Section 2), Stress factors (Section 3) and models, i.e., mathematical relationships between attributes and stress factors (Section 4).

This review is organised following the structure presented in Figure 1. It begins by summarising the results of the systematic literature review performed to investigate the most typical quality attributes used to define fish quality, including a brief description of the procedure followed to perform such a review. Secondly, it introduces the relevant stress factors affecting fish freshness. Such stress factors can be split into two groups: Pre-slaughter/slaughter conditions, and handling, storage and distribution conditions. In Section 4, we review the mathematical relationships or models that allow us to describe and predict fish quality attributes (outputs) as a function of the stress factors (inputs). As a final remark, we will discuss what we think are the main challenges for modelling fish quality and possible alternative solutions to consider in the future. As a general rule, we will denote by fish quality any positive attribute, either nutritional, organoleptic or a combination, in fresh fish, and therefore related to fish freshness. In this regard, issues not related to fresh fish were not considered in this review, such as processed fish (cooked, sterilised, etc.) or social aspects such as food security or production of added-value compounds.

2. Quality Attributes (Model Outputs)

Food quality in fresh fish and fresh fishery products is a broad and complex concept defined by a set of attributes (quality attributes) that are either nutritional, organoleptic, or a combination of both. The levels (concentrations) of some chemical compounds describe nutritional quality attributes. Examples of nutrients include vitamins, bioactive forms of oligo-elements, essential amino acids, digestible proteins or unsaturated fatty acids, among others. On the other hand, colour, texture, flavours or aroma are attributes defining the organoleptic quality of a particular food product. Organoleptic variations in fish are caused by changes in chemical, microbiological and physical properties. Sometimes, quality attributes are defined by the level of a particular biological, chemical or biochemical factor (e.g., the concentration of a given vitamin, nucleotide, enzyme, or bacteria, among others), although usually, it is the result of a combination of different factors. For example, colour is the (observable) result of a certain combination of pigments on a given food matrix that have been produced or consumed under the action of many biochemical transformations. Other quality attributes, such as freshness, are defined by the combination of nutritional and organoleptic properties which deteriorate with time. Freshness can quantitatively be described using different sensory scores, such as the Quality Index Method (QIM), or other simpler indicators, such as the shelf life dating of a given food product, either to specify the last date when the product must be sold ("sell by date"), the "high-quality" period or the date when the product must be removed from the store [18].

We have analysed this diversity of quality attributes studied in the literature, and their interconnections, by conducting a systematic bibliographic review and bibliometric analysis of the articles (in the *Web of Science Core Collection* database, *Food Science and Technology* category). Figure 2 summarises the procedure we have followed to perform the systematic literature search.

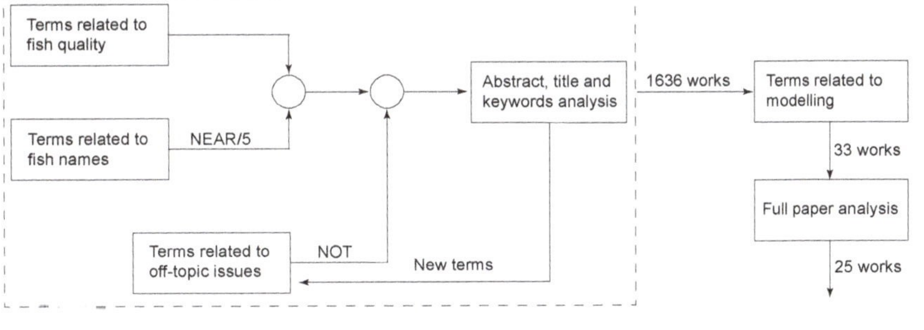

Figure 2. General scheme of the procedure followed in this manuscript to perform the systematic literature search.

Inside the dashed rectangle, we represent the iterative part of the procedure. First, we constructed two lists. One of the lists contained terms related to fish quality, such as freshness, K-value, and lipid oxidation, among others. The other one contained general terms such as seafood or shellfish and fish species names. In the search, words from both lists must appear in a major field (abstract, title or author keywords). We have found many works that included words from both lists, in which the quality terms were not related to fish. To avoid these situations, we used a maximum separation of five words (NEAR/5) between the terms in both lists. This separation was chosen by the trial and error method with the objectives of avoiding the exclusion of manuscripts within the scope of this work (shorter separations) and minimising the number of off-topic manuscripts (larger separations). Since the focus of this review is on fresh fish, we have excluded from the search (NOT) those words related to off-topic issues such as processing (sterilisation, modified atmosphere, etc.), production of added-value compounds (essential oils, gelatin, etc.), food security or other social aspects, among others. The initial search still included works not related to fresh fish quality so we identified, new terms that should be avoided, and repeated the search. The list of terms included and excluded in the systematic search is presented in Appendix A. This iterative procedure resulted in 1636 works. Details about such works and a bibliometric analysis are available online in the repository [19]. In this material, the interested reader can access 2 types of reports: a PDF file that includes the main results of the general search, and 12 interactive reports for sub-collections defined as the group of works where one of the 12 selected quality attributes is mentioned in a major field. Table 1 summarises the results of the systematic review for each of the 12 categories or groups. The table presents, for each category, the number of works, total citations, average citations per work, and the most cited works.

Table 1. Most employed quality attributes in the literature. The number of total citations per year is used to obtain the most cited articles. The terminology used for each attribute was: Lipid oxidation (fatty acid*, lipid oxidation, TBA, TBARS, thiobarbituric), Sensory analysis (QIM, QSM, sensory analysis, sensory evaluation, sensory method, TVB-N/TMA-N (TVB-N/TMA-N), Spoilage bacteria (SSO, spoilage bacteria, spoilage microorganism*), Texture properties (texture, hardness, firmness), Biogenic amines (biogenic amine*), Odour (odour, odor), Colour (colour, color, chromatism), Nutrients (nutrient*, vitamin), Water content/activity (water content, water activity) and Electrical properties (electrical properties, conductance, conductivity).

Quality Attribute	Citation Counts	No. Works	Avg. Citations per Work	Most Cited Works
Lipid oxidation	10,875	474	22.9	Herrero [12], Richards and Hultin [20], Grigorakis et al. [21]
Sensory analysis	5295	209	25.3	Ólafsdóttir et al. [22], Al Bulushi et al. [23], Olafsdottir et al. [24]
TVB-N, TMA-N	4386	185	23.7	Pacquit et al. [25], Papadopoulos et al. [26], Ruiz-Capillas and Moral [27]
Spoilage bacteria	3876	155	25.0	Al Bulushi et al. [23], Gram et al. [28], Dalgaard [29]
Texture	3587	176	20.4	Herrero [12], Olafsdottir et al. [24], Alasalvar et al. [30]
ATP degradation	3509	124	28.3	Ólafsdóttir et al. [22], Veciana-Nogués et al. [31], Jones et al. [32]
Biogenic amines	3282	112	29.3	Al Bulushi et al. [23], Veciana-Nogués et al. [31], Kim et al. [33]
Odour	3066	119	25.8	Papadopoulos et al. [26], Ramanathan and Das [34], Kawai [35]
Colour	2830	149	19.0	Pacquit et al. [25], Kuswandi et al. [36], Huang et al. [37]
Nutrients	704	62	11.3	Chakraborty and Raj [38], Moreda-Piñeiro et al. [39], Palaniappan and Vijayasundaram [40]
Water content/activity	573	32	17.9	Cakli et al. [41], Morzel et al. [42], Raju et al. [43]
Electrical properties	381	11	34.6	Olafsdottir et al. [24], Vaz-Pires et al. [44], Yao et al. [45]

Additionally, we have incorporated in the repository a PDF file that contains the information included in the interactive reports. All the reports, iterative or not, include a general analysis of the collection (including, for example, time span, collaboration index, authors per article, citations per year), and analysis of the countries, authors, articles and journals. Networks of co-citations and keyword co-occurrences were also incorporated. Moreover, the collections can be downloaded as an Excel file and the graphs and tables can be manipulated to, for example, constrain the works to those where certain words, selected by the user, are mentioned in the abstract.

Most of the manuscripts were research articles, dating back from 1949 and with citations increasing homogeneously since then, except for two jumps in publications in 1977 and 2002. Around 100 works were published in 2021, the last year considered in the review. The most cited works were written by Ólafsdóttir et al. [22] with a review on methods to evaluate fish freshness and by Ryder [46] with a method to measure ATP and its breakdown products to estimate the K_I-value. The most productive countries have been China, USA and Spain. These are also the countries where the documents with the greatest impact (in terms of average article citations) were produced. Journal of Food Science (145 documents), Food Chemistry (140 documents) and Journal of Agricultural and Food Chemistry (72 documents) are the journals with the largest number of publications. The most productive authors were from the northwest of Spain (S. P. Aubourg, Barros-Velázquez), whereas the authors with the greatest impact were Ólafsdóttir and Dalgaard from Iceland and Denmark, respectively.

After the iterative procedure was finished, we refined the search (see Figure 2) to find those works related to mathematical modelling. We read the resulting 33 manuscripts and kept those (25) that fitted within the scope of this review.

In the next sections, we define each of the attributes with references to the main works in the area. For a deeper discussion about these quality attributes we recommend some recent comprehensive reviews [6,13,14].

2.1. Lipid Oxidation

Lipid oxidation is the attribute gathering the greatest attention, both in terms of the number of citation counts and of works, with a remarkable increase of publications during the first decade of the current century remaining almost invariant during recent years. The most cited researchers on this topic are M. P. Richards and Y. Ozogul. Although fatty acids are also nutrients (described in Section 2.10), we have decided to consider them separately because of the importance of lipid oxidation. Fish constitutes the main source of polyunsaturated fatty acids (PUFA), up to 40% of long chain fatty acids [47], with quantity and composition changing with the catching period and depending on whether the fish is wild or cultured [21]. PUFA are easily oxidised into aldehydes, responsible for changes in flavour, texture and odour, known as rancidity [6,12,20,22,48], and are highly affected in most of the cases by previous fish bleeding [20]. In addition, the oxidation reaction can decrease the nutritional quality of food and certain oxidative products are potentially toxic [48]. Primary lipid peroxidation products (peroxide value being its most common measure) include hydroperoxide that is unstable and decomposes to generate various secondary products, such as aldehydes that contribute to fish rancidity [12]. The most common method to measure aldehydes is the thiobarbituric acid-reactive substance (TBARS) test [22].

The main advantage of this indicator is that the oxidation of unsaturated lipids produces alterations in smell, taste, texture, colour, and nutritional value [22]. Therefore, it provides us with a global measure of fish freshness. However, analysis of PUFA is a destructive method so the fish sample analysed cannot be commercialised.

2.2. Sensory Analysis

Despite being the second most relevant quality attribute in terms of publications (with G. Ólafsdóttir being to a large extent the most cited author), sensory analysis is the most used method to assess freshness, probably because it depends on a combination of the other quality attributes [22]. In this group, we include any form of measure or interpretation of fish freshness perceived by the senses of sight, smell, taste, or touch. They can be assessed by a trained panel or by consumers' subjective opinions about preferences. The most common method in Europe is the Quality Sensory Method (QSM) based on the Council Regulation (EC) No 2406/96 for marketing standards [49]. The output of this method is a discrete value that classifies fish in four levels of freshness (Extra, A, B, and Not Admitted). This method is based on shared characteristics in fresh fish and therefore is common to different fish species. For specific tests, the most common method is the Quality Index Method (QIM) [50]. The output of the QIM is also a discrete value $(0, 1, 2, \ldots, n)$, where lower values correspond to fresher fish. The criteria to select the value and the number of levels (n) depend on the fish species considered.

The main advantages of these methods are: (i) they are minimally invasive, and do not involve the destruction of the sample, since the sense of taste is not included; (ii) they can be used to estimate fish shelf life by agreeing from which level the food is not considered of sufficient quality to be sold to the consumer; (iii) instead of focusing on one particular feature, they provide a global evaluation of fresh fish. However, sensory analysis methods are highly subjective and depend, to a large extent, on the expertise of evaluators. The cost associated with the use of a panel of evaluators is another disadvantage. These disadvantages can be alleviated by using analytical techniques -colourimeters, electronic noses, hardness testers- instead of a panel of experts .

Another alternative is the Global Stability Index (GSI), a score gathering the influence of many attributes, sensory or not. The GSI is computed using the following general expression [51]:

$$GSI = 1 - \sum_{i=1}^{n} \alpha_i \frac{A_i - A_{i,0}}{L_i - A_{i,0}} \tag{1}$$

A_i and $A_{i,0}$ are, respectively, the values of a given attribute, for example, TVB-N or K-value, at assessment time t and at initial time $t = 0$. α_i is the weight given to each attribute, n is the total number of attributes considered, and L_i is the spoilage threshold for attribute A_i.

2.3. TVB-N/TMA-N

The formation of volatile nitrogen bases, such as trimethylamine (TMA-N), dimethylamine (DMA), or ammonia, from the reduction of trimethylamine oxide (TMAO), is a widely investigated cause of fish odour. There was a substantial increase in interest during the 10 years after 1996, being Ruiz-Capillas the most cited author. The total amount of volatile bases (TVB-N), as well as individual methylamines, have been extensively used as indicators of quality degradation in postmortem fish [52,53]. The reduction reactions are catalysed by bacteria such as *Shewanella* spp. or *Pseusomonas* spp., during fish spoilage [54]. Some authors [17,26,52] have argued that these volatile bases are poor freshness indicators for some fish species because of the low content, or even absence, of TMAO. However, the FAO specified a maximum allowable level in international trading of 10 mg of TMA-nitrogen per 100 g fish muscle [55].

2.4. Spoilage Bacteria

Research works regarding this quality attribute have increased since 1994 without any deceleration in the last years as previously described attributes. Dalgaard, Gram and Bulushi, are, markedly, the most cited authors. Fish freshness deteriorates rapidly with the growth of Gram negative psychrophilic or psychrotrophic bacteria, named spoilage bacteria (SSO), due to their ability to reduce TMAO and to produce hydrogen sulphide [28]. Common spoilage bacteria of fish at chilled temperatures are: (1) *Shewanella putrefaciens* for being H_2S-producing bacteria and with acceptability limits around 10^7 CFU/g [56] or even slightly higher $10^{7.02}$ CFU/g [57], (2) *Pseudomonas* spp. with the same acceptability limits [56–58], or larger when analysing for example *Pseudomonas psychrofila* ($10^{8.5}$ CFU/g [59]), and (3) broad measurements such as Total Viable Counts (TVC) with a limit of 10^6 CFU/g [16]. More information can be found in [28,60,61].

Since microbial growth and metabolism is the major cause of food spoilage [62], spoilage bacteria is a key indicator of fish freshness and shelf life. The main disadvantage is that the procedure to determine bacterial concentration is tedious and time consuming.

2.5. Texture Properties

Bibliometric analysis for texture attributes reveals that many works mentioned this attribute without being the focus of the work. Among the works considering texture as a relevant focus, we would like to highlight the review on fish texture [63] and a research article assessing texture, among other chemical and sensory characteristics, of sea bream [30].

Texture can be evaluated using minimally invasive techniques. However, contrary to other food matrices, such as beef, texture in fish is not usually regarded as a relevant freshness indicator, particularly when considering fresh fish. As mentioned in [64], it should be considered in combination with other indicators, such as colour and odour. Use of texture as a fish quality indicator limits to either cooked or frozen fish stored for long periods, where an increase in toughness and dryness of the tissues can be observed [63]. After cooking, the taste of fresh fish is associated with firm meat that goes to dry, crumbly with short, tough fibres for deteriorated fish. However, raw fish maximum toughness commonly occurs after 1–2 days of storage, corresponding to the minimum pH and *rigor mortis* [63]. Texture properties are typically considered in the evaluation of the QIM [65].

2.6. ATP Degradation

Although the number of citations places this attribute in the sixth position, it is the third one in average citations per work. Most of the works were published in the nineties and during the past 10 years, with G. Ólafsdóttir being the most cited author .

After fish death, ATP transforms, within the first 24-48 h, into inosine 5′-monophosphate (IMP) in three steps, producing adenosine diphosphate (ADP) and adenosine monophosphate (AMP) [66,67]. IMP degradation continues on a cascade of reactions that produces Inosine 5′-monophosphate (Ino) and hypoxanthine (Hx), which is further decomposed in other compounds such as xanthine and uric acid [52,67]. This cascade of reactions occurs in the order of days to weeks, depending on the storage temperature and bacterial concentration [68]. IMP is related to the pleasant sweet and meaty flavors in fresh fish, *umami* flavour [69,70], whereas Hx is responsible of unpleasant bitterness [71,72]. The K-value and K_I-value are two of the most widely employed indicators to evaluate freshness. They are defined as the following function of the ATP degradation products [68,73–76]:

$$\text{K-value} = \frac{\text{Ino} + \text{Hx}}{\text{ATP} + \text{ADP} + \text{AMP} + \text{IMP} + \text{Ino} + \text{Hx}} \cdot 100 \quad (2)$$

$$\text{K}_I\text{-value} = \frac{\text{Ino} + \text{Hx}}{\text{IMP} + \text{Ino} + \text{Hx}} \cdot 100 \quad (3)$$

The K-value has been also correlated with the freshness of different fish species [73]. The authors found that *very high* grade corresponded with K-value lower than 10%. *High* grade individuals had K-values lower than 20% or 30%, depending on the fish species. Fishes with K-value up to 50% correlated with *medium* grade. Finally, K-value larger than 50–70% was obtained for *low* grade samples.

The main advantages of indexes based on the degradation of ATP (K-value, K_I-value) are their reliability [66] and, as mentioned above, their direct connection with fish flavor. The main disadvantages are that the evaluation of K-value or K_I-value requires the destruction of the sample and their usefulness depends on the fish species being examined [64].

2.7. Biogenic Amines

Works studying the correlation between spoilage and biogenic amines are homogeneously increasing since the nineties, being the most cited articles by Bulushi, Ruiz-Capillas and Vecianogues. Biogenic amines here are non-volatile amines (histamine, cadaverine and putrescine) formed by decarboxylation of amino acids (histidine, lysine and ornithine, respectively). Although TMA-N and TMAO are also biogenic amines, they are not considered in this group because TMA-N is volatile and it is a result of the degradation of TMAO, therefore being one of the main contributors to the formation of TVB-N, as discussed in Section 2.3. Within non-volatile biogenic amines, histamine is the most studied due to its toxicity and allergic potential, but it is unable to correlate with the level of spoilage for different fish and conditions [23]. Cadaverine is the biogenic amine that can be used as a spoilage indicator (for example, for salmonid fish, values less than 10 mg/kg indicate good quality [77]). Putrescine, however, is not a good spoilage index because its amine, ornithine, is not present in all fish species (for example, it is missing in tuna). Alternatively, there are amine indexes combining different biogenic amines such as the amine index (AI):

$$\text{AI} = \frac{\text{putrescine} + \text{cadaverine} + \text{histamine}}{\text{Total amines}} \cdot 100$$

$$\text{Total amines} = \text{putrescine} + \text{cadaverine} + \text{histamine} + \text{tyramine} +$$

$$\text{tryptamine} + \text{methylamine} + \text{spermidine} + \text{spermine}$$

or the chemical index:

$$\text{Chemical index} = \frac{\text{putrescine} + \text{cadaverine} + \text{histamine}}{1 + \text{spermidine} + \text{spermine}}$$

For relationships between amine content and spoilage level for different fish species, the reader is referred to Table 2 in [23].

2.8. Odour

The number of citations, works, and average citations per work, regarding this attribute, are similar to ATP degradation and Biogenic amines but with a homogeneous distribution of the number of articles per year. X. Y. Huang and V. Papadopoulos are the authors with the largest number of citations regarding this attribute. Same as texture and colour, this quality attribute is usually studied in the literature together (or even correlated) with other attributes. The major aromatic compounds identified related to spoilage levels are fatty acids profiles, aldehydes, ketones, trimethyl amine (TMA), and volatile organic compounds [78]. Typically, studies focus on assessing odour following one standard sensory index (see, for example, ref. [26], for sea bass assessment). Commonly, freshness is associated with iodine shellfish and seaweed smell, and spoiled fish with muddy, putrid, faecal, pungent, smell to ammonia or ink smell in cephalopods and acidic in shrimps. Odour can be evaluated using non-invasive methods. However, using a panel of evaluators to assess this indicator is subjective and expensive. Analytical methods could be used to analyse aromatic compounds. However, in most cases, this would involve the destruction of the sample. Although, if electronic noses are properly calibrated and validated odour can be a very interesting quality indicator .

2.9. Colour

Despite being an important indicator for consumers, it is one of the attributes with the lowest number of citations and average citations per work. However, the interest of the scientific community seems to increase during the last 15 years. The most cited author in this field is A. Pacquit. Colour changes have been also used as an indicator for quality degradation in combination with other attributes or as a part of sensory analysis. Colour degradation kinetics are the result of changes in the pigments due to some biochemical transformations. For example, the oxidation of myoglobin and haemoglobin turns the flesh colour from red to brown. As pointed out by [6], some compounds present in fish, such as amines or ammonia, may react with the oxidised liquid causing serious browning.

The main advantage of using this feature to assess quality is that, as mentioned above, it is an important indicator for consumers. Besides, colour measurements can be obtained using non-invasive or minimally invasive techniques. However, fish skin is heterogeneous in many species. Therefore, the results provided by devices to measure colour, such as colourimeters, will vary depending on the regions of the skin being measured. A trained panel of experts could be used to globally evaluate the colour of fish skin. However, this alternative has the same disadvantages mentioned for sensory analysis and odour, i.e., it is expensive and subjective.

2.10. Nutrients

Fish is highly appreciated as a healthy food product [16,79,80] mainly because it is rich in nutrients such as high-quality proteins, fatty acids, and vitamins, among others. However, despite its importance, nutrients are not usually considered as a factor for determining fish freshness, except for the study of fatty acids, already considered in Section 2.1. This is the attribute with lower average citations per work, being the most cited author K. Chakraborty with 48 citations. Most of the highly cited manuscripts found in the systematic search regarding nutrients focused on different aspects [38–40,81], such as the characterisation of vitamin compounds; the use of fish oil; correlations between arsenic bioavailability and nutrient content; among others, not directly related to fish freshness. This is probably because the rate of degradation of most nutrients is slower than other indicators such as the QIM and when changes are noticeable, fish is already spoiled. Another disadvantage is that the assessment of nutrient content requires destructive methods.

2.11. Water Content/Activity

The number of citation counts and the average citations per work for this attribute are the second lowest in the list, after nutrients. This is probably because it is usually considered a stress factor and not a quality indicator itself. The most cited author is S. Cakli. Water content is related and can be described from water activity using the moisture sorption isotherm curve. Although this relationship is a non-linear function, water content increases with water activity and vice versa, and therefore both are essential quality parameters related to important textural attributes such as juiciness [12], mainly related to texture and flavour. However, as in the case of other indicators, destructive methods are used for the assessment of water content.

2.12. Electrical Properties

This attribute has gained the lowest attention in terms of the number of works. However, works focusing on this attribute are highly cited, with a review of different multi-sensors gathering most of the citations [24]. In fact, the average number of citations per work is the largest of all attributes considered in this review. Enzymatic and bacterial decomposition of proteins and lipids after fish death results in the formation of charged molecules which increase the electrical conductance (EC) of the muscle [6,82]. The loss of this kind of nutrients can be, therefore, correlated with the increase of EC. Autolytic spoilage is also responsible for cell membrane disruption, which allows the liquids to pour out increasing the EC [24,45].

The main advantage of this indicator is that it can be measured using minimally invasive techniques, avoiding the destruction of the sample. However, the correlation between EC increase and freshness degradation must be performed for the different fish species.

3. Stress Factors and Their Usual Models (Additionally, Named Secondary Models)

Food quality attributes are conditioned by the food matrix microstructure and composition as well as by several stress or environmental variables [18]. Chronologically, those factors modifying fish quality can be classified attending to the origin and slaughter conditions of the fish (pre-slaughter and slaughter) and due to handling, storage and distribution.

Fish composition and matrix microstructure depend on many factors such as the fish species, its size, age, and gender, whether it is lean or fat, fresh or saltwater fish, the fish feed, catching period (seasonal variability), geographical area or temperature of the catching waters. For example, fat, and therefore lipid oxidation, highly depends on the catching period and on whether the fish is wild or cultured [21]. Another example would be the cadaverine level which is considered a good spoilage index for wild fish, but not for aquaculture fish [23]. The slaughter procedure also affects fish quality. For example, bleeding affects lipid oxidation, preventing this oxidation in minced trout whole muscle, minced mackerel light muscle, and intact mackerel dark muscle [20]. Commonly, these factors are not modelled and, therefore, they are not considered in this review.

During handling, storage and distribution there are many external factors (stress factors) accelerating fish quality loss. Some attributes, like texture, can be affected by the just after-slaughter conditions, such as the glycolysis and *rigor mortis*, leading to gaping [63]. However, in general, outside-of-fish variables during storage and distribution are the parameters that can be manipulated to extend freshness and they are usually the focus of the mathematical models. Temperature is undoubtedly the most important and studied factor, although others such as pH [17] or CO_2, when stored under a modified atmosphere [83], have been also considered.

There are different ways of modelling the influence of temperature. Roughly, the modelling approaches can be classified into two groups: (i) models considering a direct influence of the temperature on shelf life, and (ii) models describing the effect of temperature on the degradation of biochemical compounds or on bacterial growth, which are then related to freshness. The first type consists of pure empirical input/output relationships. These models are discussed in Section 4.1. The second type consists of mechanistic-based relationships. These models do not provide a direct input/output expression, but a function of the relationship of temperature with a kinetic parameter with a major role in any of the quality attributes or outputs.

Certainly, the most common model to describe the effect of temperature, at least when the output of the model is a product of one or more biochemical reactions, is the Arrhenius model [84]:

$$K(T) = A \exp\left(\frac{-Ea}{RT}\right) \qquad \text{Arrhenius model}$$

where $K(T)$ is a degradation rate that depends on the Temperature (T) through an exponential expression. A is the pre-exponential factor, Ea is the reaction activation energy, and R is the universal gas constant.

The influence of temperature on bacterial spoilage (output) is more complex, with many models available in the literature [60], including the Arrhenius equation. Nevertheless, the most common one is the Ratkowsky or square root model to describe the change in maximum growth rate (μ_{max}) [85]:

$$\mu_{max}(T) = [b(T - T_{min})]^2 \qquad \text{Ratkowsky model}$$

being T_{min} the temperature at which growth is zero, and b the factor shaping the curvature of the function. The same functionality can be used to model the effect of temperature on the lag phase of spoilage bacteria [57].

Although the temperature is the major factor affecting fish freshness, there are other relevant stress variables such as pH (a major factor in texture [63]), water activity, salt concentration or concentration of CO_2 in packed fresh fish. The models in those cases are not so common and are of many different forms [60]. The reader is referred, for example, to the gamma concept to model the joint effect of several stress variables [86,87].

4. Models (Relationship between Model Inputs and Outputs)

The diversity of mathematical models for fish quality assessment emerges mainly from the diversity of the fish quality attributes previously described and from the complexity of the fish freshness concept. In the systematic review, we found 25 records where mathematical modelling of one or more attributes is included. These records [16,17,45,51,52,56–60,67,68,80,88–99] were revised and their main modelling information was included in the tables presented in this section. The objective of this section is to provide an overview of the most common approaches found regarding the mathematical modelling of fish quality/freshness.

In general, screened models are deterministic, lacking uncertainty analysis, and most of them are semi-empirical and described in the so-called *closed-form expression*, i.e., they are described by algebraic equations with a finite number of terms without derivatives or integrals. Fish quality usually depends on macroscopic variables that can be described using deterministic models, i.e., models without considering any random effect, thus providing the same solution for different simulations performed under the same conditions. Stochastic models, on the contrary, assume some random behaviour intrinsic in the dynamics resulting in stochastic differential equations. In that case, any realisation of the model provides a different solution. They are usually required when modelling food safety, but not for quality, where low numbers of certain variables (such as a low number of pathogenic bacteria) are decisive to assess the risk of foodborne illness [100,101]. Probabilistic models, on the other hand, lack dynamical equations or include the probabilistic part in the parameters of the dynamic. They are relevant when considering uncertainty (due to lack of information, measurement error or noise [102]) and/or variability (due to differences in the model parameters caused by, for example, changes in food matrix or spoilage bacteria strains [95]).

Attending to the type of mathematical equations, fish quality models are usually presented in their closed-form and they are based on empirical expressions used to represent certain behaviours, such as exponential growth [45,91]. When models are inspired by first principles with mechanistic or semi-mechanistic formulations, a closed-form expression may not exist, be unknown, or be too complex for practical use. In these cases, the model is directly described by differential Equations [68,96], requiring proper numerical methods for their resolution and calibration [103,104]. Models expressed in their differential form are also required when the stress variables or other model parameters (growth or degradation rates, diffusivity of a given compound, etc.) vary during storage and transportation.

When attending to the specific features of fish quality modelling, we found four different types of modelling approaches attending to their objective:

- **Shelf life soft sensors** are models that consider a direct input/output relationship. They consist of empirical functions, denoted by soft (from software) or virtual sensors. Typically, the input and the output are, respectively, temperature and shelf life.
- **Quality soft multi-sensors** are models considering a general mathematical expression that can be applied to describe more than one attribute.
- **Quality ad hoc models** are mechanistic-based models with equations specifically derived for one particular quality attribute.
- **Sensory or shelf life models** are models providing as their output a sensory score or shelf life date. However, they also require the intrinsic modelling of one or several quality indicators (such as spoilage bacterial content). To this purpose, they typically consider a quality ad hoc model. These sensors are also named *smart* when they are used not only for assessment purposes but for prediction of different degrees of fish quality as well [97].

This classification will be used to structure this section.

4.1. Shelf Life Soft Sensors (Input/Output)

In the literature, there are two common expressions to model the dependence of shelf life (SL) with temperature (T). Such expressions coincide with the ones used to model degradation rates or bacterial growth as a function of the temperature. On the one hand, sensors assume an exponential dependence of shelf life with the temperature (Shelf life decreases exponentially when increasing temperature [90]) of the form:

$$SL(T) = SL_0 \exp(-b\,T) \qquad \text{Exponential empirical shelf life model}$$

with SL_0 being the shelf life at $T = 0$ °C and b a parameter that represents the degree of influence of the temperature on the shelf life. If shelf life is highly affected by temperature, parameter b will be large, otherwise, it will be close to zero. On the other hand, the Arrhenius empirical shelf life model [56] has been also considered in the literature:

$$SL(T) = SL_{ref} \exp\left[\frac{Ea}{R}\left(\frac{1}{T_{eff}} - \frac{1}{T_{ref}}\right)\right] \quad \text{Arrhenius empirical shelf life model}$$

where now shelf life at an effective temperature T_{eff} is calculated from a reference shelf life SL_{ref} at temperature T_{ref}. Ea is the activation energy and R the universal gas constant.

There are many variations of these equations such as the school-field [98], the Exponential RRS (Relative rate of spoilage) model [105] or the square-root RRS model [85,98] that were inspired by limiting the levels of spoilage microorganisms. Shelf life can also be estimated from different quality attributes, such as the level of the spoilage bacteria. However, these are more sophisticated expressions that require modelling these attributes as described in Section 4.4.

Although shelf life is an important issue, the main disadvantage of these models is that they do not provide a measurement of the current freshness state of the fish.

As shown in Table 2, only three references considering shelf life soft sensors models were found in the systematic search. Each of these references focused on one particular fish species and the authors use one or two modelling alternatives. Therefore, there is a need for works focusing on different species. A comparative study of the different modelling approaches would be also required.

Table 2. Summary of the shelf life soft sensors models found in the literature search.

Output	Matrix	Model	References
Shelf life	Bogue	$SL(T)$ Arrhenius emp.	Taoukis et al. [56]
Shelf life	European sea bass	$SL(T)$ Exponential emp.	Limbo et al. [90]
Shelf life	Large yellow croaker	$SL(T)$ Exponential emp. & school-field	Quanyou et al. [98]

4.2. *Soft Multi-Sensors*

There are general mathematical formulations that may describe major trends in growth/increase or degradation/decay of a group of attributes. The direct advantage of this approach is that the same model structure is used for different quality indicators by adjusting the parameter values to fit the experimental behaviour of each indicator. However, the mechanisms of degradation are not considered. This results in too generic expressions that are mainly based on empirical correlations. Therefore, they cannot be applied to understanding attributes with complex dynamics (such as when the property does not increase or decrease monotonically). Besides, since the mechanisms are not considered, the predictive capabilities of these models are limited to the experimental conditions used to adjust the model parameters. The results provided by this approach are less reliable than the results obtained with the *ad hoc* models presented in the following section.

Let us denote by A_i ($i = 1, 2, \ldots m$) a given quality attribute that depends on time (t), and usually on temperature (T). m is the total number of attributes considered.

The simplest multi-sensor is based on the Weighted regression coefficients model [17]. This model relates one or several outputs (A_i) with several inputs or stress variables (S_j) using a linear expression of the form:

$$A_i = \sum_{j=1}^{w} a_{i,j} S_j + a_{i,w+1} \quad i = 1, \ldots, m \qquad \text{Weighted regression coefficients model}$$

S_j typically includes temperature, but other stress variables, such as pH, can be considered. $a_{i,j}$ are coefficients to be estimated for each attribute and each input. $a_{i,w+1}$ is the independent coefficient. Although this model is a direct input/output relationship and can be used to calculate shelf life (as in Section 4.1), the model is very general and therefore can be used to calculate many different attributes, such as the sensory freshness index [17].

Other type of multi-sensors assume that the quality attribute (A_i) behaves as a nth-order reaction as follows [51]:

$$\frac{dA_i}{dt} = K\,A_i^n \quad i = 1, \ldots, m \qquad \text{nth-order reaction model} \tag{4}$$

where K is the reaction rate, which typically is considered to depend on temperature according to the Arrhenius expression. Usually, n is considered a natural number, although it can be any positive rational number in the so-called power law models used in other contexts. nth-order reaction models are tested in some works [51], however, the usual approach is to select n so that it provides the best compromise between simplicity and performance of the model (Occam's razor principle).

Among the different expressions derived from the nth-order model, the most commonly used in the literature is the exponential model, which corresponds with a first-order reaction ($n = 1$). Considering that the reaction rate (K) remains constant during the process, the expression of the closed-form is:

$$A_i = A_{i,0}\exp(Kt) \quad i = 1, \ldots, m \qquad \text{Exponential model or first-order reaction model} \tag{5}$$

with $A_{i,0}$ being the initial condition (initial value for attribute A_i at $t = 0$).

Another expression, sufficiently general to represent the attribute dynamics, is the zeroth-order reaction ($n = 0$) model. For constant reaction rates (K), it results in a linear dependency of the attribute with time:

$$A_i = A_{i,0}Kt \quad i = 1, \ldots, m \qquad \text{Linear model or zeroth-order reaction model} \tag{6}$$

The main advantage of using the differential form, Equation (4), instead of the closed-form, Equations (5) and (6), is that it allows to consider situations where the storage or transport temperature changes.

Table 3 shows the attributes (outputs) modelled with this approach and the selected models for each case. In most cases, the growth/degradation rates depend on the temperature following the Arrhenius expression. Positive or negative values of K are used to represent, respectively, the increasing or decreasing evolution of the attributes. In general, the closed-form of the equations is used in these works so, as mentioned above, the temperature must be constant during storage and transport to obtain reliable results. K-value and TVB-N are the most typical indicators considered in this approach. As in the case of Shelf life soft sensors, only a few fish species were considered in these studies. More research is required to include other species.

We have also found, outside the systematic search, the use of linear models of the form of Equation (6) to describe the evolution of TAC, EC, K-value, and Sensory Analysis indicators in rainbow trout (*Oncorhynchus mykiss*) [106]. The authors in this work also compared the solutions obtained using either Arrhenius expression or Artificial Neural Networks (ANN) as a secondary model. Their results show that ANNs provide a better fit to experimental data than the Arrhenius expression, in particular for K-value and Sensory Analysis indicators.

Table 3. Virtual multi-sensors (same model structure for modelling different attributes). In the table TVB-N = total volatile base nitrogen, TAC = total aerobic counts, EC = electrical conductivity, GSI = global stability, SL = Shelf life, SFI = Sensory Freshness Index index, TM = Torrymeter reading, IT = Internal Temperature, ST = Superficial Temperature.

Output	Matrix	Secondary Model	Primary Model	References
TVB-N, TAC, K-value	Grass carp	$K(T)$ Arrhenius	Exponential model	Zhang et al. [91]
TVB-N, TAC, K-value, EC	Crucian carp	$K(T)$ Arrhenius	Exponential model	Yao et al. [45]
GSI (Sensory Score, TAC, TVB-N, K-value)	Bighead carp	$K(T)$ Arrhenius	Linear model	Hong et al. [51]
GSI (sensory score, K-value, TAC and TVB-N), EC	Crucian carp	$K(T)$ Arrhenius	Linear model	Zhu et al. [94]
SL, SFI	Gilt-head seabream	A^i(pH, TM, IT, ST, TVB-N)	Weighted regression coefficients.	Calanche et al. [17]

4.3. *Quality ad hoc Models*

In the systematic search, ad hoc models were found for some quality attributes, but not for texture properties, lipid oxidation/fatty acids, non-volatile biogenic amines, other nutrients, electrical conductivity, odour, colour or water activity. Lipid oxidation, despite being the most studied quality attribute in fish, was only modelled using the generic exponential model for TBA in grass carp [91]. Non-volatile biogenic amines, in particular histamine, were commonly used to model food safety [107]. However, they were not studied for describing food quality. No models for nutrient (proteins, vitamins, etc.) degradation or water activity were found, although water activity is a factor influencing texture or bacterial growth, and included as an input in those models. Regarding colour, models are mainly proposed for processed fish [108–110] but no model for colour changes in fresh fish was found in the literature. Despite odour being a relevant quality indicator by itself, it is typically used in combination with other attributes, for instance, to obtain the QIM. Electronic noses could be used to obtain reliable data that could be used to calibrate and validate models describing the evolution of odour. However, in the context of fish freshness, electronic noses are used to evaluate freshness or storage time [16]. As in the case of colour, no models were found to describe the evolution of odour.

There are, however, specific models for some of the quality attributes that are explained in detail in the next subsections. These models usually consider the mechanisms of quality degradation so the results are more reliable than those obtained with the other models described in previous sections. The main disadvantage is that such mechanisms involve, in most cases, complex phenomena. Therefore, the derivation of a mathematical model in these cases is a time consuming and complex task. Models describing the evolution of spoilage bacteria are the most commonly used ad hoc models.

4.3.1. Spoilage Bacteria Models Using Predictive Microbiology

Predictive microbiology is a field that focuses on modelling the behaviour of microorganisms, including spoilage bacteria, in different food matrices such as fresh fish. It is a broad area with established terms and community [60,88], and where three different models are usually considered: modelling microorganisms growth or inactivation dynamics (primary model), how these dynamics change with environmental stress or inputs variables (secondary models), and the implementation of these models in friendly software (tertiary models). The terminology of primary and secondary models can be useful to outline the different modelling approaches and will be used also in this review, even for models outside the predictive microbiology scope.

Primary models are diverse [111] and they focus on the dynamics of bacterial numbers (N). The most used primary models in fish spoilage bacteria are [29,112,113]:

$$
\begin{aligned}
&N = N_0 \exp\left(\mu_{max} t\right) && \text{Exponential model} \\
&N = N_0 + \frac{N_{max} - N_0}{1 + \exp\left[-\mu_{max}(t - t_i)\right]} && \text{Modified Logistic Model} \\
&N = N_0 \exp\left(\frac{\log\left(N_{max}/N_0\right)}{1 + \exp\left[\frac{4\mu}{\log\left(N_{max}/N_0\right)}(\lambda - t) + 2\right]}\right) && \text{Reparametrised Gompertz Model} \\
&\frac{dN}{dt} = \frac{a_0}{a_0 + (1 - a_0)\exp(-\mu_{max} t)} \mu_{max} N \left(1 - \frac{N}{N_{max}}\right) && \text{Reparametrised Baranyi's Model}
\end{aligned} \tag{7}
$$

with N_0 being the number of initial bacteria, N_{max} the maximum number (in the stationary phase), μ_{max} the maximum growth rate and λ the time of the lag phase. In the modified logistic model, λ is a function of the point of inflexion (t_i):

$$
\lambda = t_i - \frac{1}{\mu_{max}} \ln\left(\frac{N_{max} + N_{max}\exp(\mu_{max} t_i)}{N_{max} + N_0 \exp(\mu_{max} t_i)} - 1\right)
$$

and for simplicity in the provided equation we assume that the minimum cell number N_{min} is the initial cell number N_0. The derivative form of Baranyi's Model (closed-form solution is long and complex [114]) is presented with modifications to make lag phase zero, for $a_0 = 1$, and maximum lag (no dynamics), for $a_0 = 0$.

Works in the systematic search including modelling of spoilage bacteria are outlined in Table 4. Secondary models are only included when there is a clear description within the work. The variety of fish species considered in this case is larger than in the cases of Shelf life sensors and Soft multi-sensors. There is also a large variety of bacterial strains studied. However, most of the works found in the search consider deterministic models whereas bacterial population growth is a stochastic process. In this regard, although the mathematical structure of the model developed in [95] is deterministic, the authors estimate the variability of the model parameters using different experimental conditions and different fish samples. This variability is used to generate different combinations of parameters and each combination is used to obtain different simulation results. This approach allows us to approximate the stochastic behaviour.

4.3.2. TVB-N and TMA-N Models

Volatile nitrogenous bases (TVB-N), and its major contributor TMA-N, are widely modelled quality outputs with specific modelling approaches (in addition to the general modelling [45,91]). They are easy to measure indexes that adequately correlate to fish freshness.

Firstly, Howgate [52] pointed out that the exponential model (used in general modelling approaches [45,91]) was not descriptive of TMA-N changes since TMA-N reaches a limit, instead of increasing exponentially, because they are the sub-product of TMAO. The author suggested a logistic growth of the form:

$$
\text{TMA-N} = \frac{\text{TMA-N}_{max} - \text{TMA-N}_0}{1 + \exp\left(-K(t - t_i)\right)} \qquad \text{Modified Logistic Model}
$$

where TMA-N_0 and TMA-N_{max} are, respectively, the initial and maximum allowed concentrations of TMA-N, K is the maximum growth and t_i the point of inflexion .

On the other hand, TVB was modelled by García [97] by assuming a delay and a later production by psychrotrophic bacteria (N) with following equations:

$$
\frac{d\text{TVB}}{dt} = \left(\frac{a_0}{a_0 + (1 - a_0)\exp\left(-Kt\right)}\right) K\,N \qquad \text{Exponential Model with delay}
$$

where a_0 represents the parameter determining the duration of the delay (mathematically equivalent to the expression used by [113] for lag phase in bacterial growth, $a_0 = 1$ indicates no delay), K is the growth rate due to psychrotrophic bacteria (N). Interestingly, years after this work, a simpler exponential model was used to model TVB in rohu fish stored at 0 and 5 °C [80], claiming that TVB formation was a primary function of microbial action and suggesting the necessity to model TVB as a function of the microbial population as already carried out in the literature [97].

Table 4. Works in the systematic search, including modelling of spoilage bacteria. The following acronyms are used: TVA for total viable counts, TMAB for total mesophilic aerobic bacteria, TPAB for total psychrophilic aerobic bacteria and LAB for lactic bacteria.

Output	Matrix	Secondary Model	Primary Model	References
Pseudomonas & *Shewanella*	Bogue	$\mu_{max}(T)$ Arrhenius & Ratkowsky	Baranyi's model	Taoukis et al. [56]
Pseudomonas & *Shewanella*	Gilt-head seabream	$\mu_{max}(T)\&\lambda(T)$ Arrhenius & Ratkowsky	Mod. logistic model	Koutsoumanis and Nychas [57]
Sulphide producers & non-producers	Gilt-head seabream	$\mu_{max}(T)$ (not clearly defined)	Baranyi's model	Giuffrida et al. [92]
Pseudomonas & *Carnobacterium*	Tropical shrimp	$\mu_{max}(T)$ Arrhenius & Ratkowsky	Baranyi's model Rep. Gompertz Model	Dabadé et al. [59]
Pseudomonas & *Shewanella*	Hake	$\mu_{max}(T)$ Ratkowsky	Baranyi's model	García et al. [95]
TVC	Grass carp	–	Rep. Gompertz Model	Ying et al. [16]
Psychrotrophic counts	Cod	–	Baranyi's Model	García et al. [97]
Pseudomonas, *Enterobacteriaceae*, TMAB, TPAB & LAB	Rainbox trout	$\mu_{max}(T)$ Ratkowsky	Mod. Logistic Model	Genç and Diler [99]
Pseudomonas	Gilt-head seabream	–	Mod. Logistic Model	Correia Peres Costa et al. [58]
Biomass	Rohu fish	–	Mod. Logistic Model Gompertz Model	Prabhakar et al. [80]

4.3.3. Texture Properties

As mentioned above, no predictive models were found in the systematic search for texture properties. It must be highlighted that the works [89,93] describe the models they develop as predictive. However, as mentioned in the introduction, we use the term *predictive* to indicate the ability of the model to forecast the evolution of the quality indicators. The models developed in [89,93] are built using partial least squares regression or least-squares support vector machines to assess texture indicators using nuclear magnetic resonance (NMR) or hyperspectral imaging (HSI) measurements. In other words, these models provide a non-invasive estimation of the texture properties at the NMR or HSI measurement time, but they do not predict the future evolution of such indicators.

A predictive model to describe the viscoelastic behaviour of rohu fish (*Labeo rohita*) was developed in [115], although this work was not present in the systematic search. The authors used the modified Maxwell model to relate skin hardness and compression time for iced fish:

$$F(t) = C_0 + C\exp\left(\frac{t}{t_{rel}}\right)$$

where $F(t)$ is the force at any time, C_0 corresponds with the force at equilibrium, C is the decay force, and t_{rel} is the relaxation time. Experimental data was used to fit coefficients C_0, C and t_{rel}.

Further research is required regarding the mathematical description of the evolution of texture in fresh fish.

4.3.4. ATP Degradation

As mentioned in Section 2.6, ATP degradation occurs in a series of steps represented as:

$$\text{ATP} \rightarrow \text{ADP} \rightarrow \text{AMP} \rightarrow \text{IMP} \rightarrow \text{Ino} \rightarrow \text{Hx} \rightarrow \text{Xa} \rightarrow \text{Uric acid}$$

The first three steps occur relatively fast after slaughter so, when fish samples are analysed they contain low (or zero) concentrations of ATP, ADP and AMP. On the other hand, the degradation from Hx to Xa and uric acid is usually slow and when such products are formed, the fish is already spoiled. Therefore, in general, ad hoc models only consider the part of the scheme involving IMP, Ino and Hx. The K_I-value, Equation (3), can be obtained from these compounds.

Table 5 summarises the main features of the models derived in the different works of the systematic search. In particular, ref. [52,67] considered the reaction scheme:

$$\text{IMP} \xrightarrow{K_1} \text{Ino} \xrightarrow[K_{bac}]{K_2} \text{Hx}$$

where K_1 and K_2 are, respectively, the reaction rates for the conversion of IMP into Ino, and Ino into Hx. Bacterial conversion of Ino into Hx (K_{bac}) was also taken into account. In these works, first-order kinetics are considered. Arrhenius expressions were used to account for the dependency of reaction rates on the temperature. Bacterial growth was modelled using an exponential model of the form of Equation (7). In [52], the possibility of a loss of nucleotides by leaching (diffusion through muscle and skin) was also considered. Another interesting issue about this work is that the author presented and discussed the results obtained from data of forty-five different fish species. Reliable results were obtained for most of the considered species. In [68,96], the authors found, by fitting the models to experimental data, that alternative nucleotide degradation paths might occur in European hake (*Merluccius merluccius*). In particular, the direct conversion of IMP to Hx and other products should be considered. Leaching of nucleotides and the effect of bacteria, namely *Pseudomonas* spp. and *Shewanella* spp., on the conversion of IMP to Ino and Ino to Hx, were also considered. The standard square-root model [85] was used to represent the bacterial growth rates.

Table 5. *Ad-hoc* models found in the systematic search to describe the degradation of IMP, Ino and Hx. The K_I-value is obtained from the concentration of these components. All these models consider a cascade of first-order reactions.

Output	Matrix	Secondary Model	Primary Model	References
IMP, Ino, Hx	Rainbow trout	$K_i(T)$ Arrhenius	Exponential model, Bacterial catalysis	Howgate [67]
IMP, Ino, Hx	Forty-five species	$K_i(T)$ Arrhenius	Exponential model, Bacterial catalysis, leaching	Howgate [52]
IMP, Ino, Hx	Hake	$K_i(T)$ Arrhenius	First-order reaction model	Vilas et al. [96]
IMP, Ino, Hx	Hake	$K_i(T)$ Arrhenius	First-order reaction model, Bacterial catalysis, leaching	Vilas et al. [68]

4.4. Sensory or Shelf Life Models

Quality ad hoc models are usually a tool, more than the final aim, to assess or predict shelf life or different grades of fish quality. For example, modelling of spoilage bacteria in Table 4 is commonly used to estimate shelf life by specifying a concentration of bacterial counts at which the quality is considered not sufficient. Shelf life date, for example, is estimated using spoilage bacteria in bogue fish (*Boops boops*) and gilt-head seabream (*Sparus aurata*) for numbers greater than 7 logs ($N > 10^7$ CFU/g) [56,57] or even $N > 10^{8.5}$ CFU/g for *Pseudomonas psychrofila* in tropical shrimp [59].

There are models in the literature that are used to estimate different grades of freshness, not only shelf life or a given quality indicator. These models are summarised in Table 6. As shown in the Table, works following this approach and the number of species considered are scarce. The main challenge is to find a mathematical relationship between the outputs of the ad hoc models and different quality levels. The first attempt consisted of dividing the QIM into three indexes (QIM_S for skin, QIM_G for gills, and QIM_F for flesh), and finding their relationship with non-producers (N_w) and producers (N_b) of sulphide, present in those specific parts for the fish [92]. However, for an estimation of a final QIM index, not only models of spoilage bacteria but also of TVB-N are required [97]. In this work, a simple ANN was developed to obtain the relationship between QIM and the model variables

(TVB-N and bacterial count). A logistic model was used to describe the bacterial evolution. Other works focused on using models of spoilage bacteria to find ranges of standard sensory methods. Such methods considered fewer freshness grades than the QIM. That is the case in [16], where cod freshness, in terms of a three-level standard (SC/T 3108-1986), was correlated with the TVC value. The work by [95] used a nonlinear function of *Pseudomonas* and *Shewanella* counts to determine a four-level QSM value in European hake. This work is the only one that considers a secondary model, and therefore, it is the only one that provides the final relationship between the effect of temperature changes in the four levels of quality in QSM.

Table 6. Modelling of sensory scores using ad hoc models. QIM stands for Quality index specific for gilt-head seabream [65] or cod [50] method. S,G and F for Skin, Gills, Flesh, N_w, N_b, N_p, N_s, N_{psy} for sulphide and non-sulphide produces, *Pseudomonas*, *Shewanella* and psychrotrophic counts.

Output	Matrix	Secondary Model	Primary Model	References
QIM_S, QIM_G, QIM_F (15 levels)	Gilt-head seabream	Not clearly defined	$QIM(N_w, N_b)$	Giuffrida et al. [92]
Council Regulation(EC) No 2406/96 (1996) Standard method (4 levels)	Hake	$\mu_{max}(T)$ Ratkowsky	$SM(N_p, N_s)$	García et al. [95]
SC/T 3108-1986 Standard method (3 levels)	Cod	–	$SM(N_{TVC})$	Ying et al. [16]
QIM (23 levels)	Cod	–	$QIM(TVBN, N_{psy})$	García et al. [97]

5. Modelling Challenges and New Directions

In this review, we have described, classified and established connections for the different types of mathematical models to describe and predict fish quality. We have mainly focused on those works within the systematic search described in Appendix A, although other relevant contributions, not included in such search, were considered in this review. First of all, we should stress that the literature is larger, particularly in the case of spoilage bacteria using predictive microbiology. However, there are many comprehensive reviews regarding predictive microbiology [60,116,117], and the focus of this review was, instead, on providing connections between models that describe different fish quality attributes, and on identifying those topics that require more research attention. For example, there are still relevant quality attributes for which it was not possible to find mathematical models. The most illustrative example is lipid oxidation, the most studied attribute in the literature from the experimental point of view. Other attributes lacking modelling approaches are non-volatile biogenic amines, nutrients, odour and water activity (although this is considered as an input in several models in the literature). Other attributes, such as colour, have been modelled only for processed fish.

We need to stress that, in addition to the lack of modelling for certain quality attributes, there are major limitations in some of the models we found. Generic models, named in this review software sensors or multi-sensors, are particularly advantageous to compare results from different studies but, in this comparison, it is clearly observed that there are many inconsistencies between works. For example, both linear and exponential functions have been used to model the same attribute (see Table 3). However, whereas linear functions may approximate a short time window of an exponential model, differences between both approaches are considerable for wide prediction windows. There is a need to compare those modelling structures and detect which ones are more appropriate for the different quality outputs.

Ad-hoc predictive microbiology models, that have been studied in detail, still present limitations, mainly due to the uncertainty of the estimated parameters and the initial bacterial fish load (or model initial conditions). Microbial models require a known starting state, but measuring bacterial load takes time and involves the destruction of the sample. Some partial solutions have been considered such as (1) using the worst-case scenario [116], (2) estimating the initial conditions variability [95], or (3) estimating the numbers using indirect measurements of other variables (such as conductance measurements) that are non-invasive and fast to obtain [118]. However, taking into account that bacteria grow exponentially between lag and stationary phase, model prediction is highly affected (very sensitive) by its initial conditions. More research on this topic is required to find confidence models of spoilage bacteria.

Another challenge, still only partially addressed in the literature, is the derivation of expressions that allow the inference of sensory attributes or shelf life from the growth of spoilage bacteria. In this regard, some works derived models describing two or several attributes, for instance, shelf life and growth of *Pseudomonas*, but such attributes were only connected through the stress variables, typically temperature (see examples in Table 4). Ideally, the model should provide a final quality index, as a function of different quality attributes that depend on the stress variables, as the examples provided in Table 6.

In addition, only a few works validate the predictive capability of the models proposed, i.e., the ability of the model to describe data outside of the set used for model development and parameter estimation. Most of them use constant temperature or temperature oscillating at a high frequency, as compared with the model dynamics time-scale (using such oscillating temperature would be equivalent to using a mean constant temperature). For example, in the work by [56] two non-isothermal profiles are used for the validation. In one profile, temperature oscillates at high frequency and the change in the output signal was smaller than the experimental error of the measurement. On the other hand, the model was validated using a temperature profile with wider oscillations, that provided a change in data trend and model dynamics. Only a few works consider dynamic temperature profiles, computed using an optimal experimental design, to reduce the uncertainty of the predictive model [95].

To advance in this area, reproducibility of published works is a key aspect, particularly for ad hoc models, usually more complex and with many different mathematical structures. Currently, comparison among works is extremely difficult, not only because of the variability between fish species and conditions before and after the capture (for example in food structure [119]) but also because the proposed measure of fish quality is sometimes specifically developed for the study [118,120]. We think that for the advance of modelling of fresh fish quality, research should be focused on reproducing and predicting established sensory indexes, such as QIM [95,97,121], allowing comparison between approaches.

Finally, the community should invest in better exploitation of the available models and towards their integration into software systems for online quality prediction, as is already the case in food safety [122], or even for optimisation-based determination of the best conditions to maximise shelf life in different processes. In this regard, an ambitious objective would be the derivation of a digital twin [123] for fresh fish degradation. Let us here use an illustrative example of the potential of these models in a study developed in our group, namely [124]. In this work, existing models were used to find the best active package configuration (including the type of packaging and concentration of antimicrobial) that maximises food quality while ensuring food safety. In addition, the model was used to predict, at any moment, the expected food quality for the expected stress variables along the food chain.

Author Contributions: M.R.G. and C.V. designed the original idea, drafted and review the manuscript. J.A.F.-R. conducted the bibliometric analysis. All authors have read and agreed to the published version of the manuscript.

Funding: We thanks the grants RTI2018-093560-J-I00 funded by MCIN/AEI/10.13039/501100011033, "ERDF A way of making Europe", the CSIC grant 20213AT001 and Project Pescando4.0 (ref. 202178002) funded by NextGenerationEU funds in the framework of the *Plan de Recuperación, Transformación y Resiliencia (PRTR)* managed by the *Ministerio de Agricultura, Pesca y Alimentación* through the RD 685/2021.

Institutional Review Board Statement: Not applicable.

Informed Consent Statement: Not applicable.

Data Availability Statement: All bibliometric analysis is shared in the repository [19], with link doi:10.5281/zenodo.6414360.

Acknowledgments: Authors M.R.G. and C.V. would like to express their gratitude and dedicate this contribution to Antonio A. Alonso, the researcher that inspired this work, which unfortunately left us prematurely. We thank their endless support and continuous guidance.

Conflicts of Interest: The authors declare no conflict of interest.

Appendix A. Systematic Literature Search

Table A1. Terms of the search performed on the web of science. The search was limited to the *Web of Science Core Collection* database and to the *Food Science and Technology* category. All years until end 2021 are considered.

Set	Search	Records
#1	(**TI** OR **AB** OR **AK**) = (("quality" OR "freshness" OR "shelf life" OR "shelf-life" OR "K-value" OR "KI-value" OR "ATP" OR "adenosine triphosphate" OR "IMP" OR "inosine monophosphate" OR "hypoxanthine" OR "colour" OR "color" OR "chromatism" OR "fatty acid*" OR "lipid oxidation" OR "tba" OR "electrical properties" OR "electrical conductance" OR "electrical conductivity" OR "texture" OR "hardness" OR "firmness" OR "odour" OR "odor" OR "nutrient*" OR "vitamin" OR "biogenic amine*" OR "water content" OR "water activity" OR "tvb-n" OR "tma-n" OR "qim" OR "qsm" OR "sensory analysis" OR "sensory evaluation" OR "sensory method" OR "sso" OR "spoilage bacteria" OR "spoilage microorganism*") NEAR/5 ("fish" OR "fishes" OR "shellfish" OR "seafood*" OR "albacore" OR "amberjack" OR "anchovy" OR "angler" OR "barbel" OR "barracuda*" OR "sea bass" OR "beluga" OR "bigeye" OR "blackfish" OR "bluefish" OR "blue runner" OR "blue shark" OR "branzino" OR "seabream" OR "sea bream" OR "butterfish" OR "carp" OR "catfish" OR "catshark" OR "comber" OR "conger" OR "cutlassfish" OR "danubian wels" OR "dogfish" OR "eel" OR "eels" OR "flounder" OR "flying fish" OR "forkbeard" OR "garfish" OR "garrick" OR "guitarfish" OR "gunard" OR "haddock" OR "hake" OR "halibut" OR "hammerhead" OR "herring" OR "icefish" OR "John dory" OR "lamprey" OR "lanternfish" OR "leerfish" OR "little tunny" OR "mackerel" OR "mahi mahi" OR "marlin" OR "megrim" OR "melva" OR "monkfish" OR "moonfish" OR "needlefish" OR "pandoras" OR "panga" OR "pangasius" OR "parrotfish" OR "parrot fish" OR "perch" OR "pike fish" OR "pilchard" OR "pilotfish" OR "pilot fish" OR "plaice" OR "pollack" OR "pollock" OR "ponyfish" OR "porbeagle" OR "rainbow trout" OR "cownose ray" OR "devilray" OR "butterfly ray" OR "softnose skate" OR "legskate" OR "sawfish" OR "ribbonfish" OR "rockfish" OR "rosefish" OR "sablefish" OR "sailfish" OR "salmon" OR "sardine" OR "sardinella" OR "scabbardfish" OR "scorpionfish" OR "sheatfish" OR "shi drum" OR "sillago" OR "skipjack" OR "smooth hound" OR "smooth-hound" OR "spearfish" OR "St Peter's fish" OR "stargazer" OR "stingray" OR "sturgeon" OR "surgeon fish" OR "swordfish" OR "tilapia" OR "threadfin" OR "triggerfish" OR "trout" OR "tubefish" OR "tuna" OR "turbot" OR "walleye" OR "whitebait" OR "whiting" OR "yellowtail" OR "octopus" OR "squid*" OR "crab" OR "lobster*" OR "prawn*" OR "shrimp*" OR "cuttlefish*" OR "crayfish*" OR "langoustine*" OR "scampi*" OR "urchin"))	6429

Table A1. *Cont.*

Set	Search	Records
#2	(**TI** OR **AB** OR **AK**) = ("high pressure" OR "atmospheric cold plasma" OR "modified atmosphere*" OR "sterilization" OR "sterilisation" OR "frozen" OR "thawed" OR "pH-shift processing" OR "cross-processing" OR "fillet*" OR "slice*" OR "fish oil*" OR "quality of water" OR "water quality" OR "frying" OR "garlic" OR "canned fish*" OR "surimi" OR "fish sauce" OR "nugget" OR "chitosan" OR "x-ray*" OR "farming" OR "farm-level" OR "antibacterial" OR "antimicrobial" OR "electronarcosis" OR "immun*" OR "Fluorescence in situ hybridization" OR "cooking" OR "microplastic*" OR "embryo*" OR "rice" OR "drying" OR "dried" OR "nursery" OR "fishmeal" OR "plastic" OR "irradiation" OR "larva*" OR "feed*" OR "coating" OR "diet" OR "dietary" OR "reproductive" OR "ibuprofen" OR "fertilizer*" OR "valorization" OR "catch damage" OR "thermal process*" OR "non-thermal process*" OR "non thermal process*" OR "children" OR "gear design" OR "additive" OR "mimicry" OR "seed quality" OR "sodium alginate" OR "edible film*" OR "TYRP1 gene*" OR "transgene" OR "antioxida* peptide*" OR "antioxida* capacity" OR "antioxida* solution" OR "antioxida* effect*" OR "anti-oxida* activity" OR "antioxida* activity" OR "rearing" OR "algae" OR "collagen expression" OR "genomic*" OR "proteomic*" OR "s-potential*" OR "fish meal" OR "synthesis" OR "egg quality" OR "carotenogenesis" OR "nutrient requirement*" OR "maternal" OR "oily fish" OR "gelatin" OR "polychlorinated" OR "nutrition of salmonoid" OR "fish retina" OR "color picture*" OR "elderly" OR "inheritance of color" OR "short read mapping program" OR "tea polyphenol" OR "fish consumption" OR "inter-specific hybrids" OR "extracellular lipase" OR "pathology" OR "metabolic polymorphisms" OR "transgenic" OR "eco-label" OR "lethality" OR "micro-squid" OR "emulsion*" OR "vegetable production" OR "source of nutrient*" OR "hydrolyzate*" OR "epiphitic" OR "bilirubin" OR "essential oil*" OR "histology" OR "egg-yolk" OR "chd" OR "silver toxicity" OR "biomanipulation" OR "hormonal-control" OR "protein crosslinking" OR "food security" OR "high hydrostatic pressure" OR "red tide*" OR "intake" OR "digestibility" OR "nutrient absorption" OR "nutrition" OR "docking" OR "egg" OR "diet" OR "nutrient recycling" OR "farm effluents" OR "fish behavior" OR "smoked" OR "chromatophore*" OR "hydroponic" OR "recovery of fish" OR "fish recovery" OR "fish-odor syndrome" OR "fish-odour syndrome" OR "water chemistry" OR "natural preservatives" OR "mince" OR "epiphytic" OR "phosphorus" OR "omega3" OR "hemolysate" OR "hemolysis" OR "availability of nutrients" OR "biosynthesis" OR "health" OR "globalization" OR "inhibition of polyphenoloxidase" OR "species identification" OR "virus" OR "norovirus" OR "ferment*" OR "bread" OR "plant extract*" OR "nanoencapsulate*" OR "starch" OR "edible compound*" OR "edible natural compound*" OR "fish burger*" OR "wafer*" OR "bromelain" OR "fish ball*" OR "fish cake*" OR "dehydrat*" OR "grill*" OR "4-hexylresorcinol" OR "freezing" OR "freezing-point" OR "restructure*" OR "squid oil" OR "molecular distilation")	332,851
#3	#1 NOT #2	1636
#4	(**TI** OR **AB** OR **AK**) = ("mathematical model*" OR "predictive model*" OR "dynamic model*" OR "growth model*" OR "predictive microbiology" OR "model")	922,980
#5	#3 AND #4	33

The use of the term ID (keywords plus) in our search caused the selection of many spurious publications. Therefore, we only considered *title* (TI), *abstract* (AB), and *author keywords* (AK) in the search. In the set search #1, we have considered two blocks. The first one contains terms related to quality indicators (for example, quality, freshness, K-value, etc.). The second one was used to limit the search to fish products. On the other hand, set search #2 was included to avoid those works that were not related to fresh fish, which is the focus of this review.

References

1. Raak, N.; Symmank, C.; Zahn, S.; Aschemann-Witzel, J.; Rohm, H. Processing- and product-related causes for food waste and implications for the food supply chain. *Waste Manag.* **2017**, *61*, 461–472. [CrossRef] [PubMed]
2. Corradini, M.G. Shelf Life of Food Products: From Open Labeling to Real-Time Measurements. *Annu. Rev. Food Sci. Technol.* **2018**, *9*, 251–269. [CrossRef] [PubMed]
3. Zöller, S.; Wachtel, M.; Knapp, F.; Steinmetz, R. Going all the way—Detecting and transmitting events with wireless sensor networks in logistics. In Proceedings of the 38th Annual IEEE Conference on Local Computer Networks—Workshops, Sydney, Australia, 21–24 October 2013; pp. 39–47. [CrossRef]
4. Koutsoumanis, K.; Allende, A.; Alvarez-Ordóñez, A.; Bolton, D.; Chemaly, M.; Davies, R.; De Cesare, A.; Herman, L.; Hilbert, F.; Lindqvist, R.; et al. The use of the so-called 'tubs' for transporting and storing fresh fishery products. *EFSA J.* **2020**, *18*, 1–123. [CrossRef]
5. Koutsoumanis, K.; Allende, A.; Alvarez-Ordóñez, A.; Bolton, D.; Chemaly, M.; Davies, R.; De Cesare, A.; Herman, L.; Hilbert, F.; Lindqvist, R.; et al. The use of the so-called 'superchilling' technique for the transport of fresh fishery products. *EFSA J.* **2021**, *19*. [CrossRef]
6. Wu, L.; Pu, H.; Sun, D.W. Novel techniques for evaluating freshness quality attributes of fish: A review of recent developments. *Trends Food Sci. Technol.* **2019**, *83*, 259–273. [CrossRef]
7. Venugopal, V. Biosensors in fish production and quality control. *Biosens. Bioelectron.* **2002**, *17*, 147–157. [CrossRef]

8. Olafsdottir, G.; Jonsdottir, R.; Lauzon, H.L.; Luten, J.; Kristbergsson, K. Characterization of volatile compounds in chilled cod (*Gadus morhua*) fillets by gas chromatography and detection of quality indicators by an electronic nose. *J. Agric. Food Chem.* **2005**, *53*, 10140–10147. [CrossRef]
9. Li, J.; Feng, H.; Liu, W.; Gao, Y.; Hui, G. Design of A Portable Electronic Nose system and Application in K Value Prediction for Large Yellow Croaker (*Pseudosciaena crocea*). *Food Anal. Methods* **2016**, *9*, 2943–2951. [CrossRef]
10. Wu, D.; Sun, D.W.; He, Y. Novel non-invasive distribution measurement of texture profile analysis (TPA) in salmon fillet by using visible and near infrared hyperspectral imaging. *Food Chem.* **2014**, *145*, 417–426. [CrossRef] [PubMed]
11. Qu, J.H.; Sun, D.W.; Cheng, J.H.; Pu, H. Mapping moisture contents in grass carp (*Ctenopharyngodon idella*) slices under different freeze drying periods by Vis-NIR hyperspectral imaging. *LWT* **2017**, *75*, 529–536. [CrossRef]
12. Herrero, A.M. Raman spectroscopy a promising technique for quality assessment of meat and fish: A review. *Food Chem.* **2008**, *107*, 1642–1651. [CrossRef]
13. Prabhakar, P.K.; Vatsa, S.; Srivastav, P.P.; Pathak, S.S. A comprehensive review on freshness of fish and assessment: Analytical methods and recent innovations. *Food Res. Int.* **2020**, *133*, 109157. [CrossRef] [PubMed]
14. Franceschelli, L.; Berardinelli, A.; Dabbou, S.; Ragni, L.; Tartagni, M. Sensing technology for fish freshness and safety: A review. *Sensors* **2021**, *21*, 1373. [CrossRef] [PubMed]
15. Chantarachoti, J.; Oliveira, A.C.; Himelbloom, B.H.; Crapo, C.A.; McLachlan, D.G. Portable electronic nose for detection of spoiling alaska pink salmon (*Oncorhynchus gorbuscha*). *J. Food Sci.* **2006**, *71*, 414–421. [CrossRef]
16. Ying, X.; Zinnai, A.; Venturi, F.; Sanmartin, C.; Deng, S. Freshness evaluation of grass carp (Ctenopharyngodon idellus) by electronic nose. *J. Food Meas. Charact.* **2017**, *11*, 1026–1034. [CrossRef]
17. Calanche, J.; Pedrós, S.; Roncalés, P.; Beltrán, J.A. Design of predictive tools to estimate freshness index in farmed sea bream (*Sparus aurata*) stored in ice. *Foods* **2020**, *9*, 69. [CrossRef] [PubMed]
18. Martins, R.C.; Lopes, V.V.; Vicente, A.A.; Teixeira, J.A. Computational shelf-life dating: Complex systems approaches to food quality and safety. *Food Bioprocess Technol.* **2008**, *1*, 207–222. [CrossRef]
19. Férez-Rubio, J.; García, M.R.; Vilas, C. A systematic review on fish freshness and quality indicators. *Zenodo* **2022**, *6414360*. [CrossRef]
20. Richards, M.P.; Hultin, H.O. Contributions of blood and blood components to lipid oxidation in fish muscle. *J. Agric. Food Chem.* **2002**, *50*, 555–564. [CrossRef] [PubMed]
21. Grigorakis, K.; Alexis, M.N.; Anthony Taylor, K.D.; Hole, M. Comparison of wild and cultured gilthead sea bream (*Sparus aurata*); composition, appearance and seasonal variations. *Int. J. Food Sci. Technol.* **2002**, *37*, 477–484. [CrossRef]
22. Ólafsdóttir, G.; Martinsdóttir, E.; Oehlenschläger, J.; Dalgaard, P.; Jensen, B.; Undeland, I.; Mackie, I.M.; Henehan, G.; Nielsen, J.; Nilsen, H. Methods to evaluate fish freshness in research and industry. *Trends Food Sci. Technol.* **1997**, *8*, 258–265. [CrossRef]
23. Al Bulushi, I.; Poole, S.; Deeth, H.C.; Dykes, G.A. Biogenic amines in fish: Roles in intoxication, spoilage, and nitrosamine formation—A review. *Crit. Rev. Food Sci. Nutr.* **2009**, *49*, 369–377. [CrossRef] [PubMed]
24. Olafsdottir, G.; Nesvadba, P.; Di Natale, C.; Careche, M.; Oehlenschläger, J.; Tryggvadóttir, S.V.; Schubring, R.; Kroeger, M.; Heia, K.; Esaiassen, M.; et al. Multisensor for fish quality determination. *Trends Food Sci. Technol.* **2004**, *15*, 86–93. [CrossRef]
25. Pacquit, A.; Frisby, J.; Diamond, D.; Lau, K.T.; Farrell, A.; Quilty, B.; Diamond, D. Development of a smart packaging for the monitoring of fish spoilage. *Food Chem.* **2007**, *102*, 466–470. [CrossRef]
26. Papadopoulos, V.; Chouliara, I.; Badeka, A.; Savvaidis, I.N.; Kontominas, M.G. Effect of gutting on microbiological, chemical, and sensory properties of aquacultured sea bass (*Dicentrarchus labrax*) stored in ice. *Food Microbiol.* **2003**, *20*, 411–420. [CrossRef]
27. Ruiz-Capillas, C.; Moral, A. Sensory and biochemical aspects of quality of whole bigeye tuna (*Thunnus obesus*) during bulk storage in controlled atmospheres. *Food Chem.* **2005**, *89*, 347–354. [CrossRef]
28. Gram, L.; Trolle, G.; Huss, H.H. Detection of specific spoilage bacteria from fish stored at low (0 °C) and high (20 °C) temperatures. *Int. J. Food Microbiol.* **1987**, *4*, 65–72. [CrossRef]
29. Dalgaard, P. Modelling of microbial activity and prediction of shelf life for packed fresh fish. *Int. J. Food Microbiol.* **1995**, *26*, 305–317. [CrossRef]
30. Alasalvar, C.; Taylor, K.D.; Öksüz, A.; Garthwaite, T.; Alexis, M.N.; Grigorakis, K. Freshness assessment of cultured sea bream (*Sparus aurata*) by chemical, physical and sensory methods. *Food Chem.* **2001**, *72*, 33–40. [CrossRef]
31. Veciana-Nogués, M.T.; Mariné-Font, A.; Vidal-Carou, M.C. Biogenic Amines as Hygienic Quality Indicators of Tuna. Relationships with Microbial Counts, ATP-Related Compounds, Volatile Amines, and Organoleptic Changes. *J. Agric. Food Chem.* **1997**, *45*, 2036–2041. [CrossRef]
32. Jones, N.R.; Murray, J.; Livingston, E.I.; Murray, C.K. Rapid estimations of hypoxanthine concentrations as indices of the freshness of chill-stored fish. *J. Sci. Food Agric.* **1964**, *15*, 763–774. [CrossRef]
33. Kim, M.K.; Mah, J.H.; Hwang, H.J. Biogenic amine formation and bacterial contribution in fish, squid and shellfish. *Food Chem.* **2009**, *116*, 87–95. [CrossRef]
34. Ramanathan, L.; Das, N.P. Studies on the Control of Lipid Oxidation in Ground Fish by Some Polyphenolic Natural Products. *J. Agric. Food Chem.* **1992**, *40*, 17–21. [CrossRef]
35. Kawai, T. Fish Flavor. *Crit. Rev. Food Sci. Nutr.* **1996**, *36*, 257–298. [CrossRef]
36. Kuswandi, B.; Jayus; Restyana, A.; Abdullah, A.; Heng, L.Y.; Ahmad, M. A novel colorimetric food package label for fish spoilage based on polyaniline film. *Food Control* **2012**, *25*, 184–189. [CrossRef]

37. Huang, X.; Xin, J.; Zhao, J. A novel technique for rapid evaluation of fish freshness using colorimetric sensor array. *J. Food Eng.* **2011**, *105*, 632–637. [CrossRef]
38. Chakraborty, K.; Raj, R.P. An extra-cellular alkaline metallolipase from Bacillus licheniformisMTCC 6824: Purification and biochemical characterization. *Food Chem.* **2008**, *109*, 727–736. [CrossRef] [PubMed]
39. Moreda-Piñeiro, J.; Alonso-Rodríguez, E.; Romarís-Hortas, V.; Moreda-Piñeiro, A.; López-Mahía, P.; Muniategui-Lorenzo, S.; Prada-Rodríguez, D.; Bermejo-Barrera, P. Assessment of the bioavailability of toxic and non-toxic arsenic species in seafood samples. *Food Chem.* **2012**, *130*, 552–560. [CrossRef]
40. Palaniappan, P.R.; Vijayasundaram, V. Fourier transform infrared study of protein secondary structural changes in the muscle of Labeo rohita due to arsenic intoxication. *Food Chem. Toxicol.* **2008**, *46*, 3534–3539. [CrossRef] [PubMed]
41. Cakli, S.; Kilinc, B.; Cadun, A.; Dincer, T.; Tolasa, S. Quality differences of whole ungutted sea bream (*Sparus aurata*) and sea bass (*Dicentrarchus labrax*) while stored in ice. *Food Control* **2007**, *18*, 391–397. [CrossRef]
42. Morzel, M.; Sohier, D.; Van De Vis, H. Evaluation of slaughtering methods for turbot with respect to animal welfare and flesh quality. *J. Sci. Food Agric.* **2003**, *83*, 19–28. [CrossRef]
43. Raju, C.V.; Shamasundar, B.A.; Udupa, K.S. The use of nisin as a preservative in fish sausage stored at ambient (28 ± 2 °C) and refrigerated (6 ± 2 °C) temperatures. *Int. J. Food Sci. Technol.* **2003**, *38*, 171–185. [CrossRef]
44. Vaz-Pires, P.; Seixas, P.; Mota, M.; Lapa-Guimarães, J.; Pickova, J.; Lindo, A.; Silva, T. Sensory, microbiological, physical and chemical properties of cuttlefish (*Sepia officinalis*) and broadtail shortfin squid (*Illex coindetii*) stored in ice. *LWT - Food Sci. Technol.* **2008**, *41*, 1655–1664. [CrossRef]
45. Yao, L.; Luo, Y.; Sun, Y.; Shen, H. Establishment of kinetic models based on electrical conductivity and freshness indictors for the forecasting of crucian carp (*Carassius carassius*) freshness. *J. Food Eng.* **2011**, *107*, 147–151. [CrossRef]
46. Ryder, J.M. Determination of Adenosine Triphosphate and Its Breakdown Products in Fish Muscle by High-Performance Liquid Chromatography. *J. Agric. Food Chem.* **1985**, *33*, 678–680. [CrossRef]
47. Secci, G.; Parisi, G. From farm to fork: Lipid oxidation in fish products. A review. *Ital. J. Anim. Sci.* **2016**, *15*, 124–136. [CrossRef]
48. Kanner, J. Oxidative processes in meat and meat products: Quality implications. *Meat Sci.* **1994**, *36*, 169–189. [CrossRef]
49. (EC) Regulation Council. Laying down common marketing standards foe certain fishery products. *Off. J. Eur. Communities* **1996**, *334*, L–334/2.
50. Luten, J.B.; Martinsdottir, E. QIM: A European tool for fish freshness evaluation in the fishery chain. In *Methods to Determine the Freshness of Fish in Research and Industry, Proceedings of the Final Meeting of the Concerted Action 'Evaluation of Fish Freshness' AIR3CT94 2283, Nantes, France, 12–14 November 1997*; Elsevier: Amsterdam, The Netherlands, 1997; pp. 287–296.
51. Hong, H.; Luo, Y.; Zhu, S.; Shen, H. Application of the general stability index method to predict quality deterioration in bighead carp (*Aristichthys nobilis*) heads during storage at different temperatures. *J. Food Eng.* **2012**, *113*, 554–558. [CrossRef]
52. Howgate, P. A review of the kinetics of degradation of inosine monophosphate in some species of fish during chilled storage. *Int. J. Food Sci. Technol.* **2006**, *41*, 341–353. [CrossRef]
53. Dai, Q.; Cheng, J.H.; Sun, D.W.; Zhu, Z.; Pu, H. Prediction of total volatile basic nitrogen contents using wavelet features from visible/near-infrared hyperspectral images of prawn (*Metapenaeus ensis*). *Food Chem.* **2016**, *197*, 257–265. [CrossRef] [PubMed]
54. Parlapani, F.F.; Mallouchos, A.; Haroutounian, S.A.; Boziaris, I.S. Microbiological spoilage and investigation of volatile profile during storage of sea bream fillets under various conditions. *Int. J. Food Microbiol.* **2014**, *189*, 153–163. [CrossRef] [PubMed]
55. Johnston, W.A.; Nicholson, F.J.; Roger, A.; Stroud, G.D. *Freezing and Refrigerated Storage in Fisheries*; Food & Agriculture Org.: Rome, Italy, 1994.
56. Taoukis, P.S.; Koutsoumanis, K.; Nychas, G.J. Use of time-temperature integrators and predictive modelling for shelf life control of chilled fish under dynamic storage conditions. *Int. J. Food Microbiol.* **1999**, *53*, 21–31. [CrossRef]
57. Koutsoumanis, K.; Nychas, G.J.E. Application of a systematic experimental procedure to develop a microbial model for rapid fish shelf life predictions. *Int. J. Food Microbiol.* **2000**, *60*, 171–184. [CrossRef]
58. Correia Peres Costa, J.C.; Floriano, B.; Bascón Villegas, I.M.; Rodríguez-Ruiz, J.P.; Posada-Izquierdo, G.D.; Zurera, G.; Pérez-Rodríguez, F. Study of the microbiological quality, prevalence of foodborne pathogens and product shelf-life of Gilthead sea bream (*Sparus aurata*) and Sea bass (*Dicentrarchus labrax*) from aquaculture in estuarine ecosystems of Andalusia (Spain). *Food Microbiol.* **2020**, *90*, 103498. [CrossRef] [PubMed]
59. Dabadé, D.S.; Azokpota, P.; Nout, M.J.; Hounhouigan, D.J.; Zwietering, M.H.; Den Besten, H.M. Prediction of spoilage of tropical shrimp (*Penaeus notialis*) under dynamic temperature regimes. *Int. J. Food Microbiol.* **2015**, *210*, 121–130. [CrossRef] [PubMed]
60. McMeekin, T.A.; Ross, T.; Olley, J. Application of predictive microbiology to assure the quality and safety of fish and fish products. *Int. J. Food Microbiol.* **1992**, *15*, 13–32. [CrossRef]
61. Davies, A.R. Modified-atmosphere packaging of fish and fish products. *Fish Process. Technol.* **1997**, *33*, 200–223. [CrossRef]
62. Gram, L.; Dalgaard, P. Fish spoilage bacteria—Problems and solutions. *Curr. Opin. Biotechnol.* **2002**, *13*, 262–266. [CrossRef]
63. Dunajski, E. Texture of Fish Muscle. *J. Texture Stud.* **1980**, *10*, 301–318. [CrossRef]
64. Oehlenschläger, J. Seafood Quality Assessment. In *Seafood Processing*; John Wiley & Sons, Ltd.: Hoboken, NJ, USA, 2014; Chapter 14, pp. 359–386. [CrossRef]
65. Huidobro, A.; Pastor, A.; Tejada, M. Quality index method developed for raw gilthead seabream (*Sparus aurata*). *J. Food Sci.* **2000**, *65*, 1202–1205. [CrossRef]
66. Lakshmanam, P.T.; Gopakumar, K. K-value, an index for estimating fish freshness and quality. *Curr. Sci.* **1999**, *76*, 400–404.

67. Howgate, P. Kinetics of degradation of adenosine triphosphate in chill-stored rainbow trout (*Oncorhynchus mykiss*). *Int. J. Food Sci. Technol.* **2005**, *40*, 579–588. [CrossRef]
68. Vilas, C.; Alonso, A.A.; Herrera, J.R.; Bernárdez, M.; García, M.R. A mathematical model to predict early quality attributes in hake during storage at low temperature. *J. Food Eng.* **2018**, *222*, 11–19. [CrossRef]
69. Kuda, T.; Fujita, M.; Goto, H.; Yano, T. Effects of retort conditions on ATP-related compounds in pouched fish muscle. *LWT-Food Sci. Technol.* **2008**, *41*, 469–473. [CrossRef]
70. Hong, H.; Regenstein, J.M.; Luo, Y. The importance of ATP-related compounds for the freshness and flavor of post-mortem fish and shellfish muscle: A review. *Crit. Rev. Food Sci. Nutr.* **2017**, *57*, 1787–1798. [CrossRef]
71. Haard, N. The role of enzymes in determining seafood color, flavour and texture. In *Safety and Quality Issues in Fish Processing*; Bremmer, H.A., Ed.; Woodhead Publishing Limited: Sawston, UK, 2002; pp. 220–253. [CrossRef]
72. Li, K.; Luo, Y.; Shen, H. Postmortem Changes of Crucian Carp (*Carassius auratus*) During Storage in Ice. *Int. J. Food Prop.* **2015**, *18*, 205–212. [CrossRef]
73. Saito, T.; ichi Arai, K.; Matsuyoshi, M. A New Method for Estimating the Freshness of Fish. *Nippon Suisan Gakkaishi* **1959**, *24*, 749–750. [CrossRef]
74. Karube, I.; Matsuoka, H.; Suzuki, S.; Watanabe, E.; Toyama, K. Determination of Fish Freshness with an Enzyme Sensor System. *J. Agric. Food Chem.* **1984**, *32*, 314–319. [CrossRef]
75. Surette, M.E.; Gill, T.A.; LeBlanc, P.J. Biochemical basis of postmortem nucleotide catabolism in cod (*Gadus morhua*) and its relationship to spoilage. *J. Agric. Food Chem.* **1988**, *36*, 19–22. [CrossRef]
76. Zaragozá, P.; Fuentes, A.; Fernández-Segovia, I.; Vivancos, J.L.; Rizo, A.; Ros-Lis, J.V.; Barat, J.M.; Martínez-Máñez, R. Evaluation of sea bream (*Sparus aurata*) shelf life using an optoelectronic nose. *Food Chem.* **2013**, *138*, 1374–1380. [CrossRef]
77. Yamanaka, H.; Shiomi, K.; Kikuchi, T. Cadaverine as a Potential Index for Decomposition of Salmonoid Fishes. *Food Hyg. Saf. Sci. (Shokuhin Eiseigaku Zasshi)* **1989**, *30*, 170. [CrossRef]
78. Loutfi, A.; Coradeschi, S.; Mani, G.K.; Shankar, P.; Rayappan, J.B.B. Electronic noses for food quality: A review. *J. Food Eng.* **2015**, *144*, 103–111. [CrossRef]
79. Mabuchi, R.; Adachi, M.; Ishimaru, A.; Zhao, H.; Kikutani, H.; Tanimoto, S. Changes in Metabolic Profiles of Yellowtail (*Seriola quinqueradiata*) Muscle during Cold Storage as a Freshness Evaluation Tool Based on GC-MS Metabolomics. *Foods* **2019**, *8*, 511. [CrossRef] [PubMed]
80. Prabhakar, P.K.; Srivastav, P.P.; Pathak, S.S.; Das, K. Mathematical Modeling of Total Volatile Basic Nitrogen and Microbial Biomass in Stored Rohu (*Labeo rohita*) Fish. *Front. Sustain. Food Syst.* **2021**, *5*, 1–14. [CrossRef]
81. Watanabe, F.; Katsura, H.; Takenaka, S.; Enomoto, T.; Miyamoto, E.; Nakatsuka, T.; Nakano, Y. Characterization of vitamin B 12 compounds from edible shellfish, clam, oyster, and mussel. *Int. J. Food Sci. Nutr.* **2001**, *52*, 263–268. [CrossRef] [PubMed]
82. Ekanem, E.O.; Achinewhu, S.C. Effects of shucking method on opening, meat yield and selected quality parameters of West African clam, Galatea paradoxa (Born). *J. Food Process. Preserv.* **2000**, *24*, 365–377. [CrossRef]
83. Antunes-Rohling, A.; Artaiz, Á.; Calero, S.; Halaihel, N.; Guillén, S.; Raso, J.; Álvarez, I.; Cebrián, G. Modelling microbial growth in modified-atmosphere-packed hake (*Merluccius merluccius*) fillets stored at different temperatures. *Food Res. Int.* **2019**, *122*, 506–516. [CrossRef] [PubMed]
84. Winkelmann, A. Den Einfluss der Temperatur auf die Verdampfung. *Z. Anal. Chem.* **1889**, *28*, 329. [CrossRef]
85. Ratkowsky, D.A.; Olley, J.; McMeekin, T.A.; Ball, A. Relationship between temperature and growth rate of bacterial cultures. *J. Bacteriol.* **1982**, *149*, 1–5. [CrossRef] [PubMed]
86. McMeekin, T.A.; Chandler, R.E.; Doe, P.E.; Garland, C.D.; Olley, J.; Putro, S.; Ratkowsky, D.A. Model for combined effect of temperature and salt concentration/water activity on the growth rate of Staphylococcus xylosus. *J. Appl. Bacteriol.* **1987**, *62*, 543–550. [CrossRef] [PubMed]
87. Zwietering, M.H.; Wijtzes, T.; De Wit, J.C.; Riet, K.V.T. A decision support system for prediction of the microbial spoilage in foods. *J. Food Prot.* **1992**, *55*, 973–979. [CrossRef]
88. Dalgaard, P.; Buch, P.; Silberg, S. Seafood Spoilage Predictor—Development and distribution of a product specific application software. *Int. J. Food Microbiol.* **2002**, *73*, 343–349. [CrossRef]
89. Bro, R.; Van den Berg, F.; Thybo, A.; Andersen, C.M.; Jørgensen, B.M.; Andersen, H. Multivariate data analysis as a tool in advanced quality monitoring in the food production chain. *Trends Food Sci. Technol.* **2002**, *13*, 235–244. [CrossRef]
90. Limbo, S.; Sinelli, N.; Torri, L.; Riva, M. Freshness decay and shelf life predictive modelling of European sea bass (*Dicentrarchus labrax*) applying chemical methods and electronic nose. *LWT - Food Sci. Technol.* **2009**, *42*, 977–984. [CrossRef]
91. Zhang, L.; Li, X.; Lu, W.; Shen, H.; Luo, Y. Quality predictive models of grass carp (*Ctenopharyngodon idellus*) at different temperatures during storage. *Food Control* **2011**, *22*, 1197–1202. [CrossRef]
92. Giuffrida, A.; Valenti, D.; Giarratana, F.; Ziino, G.; Panebianco, A. A new approach to modelling the shelf life of Gilthead seabream (*Sparus aurata*). *Int. J. Food Sci. Technol.* **2013**, *48*, 1235–1242. [CrossRef]
93. Dai, Q.; Cheng, J.H.; Sun, D.W.; Zeng, X.A. Potential of hyperspectral imaging for non-invasive determination of mechanical properties of prawn (*Metapenaeus ensis*). *J. Food Eng.* **2014**, *136*, 64–72. [CrossRef]
94. Zhu, S.; Luo, Y.; Feng, L.; Bao, Y. Establishment of kinetic models based on electrical conductivity and global stability index for predicting the quality of allogynogenetic crucian carps (*Carassius auratus gibelio*) during chilling storage. *J. Food Process. Preserv.* **2015**, *39*, 167–174. [CrossRef]

95. García, M.R.; Vilas, C.; Herrera, J.R.; Bernárdez, M.; Balsa-Canto, E.; Alonso, A.A. Quality and shelf-life prediction for retail fresh hake (*Merluccius merluccius*). *Int. J. Food Microbiol.* **2015**, *208*, 65–74. [CrossRef]
96. Vilas, C.; Alonso, A.A.; Herrera, J.R.; García-Blanco, A.; García, M.R. A model for the biochemical degradation of inosine monophosphate in hake (*Merluccius merluccius*). *J. Food Eng.* **2017**, *200*, 95–101. [CrossRef]
97. García, M.R.; Cabo, M.L.; Herrera, J.R.; Ramilo-Fernández, G.; Alonso, A.A.; Balsa-Canto, E. Smart sensor to predict retail fresh fish quality under ice storage. *J. Food Eng.* **2017**, *197*, 87–97. [CrossRef]
98. Guo, Q.; Li, B.; Jiang, C. Shelf life prediction and Bacterial flora for the fresh and lightly salted Pseudosciaena crocea stored at different temperatures. *Emir. J. Food Agric.* **2018**, *30*, 39–48. [CrossRef]
99. Genç, I.Y.; Diler, A. Development of Shelf Life Prediction Model in Rainbow Trout Stored at Different Temperatures. *J. Aquat. Food Prod. Technol.* **2019**, *28*, 1027–1036. [CrossRef]
100. Shorten, P.R.; Soboleva, T.K.; Pleasants, A.B.; Membré, J.M. A risk assessment approach applied to the growth of Erwinia carotovora in vegetable juice for variable temperature conditions. *Int. J. Food Microbiol.* **2006**, *109*, 60–70. [CrossRef]
101. García, M.R.; Vázquez, J.A.; Teixeira, I.G.; Alonso, A.A. Stochastic individual-based modeling of bacterial growth and division using flow cytometry. *Front. Microbiol.* **2018**, *8*, 2626. [CrossRef]
102. Garre, A.; Egea, J.A.; Esnoz, A.; Palop, A.; Fernandez, P.S. Tail or artefact? Illustration of the impact that uncertainty of the serial dilution and cell enumeration methods has on microbial inactivation. *Food Res. Int.* **2019**, *119*, 76–83. [CrossRef]
103. Balsa-Canto, E.; Alonso, A.A.; Arias-Méndez, A.; García, M.R.; López-Núñez, A.; Mosquera-Fernández, M.; Vázquez, C.; Vilas, C. *Modeling and Optimization Techniques with Applications in Food Processes, Bio-Processes and Bio-Systems*; Springer: Berlin/Heidelberg, Germany, 2016; Volume 9, pp. 187–216. [CrossRef]
104. Vilas, C.; Arias-Méndez, A.; García, M.R.; Alonso, A.A.; Balsa-Canto, E. Toward predictive food process models: A protocol for parameter estimation. *Crit. Rev. Food Sci. Nutr.* **2018**, *58*, 436–449. [CrossRef]
105. Dalgaard, P. Predictive microbiological modelling and seafood quality. *Seaf. Prod. Consum. Integr. Approach Qual.* **1997**, *38*, 431–443.
106. Liu, X.; Jiang, Y.; Shen, S.; Luo, Y.; Gao, L. Comparison of Arrhenius model and artificial neuronal network for the quality prediction of rainbow trout (*Oncorhynchus mykiss*) fillets during storage at different temperatures. *LWT - Food Sci. Technol.* **2015**, *60*, 142–147. [CrossRef]
107. Emborg, J.; Dalgaard, P. Modelling the effect of temperature, carbon dioxide, water activity and pH on growth and histamine formation by Morganella psychrotolerans. *Int. J. Food Microbiol.* **2008**, *128*, 226–233. [CrossRef]
108. Banga, J.R.; Alonso, A.A.; Gallardo, J.M.; Pérez-Martín, R.I. Kinetics of thermal degradation of thiamine and surface colour in canned tuna. *Z. Fur Lebensm.-Unters. Und Forsch.* **1993**, *197*, 131–137. [CrossRef]
109. Kong, F.; Tang, J.; Rasco, B.; Crapo, C. Kinetics of salmon quality changes during thermal processing. *J. Food Eng.* **2007**, *83*, 510–520. [CrossRef]
110. Scherer, E.; Sandoval, A.; Barreiro, J.A. Kinetics of heat-induced color change of a tuna-vegetable mixture. *Interciencia* **2009**, *34*, 888–892.
111. Van Impe, J.F.; Poschet, F.; Geeraerd, A.H.; Vereecken, K.M. Towards a novel class of predictive microbial growth models. *Int. J. Food Microbiol.* **2005**, *100*, 97–105. [CrossRef] [PubMed]
112. Zwietering, M.H.; Jongenburger, I.; Rombouts, F.M.; Va not Riet, K. Modeling of the bacterial growth curve. *Appl. Environ. Microbiol.* **1990**, *56*, 1875–1881. [CrossRef] [PubMed]
113. Baranyi, J.; Roberts, T.A. A dynamic approach to predicting bacterial growth in food. *Int. J. Food Microbiol.* **1994**, *23*, 277–294. [CrossRef]
114. McKellar, R.C.; Lu, X. Primary models. *Model. Microb. Responses Food* **2003**, 21–62. [CrossRef]
115. Jain, D.; Pathare, P.B.; Manikantan, M.R. Evaluation of texture parameters of Rohu fish (*Labeo rohita*) during iced storage. *J. Food Eng.* **2007**, *81*, 336–340. [CrossRef]
116. Ross, T.; McMeekin, T.A. Predictive microbiology. *Int. J. Food Microbiol.* **1994**, *23*, 241–264. [CrossRef]
117. Koutsoumanis, K. Predictive Modeling of the Shelf Life of Fish under Nonisothermal Conditions. *Appl. Environ. Microbiol.* **2001**, *67*, 1821–1829. [CrossRef] [PubMed]
118. Koutsoumanis, K.; Giannakourou, M.C.; Taoukis, P.S.; Nychas, G.J. Application of shelf life decision system (SLDS) to marine cultured fish quality. *Int. J. Food Microbiol.* **2002**, *73*, 375–382. [CrossRef]
119. Wilson, P.D.; Brocklehurst, T.F.; Arino, S.; Thuault, D.; Jakobsen, M.; Lange, M.; Farkas, J.; Wimpenny, J.W.; Van Impe, J.F. Modelling microbial growth in structured foods: Towards a unified approach. *Int. J. Food Microbiol.* **2002**, *73*, 275–289. [CrossRef]
120. Giannakourou, M.C.; Koutsoumanis, K.; Nychas, G.J.; Taoukis, P.S. Field evaluation of the application of time temperature integrators for monitoring fish quality in the chill chain. *Int. J. Food Microbiol.* **2005**, *102*, 323–336. [CrossRef]
121. Nuin, M.; Alfaro, B.; Cruz, Z.; Argarate, N.; George, S.; Le Marc, Y.; Olley, J.; Pin, C. Modelling spoilage of fresh turbot and evaluation of a time-temperature integrator (TTI) label under fluctuating temperature. *Int. J. Food Microbiol.* **2008**, *127*, 193–199. [CrossRef]
122. Possas, A.; Valero, A.; Pérez-Rodríguez, F. New software solutions for microbiological food safety assessment and management. *Curr. Opin. Food Sci.* **2022**, *44*, 100814. [CrossRef]

123. de Prada, C.; Galán-Casado, S.; Pitarch, J.L.; Sarabia, D.; Galán, A.; Gutiérrez, G. Gemelos Digitales en la Industria de Procesos. *Rev. Iberoam. Autom. Inform. Ind.* 2022, *in press*. [CrossRef]
124. Vilas, C.; Mauricio-Iglesias, M.; García, M.R. Model-based design of smart active packaging systems with antimicrobial activity. *Food Packag. Shelf Life* **2020**, *24*, 100446. [CrossRef]

foods

MDPI

Article

Dynamic Modeling of the Impact of Temperature Changes on CO_2 Production during Milk Fermentation in Batch Bioreactors

Jožef Ritonja [1,*], Andreja Goršek [2], Darja Pečar [2], Tatjana Petek [1] and Boštjan Polajžer [1]

1 Faculty of Electrical Engineering and Computer Science, University of Maribor, Koroška Cesta 46, 2000 Maribor, Slovenia; tatjana.petek@um.si (T.P.); bostjan.polajzer@um.si (B.P.)

2 Faculty of Chemistry and Chemical Engineering, University of Maribor, Smetanova Ulica 17, 2000 Maribor, Slovenia; andreja.gorsek@um.si (A.G.); darja.pecar@um.si (D.P.)

* Correspondence: jozef.ritonja@um.si; Tel.: +386-2-220-7074

Abstract: Knowledge of the mathematical models of the fermentation processes is indispensable for their simulation and optimization and for the design and synthesis of the applicable control systems. The paper focuses on determining a dynamic mathematical model of the milk fermentation process taking place in a batch bioreactor. Models in the literature describe milk fermentation in batch bioreactors as an autonomous system. They do not enable the analysis of the effect of temperature changes on the metabolism during fermentation. In the presented extensive multidisciplinary study, we have developed a new mathematical model that considers the impact of temperature changes on the dynamics of the CO_2 produced during fermentation in the batch bioreactor. Based on laboratory tests and theoretical analysis, the appropriate structure of the temperature-considered dynamic model was first determined. Next, the model parameters of the fermentation process in the laboratory bioreactor were identified by means of particle swarm optimization. Finally, the experiments with the laboratory batch bioreactor were compared with the simulations to verify the derived mathematical model. The developed model proved to be very suitable for simulations, and, above all, it enables the design and synthesis of a control system for batch bioreactors.

Keywords: bioprocess engineering; fermentation process; batch bioreactors; dynamical non-linear mathematical model; model identification; particle swarm optimization; simulation

Citation: Ritonja, J.; Goršek, A.; Pečar, D.; Petek, T.; Polajžer, B. Dynamic Modeling of the Impact of Temperature Changes on CO_2 Production during Milk Fermentation in Batch Bioreactors. *Foods* **2021**, *10*, 1809. https://doi.org/10.3390/foods10081809

Academic Editors: Carlos Vilas, Míriam R. García and Jose A. Egea

Received: 18 June 2021
Accepted: 2 August 2021
Published: 5 August 2021

1. Introduction

1.1. Basic Terms and Topic Relevance

Biotechnology is an engineering and science discipline with enormous importance and potential [1]. The world biotechnology market was estimated at USD 369.62 billion in 2016 [2] and USD 414.50 billion in 2017 [3]. For comparison, it was four times larger than the worldwide production of electric motors (USD 96.97 billion in 2017) [4]. The global biotechnology market is projected to reach USD 727.1 billion by 2025 (the global electric motor market is estimated to reach USD 136.5 billion by 2025). Additionally, in many cases, waste from agriculture or renewable raw material is used as source material in the industrial applications of biotechnology. This is the reason that biotechnology is also essential—to lower dependence on fossil fuels and to reduce greenhouse gas emissions. It is expected that the usage of biotechnology for industrial systems can reduce energy consumption by 20%, water consumption by 75%, and carbon dioxide pollution by 50% [5].

Bioprocess engineering is a biotechnology sub-discipline that is responsible for transferring the discoveries of science results into practical processes, systems, or products that can serve the needs of society [6]. Although the production of biopharmaceuticals is the most visible, bioprocess engineering also has a dominant position in the existing fermentation industries responsible for ethanol production (used for beverages, fuel), lactic acid (for milk products), carbon dioxide, hydrogen gas, butanediol (in pharmaceuticals and cosmetics applications), propanediol (for production of biopolymers), succinic acid (in the chemical, pharmaceutical, food and agricultural industries), and aspartic acid (for production of polymers) [7].

Industrial fermentation is the primary bioengineering process. It represents a planned use of microorganisms (bacteria, yeasts, molds, or algae) or cells (animal or plant cells) to make products advantageous to humans [8].

Industrial fermentation is executed in bioreactors. Bioreactors are classified based on their construction and, consequently, their mode of operation. There are three types of bioreactor: batch bioreactors, fed-batch bioreactors, and continuous bioreactors [9]. The characteristic of batch bioreactors is that, during the implemented biological process, the bioreactor's content has no contact with external substances or organisms. That means it is closed during the operation, and no inlet or outlet from the bioreactor is possible. Such an operation mode allows non-complicated construction of batch bioreactors, which is seen in low production costs and simple maintenance. This is the reason for the prevalence of batch bioreactors. In fed-batch bioreactors, it is possible to add substances during the execution of the fermentation process. All products wait in the bioreactor till the conclusion of the biological process. Continuous bioreactors (or flow bioreactors) enable the inlet and outlet of substances or organisms into/from the reactor as a flow.

1.2. Problem Identification and Aim of the Study

A simple mode of operation and an associated undemanding and inexpensive construction and maintenance represent the great advantage of batch bioreactors. However, their great disadvantage is that we cannot add or remove external substances during the operation, and consequently, we cannot control the dynamics of the metabolism. The microorganism, substrate, and fermentation product concentrations change during the fermentation and, in the current batch bioreactors, depend only on the initial concentrations. Although it is not possible to add or remove individual substances during the operation of batch bioreactors, we can change the temperature during the execution of the fermentation in the modern batch bioreactor. This fact led us to the assumption that we could influence the metabolism during the fermentation process in batch bioreactors by changing the inner temperature. To analyze the influence of the changing temperature and to develop a control system that would ensure that the metabolism will follow the reference, we need an appropriate mathematical model. The development of the dynamic mathematical model of milk fermentation in a batch bioreactor, which would describe the influence of the temperature changes on CO_2 production, was the main goal of our study.

1.3. Related Works and Literature Review

Over recent decades, enormous development has occurred in the fields of Design, Synthesis, and Implementation of Automatic Control for Bioprocess Engineering. A thorough review of the relevant literature in the field of Control of Fermentation Processes in Bioreactors is performed in [10]. Interestingly, however, in most studies in the field of Control of Fermentation Processes, a fundamental kinetic mathematical model [11] is still used, both for theoretical and numerical analysis [12–14]. The derivation of a parametric mathematical model of the fermentation process, which also describes the influence of temperature changes on the time course of concentrations of microorganisms, substrate, and fermentation product, is a relatively new field. Although there are many studies showing the influence of temperature on the fermentation processes of different species, these are limited mainly to the observation and numerical evaluation of the influence of different (constant) temperatures on the course of the fermentation process and do not result in deriving a parametric mathematical model [15–17].

Mathematical modeling of the fermentation process in bioreactors is highly current, as shown by numerous new publications. Reference [18] shows the implementation of the computational fluid dynamics model of a bioreactor. In [19], the stability analysis of a fundamental kinetic mathematical model of a continuous bioreactor is shown. Reference [20] shows how mathematical modeling can be usefully applied in designing and optimizing bioreactors.

The authors of this paper started working on a parametric dynamical mathematical model that describes the impact of temperature variation on the fermentation dynamics in the year 2018. The influence of temperature change on fermentation dynamics was experimentally observed and described first in [21]. In [22], the authors introduce a new supplementary transfer function that considers the impact of temperature variation on a bioprocess. A conventional control system was developed, and a tuning method was proposed on the basis of this model. In [23], the derived mathematical model was used for the optimization of the control system. In all cases, a mathematical model was derived for probiotic beverages' fermentation.

The mathematical model presented in [21–23] is a "hybrid" mathematical model consisting of an autonomous non-linear model and a non-autonomous linear model connected in parallel. The non-linear model describes the influence of initial conditions on the course of fermentation variables, and the linear model is used to model the impact of temperature changes on their responses. The derived model proved to be a very accurate description of what happened during the fermentation process. This model allows efficient analysis of the fermentation process, but it is demanding to design and synthesize control systems due to its complicated structure. In order to increase the compactness of the model and, thus, increase the suitability of the model for the needs of design and synthesis of control systems, we developed a temperature-considered dynamic mathematical model of the milk fermentation process, which is uniform and does not consist of two parallel sub-models. The derived model is presented in this article.

1.4. Paper Contributions

The premise that the control of the metabolism during the fermentation is possible by changing the temperature in the bioreactor leads to the conclusion that, also for batch bioreactors, we can develop a control system, which will control the fermentation process during the operation. The paper's first contribution is the confirmation of this premise by experiments on a laboratory batch bioreactor. The logical continuation of this finding is the intention to determine the mathematical model that will allow the design and synthesis of the control system. The paper's second contribution is the derivation of the appropriate mathematical model. The developed model is required to accurately describe the phenomenon during the fermentation process, and, at the same time, the model must be structurally suitable for the development of the control system.

2. Materials and Methods

In our study, we focused on the dynamic modeling of the fermentation process in batch bioreactors. Therefore, the subsequent chapters describe in more detail the considered fermentation process and the batch bioreactor, along with the necessary equipment for measurements.

2.1. Fermentation Process

The presented study considers milk fermentation with kefir grains. Traditionally, kefir is produced by inoculating kefir grains, which are a mixture of proteins, polysaccharides, mesophilic, homofermentative, and heterofermentative lactic acid streptococci, thermophilic and mesophilic lactobacilli, acetic acid bacteria, and yeast. Fermentation of the milk by the inoculum proceeds for ca. 24 h, during which time homofermentative lactic acid streptococci grow rapidly, initially causing a drop in pH. This low pH favors the growth of lactobacilli but causes the streptococci number to decline. The presence of yeasts in the mixture, together with fermentation temperature (21–23 °C), encourages the growth of aroma-producing heterofermentative streptococci. As fermentation proceeds, the growth of lactic acid bacteria is favored over the growth of yeasts and acetic acid bacteria.

Before the fermentation, kefir grains (40 g) were activated for 5 successive days, washed daily with cold water and put into 500 mL of fresh pasteurized whole fat milk at room temperature. To start the fermentation, 500 mL of fresh pasteurized whole fat milk was preheated in the fermenter to the desired temperature and then inoculated with 40 g of active kefir grains. For the presented fermentation, the full activated (5 days activation) kefir grains were used. Different fermentation processes were obtained by means of differently activated kefir grains.

During the fermentation, carbon dioxide, acetic acid, diacetyl, acetaldehyde, ethyl alcohol, and several other substances are formed, and these give the products their characteristic fresh taste and aroma.

Milk fermentation with kefir grains propagation is an inherently very complex process because of the specific nature of the microbial metabolism, as well as the non-linearity of its kinetics. Therefore, monitoring and control are extremely important to develop models that are able to provide an accurate description of the progress of the process.

2.2. *Laboratory Equipment*

We needed an appropriate laboratory system to determine how the temperature changes affect the time responses of the substance concentrations of the fermentation process. For this purpose, a laboratory system was built, which enabled controlled temperature changes in the bioreactor and dynamic measurement of concentrations of substances during the fermentation process. The laboratory system consists of a batch bioreactor with a heating/cooling thermal system, a measurement system, and a data acquisition system. All parts are described below.

2.2.1. Batch Bioreactor

The fermentation process was primarily analyzed in the computer-controlled laboratory batch bioreactor RC1e from Mettler Toledo (Greifensee, Switzerland). The laboratory bioreactor has additional equipment for the identification of model parameters. The studied batch bioreactor is described in detail in [10,22,23].

2.2.2. Heating/Cooling System

For the identification of the parameters of the mathematical model, an integrated heating/cooling system was used, which controls the temperature of the bioreactor's mixture. Silicone oil represents the heat transfer agent. The oil is circulated through the bioreactor's double jacket in a closed circulation system. The temperature control system keeps the temperature of the bioreactor's contents at the reference value [23]. Compared to the dynamics of the fermentation process, the response of the heating/cooling system is very fast. The identified time constant of the heating/cooling system is approximately 6 min.

2.2.3. Dissolved Carbon Dioxide Measurement System

In the studied fermentation process, the carbon dioxide (CO_2) dissolved in the bioreactor's medium is the fermentation product. CO_2 is a product of the cellular metabolism of microorganisms. Assuming a growth medium with a sufficient carbon source, the measured CO_2 concentration profile could also be the indicator of the fermentation progress. Thus, dissolved CO_2 was chosen as the output and identified as an observable variable. The distribution of the CO_2 in the bioreactor medium is very homogeneous. The sensors for the measurement of CO_2 concentration are reliable and accurate. The multi-parameter measuring device SevenMulti from Mettler Toledo with modular expansion possibilities was used as a basic unit. For the monitoring of dissolved CO_2 concentration in liquid media, the SevenMulti apparatus was connected to the ISE51B ion-selective electrode. Electrode potential response to CO_2 concentration is, in a semilogarithmic scale, a straight line over two decades of the concentration ($5{\cdot}10^{-4}$ g/L to $2{\cdot}10^{-2}$ g/L). A change in temperature causes the electrode response to shift and change slope (temperature variation 5 °C change, slope approx. 1.7%). The detailed principles of the measurement system are described in [10].

2.2.4. Data Acquisition Equipment

For the connection of the SevenMulti basic device and the ion-selective electrode sensor, an expansion module was added to the basic unit. The analog 1st order low pass filter for the elimination of sensor noise is integrated into the expansion module. For the transfer of measured signals from the SevenMulti basic unit to PC, the basic device was equipped with a USB communication module. For the transfer of the measured temperature signal and the reference temperature signal from the heating/cooling system to PC, a dSpace 1103 PPC controller board was utilized. The controller is equipped with 16 bit A/D and D/A converters as well as serial and CAN interfaces [10].

For the comprehensive measurement of several quantities over a long time period and for the necessary signal processing, software LabX direct pH 2.3, Mettler Toledo (Greifensee, Switzerland) was installed on PC. This is professional equipment used for data logging and data analysis. The selected sampling time was 10 min. This was sufficient due to the slow dynamics of the fermentation process. During the performing the experiments, the sampling time was changed and adjusted to the dynamics of the measured signal. Measured data was saved into Excel, Microsoft Office 365 (USA) documents, transferred into MATLAB R2021a, MathWorks (USA), and processed using MATLAB with its Optimization toolbox functions [10]. A block diagram of the batch bioreactor with measurement system is displayed in Figure 1.

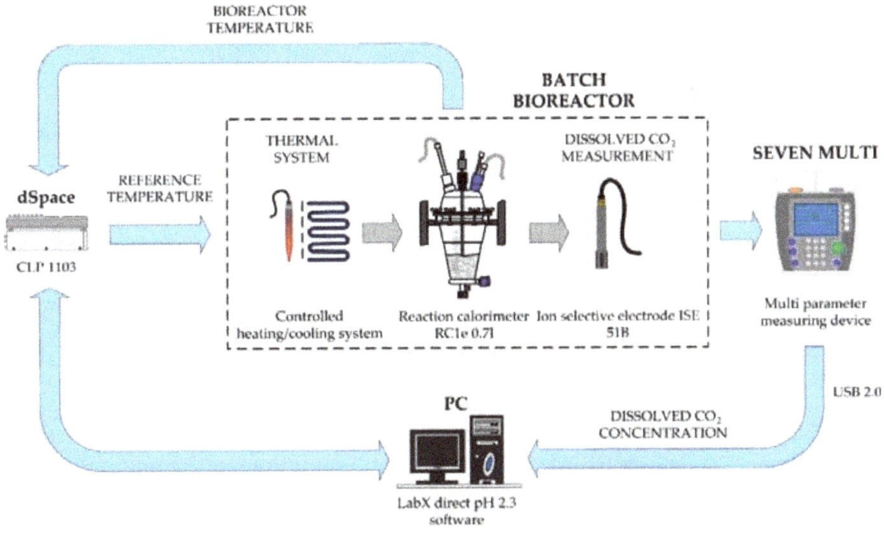

Figure 1. Laboratory system for experiments and identification of the mathematical model.

2.3. *Fundamental Mathematical Model of the Fermentation Process in Batch Bioreactor*

Fermentation is a process whereby microorganisms induce a substrate to break down into a fermentation product. Microorganisms, substrates, and fermentation products are present in all fermentation.

The fermentation process is a complex non-linear system with mainly unknown structures and unknown time-varying parameters. In the fermentation process, a multitude of biochemical reactions occurs, representing the kinetics inside a microorganism, in addition to physical transfer rates [9]. The adaptation of the microorganisms to the environment through mutations represents an additional challenge. These are the reasons for the absence of mathematical models, useful from an engineering viewpoint, which would take into account the numerous factors which influence the growth of microorganisms and the execution of the fermentation process.

However, we can find in the literature various mathematical models of different complexity describing the dynamics of fermentation processes in batch bioreactors. These models are mostly based on the mass balances of substances that occur in fermentation. A fundamental kinetic model is written in the state-space form of 3rd order (1)–(3) [10,23],

$$\dot{x}_1(t) = \frac{\mu_\mathrm{m}\left(1 - \frac{1}{P_\mathrm{i}}x_3(t)\right)x_2(t)}{S_\mathrm{m} + x_2(t) + \frac{1}{S_\mathrm{i}}(x_2(t))^2}x_1(t) \tag{1}$$

$$\dot{x}_2(t) = -\frac{\mu_\mathrm{m}\left(1 - \frac{1}{P_\mathrm{i}}x_3(t)\right)x_2(t)}{S_\mathrm{m} + x_2(t) + \frac{1}{S_\mathrm{i}}(x_2(t))^2}x_1(t) \tag{2}$$

$$\dot{x}_3(t) = \left(\alpha\frac{\mu_\mathrm{m}\left(1 - \frac{1}{P_\mathrm{i}}x_3(t)\right)x_2(t)}{S_\mathrm{m} + x_2(t) + \frac{1}{S_\mathrm{i}}(x_2(t))^2} + \beta\right)x_1(t) \tag{3}$$

where the state space variables and the model parameters are:

$x_1(t)$—the microorganisms' concentration (g/L),
$x_2(t)$—the substrate's concentration (g/L),
$x_3(t)$—the fermentation product's concentration (g/L)
μ_m—the maximum microorganisms' growth rate (h^{-1}),
P_i—the product inhibition constant (g/L),
S_m—the substrate saturation constant (g/L),
S_i—the substrate inhibition constant (g/L),
A—the parameter that describes the relation between product yield and microorganism growth, and
β—the growth independent constant (h^{-1}).

The mathematical model presented in (1)–(3) is autonomous. The model has no input variable. This corresponds to the actual realization of batch bioreactors because they do not have an input quantity to be used for control of the fermentation process. All three substances of the bioprocess are placed in the bioreactor at the start of the fermentation. During the fermentation, it is not possible to add to or remove any of them. The dynamics of the fermentation process depend only on the quantity and quality of substances used and the type of batch bioreactor. Accordingly, the transients of the model are obtained as the response to the initial values of the model variables and depend on the parameters of the mathematical model.

During the fermentation process, the quantity of the microorganisms and product increase, and the amount of substrate decreases. The quantities of substances are measured and applied in the model as concentrations.

In the case of an autonomous fermentation process (the fermentation process that depends only on initial concentrations, where the temperature is constant during the whole fermentation process), all parameters of the fundamental kinetic mathematical model are constant throughout the entire duration of the fermentation process.

The presented fundamental kinetic mathematical model of the fermentation process enables simple and efficient simulation and analysis of the time courses of the concentrations of microorganisms, substrate, and product in the case of different initial concentrations. The main disadvantage of this model is its inability to consider the effect of temperature change on the fermentation process, which is necessary for control system purposes.

This model does not allow the evaluation of the impact of temperature changes on the time courses of concentrations of individual substances during the fermentation process, which is necessary for the design and synthesis of suitable control systems. Therefore, it is necessary to derive a new mathematical model of the fermentation process in the batch bioreactor, which will evaluate the influence of temperature change.

3. Results

3.1. Experimental Analysis of the Fermentation Process

3.1.1. Responses of the Autonomous Process

First, an analysis was carried out of how the diverse constant temperatures affect the time course of the CO_2 production. The same initial values for all substances of the fermentation process were used in all experiments. The bioreactor has a constant temperature during the whole fermentation process. The obtained results were expected, meaning that a more responsive fermentation course and a greater end amount of the CO_2 concentration were achieved at a higher constant temperature. The equal findings were obtained in experiments with different fermentation processes. Figure 2 shows the measured time courses of the CO_2 for the various constant temperatures of the studied bioreactor.

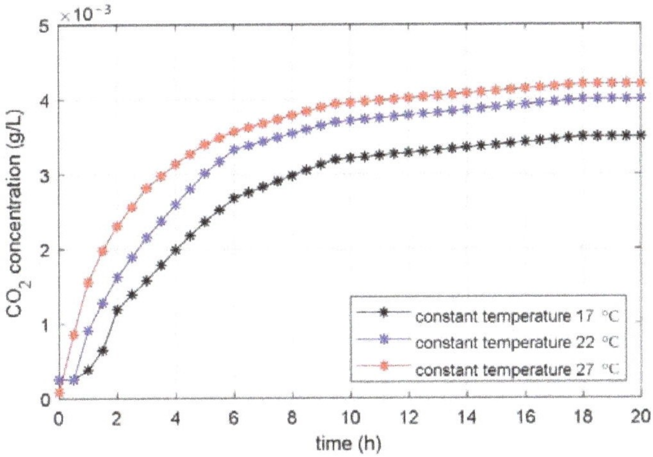

Figure 2. Measured time courses of the CO_2 concentration during the fermentation processes with different bioreactor temperatures.

3.1.2. Responses of the Non-Autonomous Process

In the second phase, the impact of the temperature changes during the fermentation was studied. In this phase, many tests were also carried out (different amplitudes, slopes, signs, and different moments of temperature change). It was visible from all experiments that an increase in the temperature speeds up the fermentation, and a temperature decrease decelerates it. The influence of the step change of the temperature on the time course of the CO_2 concentration is presented in Figure 3. The figure shows the transient response of the CO_2 concentration when temperature changes from 22 °C to 27 °C. The time course of the CO_2 for the constant temperature (T = 22 °C) during the entire fermentation was also added to this graph.

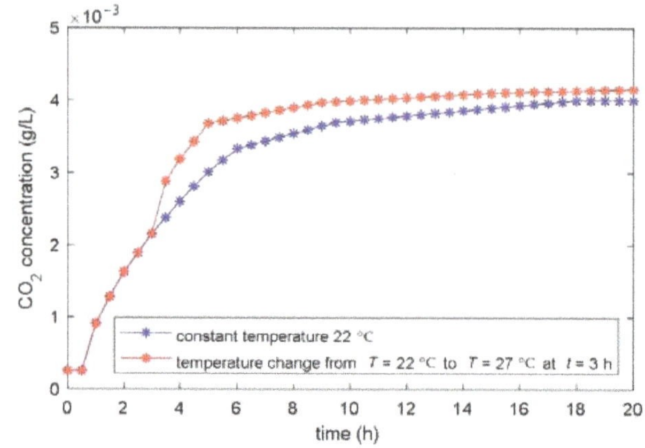

Figure 3. Measured time courses of the CO_2 concentration during the fermentation processes with constant and changeable bioreactor temperatures.

3.2. Analysis of Correlation between Parameters and Responses of Mathematical Model

The aim of the paper was to find a non-autonomous mathematical model whose input variable would be the desired temperature of the heating/cooling system, and the model's state variables would be the concentrations of microorganisms, substrate, and fermentation product. The starting point for the development of such a model was the idea that the temperature in the bioreactor affects the parameters of the associated mathematical model. As a basis for the development of a new mathematical model, the presented fundamental kinetic mathematical model (1–3) was used. The correlation between the fermentation process and the parameters of the fundamental kinetic mathematical model (1–3) was determined using systematic and extensive simulations. It has been shown that it is significant to analyze the impact of the model's parameters on the time course of the CO_2 concentration of the fermentation product (variable $x_3(t)$ in (1)–(3)). Figures 4–9 show the influence of all mathematical model parameters (1)–(3) on the time course of the CO_2 concentration, separately. To obtain these results, modified values of the parameters of the mathematical model were presumed. Constant values of the model's parameters were used during the particular simulation of the fermentation process. Results are shown in the following Figures:

- the impact of the maximum microorganisms' growth rate μ_m on the CO_2 concentration $x_3(t)$ is shown in Figure 4,
- the impact of the product inhibition constant P_i on the CO_2 concentration $x_3(t)$ is shown in Figure 5,
- the impact of the substrate saturation constant S_m on the CO_2 concentration $x_3(t)$ is shown in Figure 6,
- the impact of the substrate inhibition constant S_i on the CO_2 concentration $x_3(t)$ is shown in Figure 7,
- the impact of the parameter of the product yield in consequence of microorganism growth α on the CO_2 concentration $x_3(t)$ is shown in Figure 8,
- the impact of the independent product yield β on the CO_2 concentration $x_3(t)$ is shown in Figure 9.

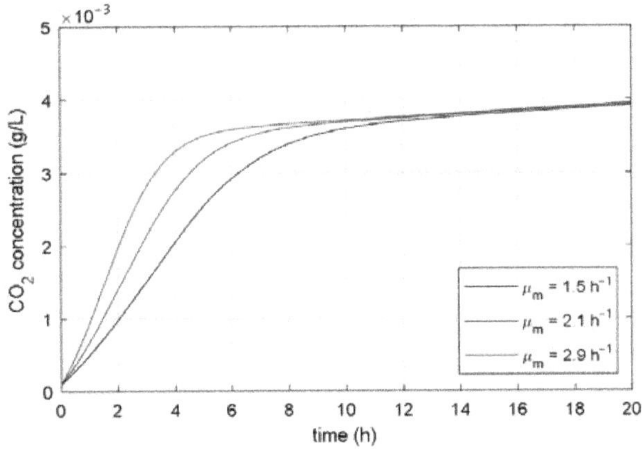

Figure 4. The impact of the maximum microorganisms' growth rate μ_m on the time course of the CO_2 concentration.

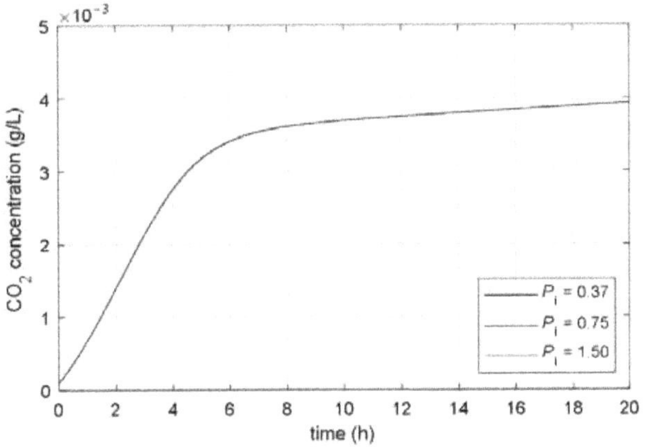

Figure 5. The impact of the product inhibition constant P_i on the time course of the CO_2 concentration.

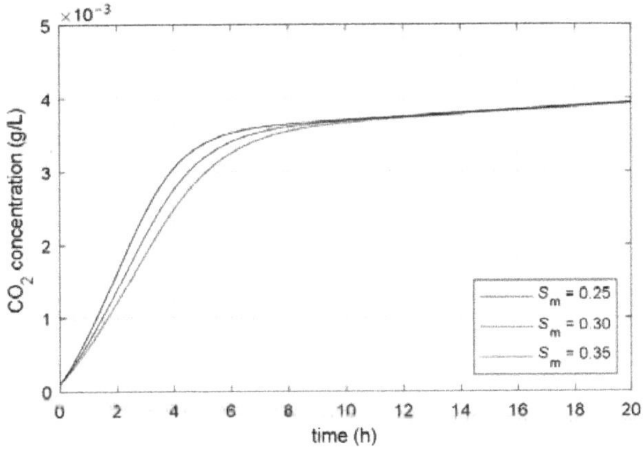

Figure 6. The impact of the substrate saturation constant S_m on the time course of the CO_2 concentration.

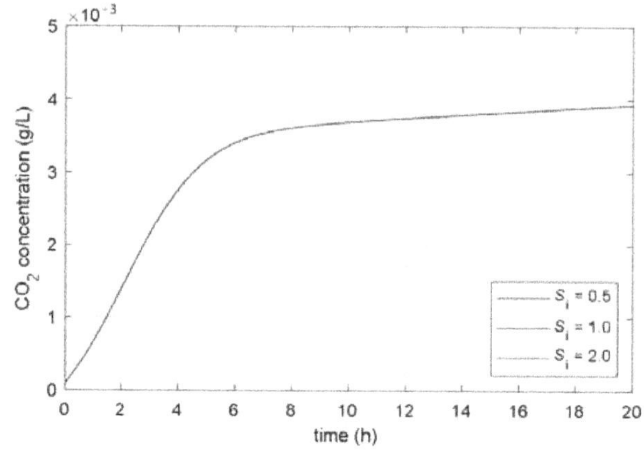

Figure 7. The impact of the substrate inhibition constant S_i on the time course of the CO_2 concentration.

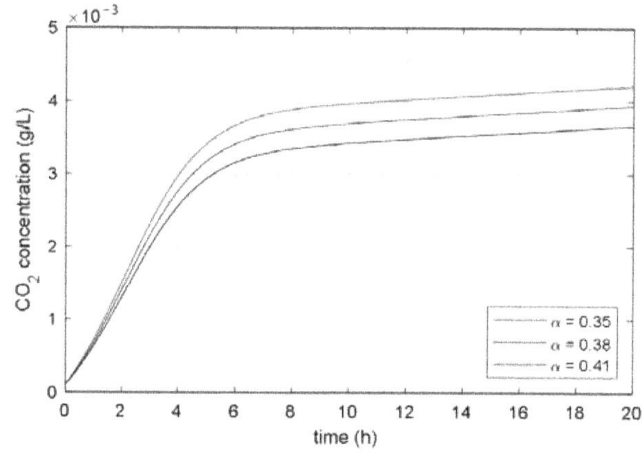

Figure 8. The impact of the parameter α on the time course of the CO_2 concentration.

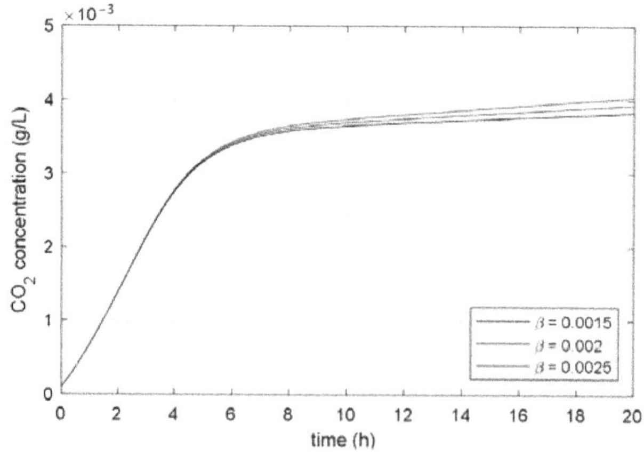

Figure 9. The impact of the parameter β on the time course of the CO_2 concentration.

The goal of the present work was to find the connection between the parameters of the mathematical model and the dynamics of the fermentation process. We were interested in discovering which parameter of the mathematical model had a high impact on the fermentation process dynamics. Based on mathematical models of different fermentation processes in different batch bioreactors, we carried out an extensive systematic numerical analysis. The analyzed fermentation processes have similar impacts of parameter changes on fermentation process dynamics. The main conclusions of all simulations were:

- the impact of the product inhibition constant P_i and the substrate inhibition constant S_i on the fermentation process transient and steady-state is very small,
- the substrate saturation constant S_m has a very small impact on the fermentation process steady-state and a small impact on the fermentation process transient,
- the parameter that describes the product yield that is independent of the microorganisms' growth β has a very small impact on the fermentation process transient and a small impact on the fermentation process steady state,
- the maximum microorganisms' growth rate μ_m has a significant impact on the fermentation process transient and very small impact on the fermentation process steady-state,
- the parameter that describes the relation between product yield and microorganisms' growth α has a significant impact on the fermentation process steady-state and very small impact on the fermentation process transient.

Determination of parameters that have an impact on the dynamics and statics of the fermentation process is also possible using sensitivity analysis [24–26]. Sensitivity analysis allocates model uncertainty to the various sources of uncertainty, facilitating the targeted reduction of output uncertainty. Sensitivity analysis methods have provided numerous interesting results in a wide variety of different applications. Its advantage is visible especially in systems with a large number of parameters.

3.3. Temperature-Considered Model of the Fermentation Process in Batch Bioreactor

From the comparison of the experimental and simulation results, we can see that the temperature changes have a similar effect on the experimentally obtained courses of the fermentation product (Figure 2) as changing the parameters μ_m and α on the numerically calculated courses (Figures 4 and 8). Based on the results in Sections 3.1 and 3.2, we can conclude that the temperature change significantly affects parameters μ_m and α, while it does not have a large effect on other parameters. It can be seen from Figure 4 that the maximum microorganisms' growth rate μ_m affects the fermentation speed but does not affect the steady-state value of the fermentation product achieved at the end of fermentation. In contrast, the parameter α affects the steady-state value of the fermentation product and does not significantly affect the transient phenomenon, as shown in Figure 8.

This means that in the case of variable temperature in the bioreactor, the parameters μ_m and α will no longer be constant, but their values will change according to the temperature in the bioreactor. Therefore, we added a differential equation to the fundamental mathematical model of the fermentation process, with which we calculate the temperature in the batch bioreactor. The temperature in the bioreactor is described by a new variable $x_4(t)$. Instead of constant parameters μ_m and α, we introduce temperature-dependent parameters $\mu_{m\vartheta}(t)$ and $\alpha_\vartheta(t)$. In [27], it was shown that there is a static relationship between temperature and the parameters of the mathematical model. Since the temperature during the fermentation varies in a relatively small range (too high or too low temperatures can damage the microorganisms), we used a linear dependence between temperature deviation and model parameters deviation. Consequently, we expressed temperature-dependent parameters $\mu_{m\vartheta}(t)$ and $\alpha_\vartheta(l)$ with following static functions:

$$\mu_{m\vartheta}(t) = \mu_m(1 + k_{\mu m}(x_4(t) - \vartheta_0)) \tag{4}$$

$$\alpha_\vartheta(t) = \alpha\,(1 + k_\alpha(x_4(t) - \vartheta_0)) \tag{5}$$

where the new parameters in the linear static equations are:

$x_4(t)$—the temperature in the bioreactor (°C),
ϑ_0—the temperature of the bioreactor's contents at the beginning of the fermentation process (°C), where normally ϑ_0 is equal to the outside temperature,
$k_{\mu m}$—the coefficient that outlined the effect of the temperature changing on the maximum microorganisms' growth rate μ_m $(°C)^{-1}$,
k_α—the coefficient that describes the impact of the temperature change on the parameter that describes the relation between product yield and microorganism growth $(°C)^{-1}$,
$\mu_{m\vartheta}(t)$—the temperature-dependent maximum microorganisms' growth rate (h^{-1}), and
$\alpha_\vartheta(t)$—the temperature-dependent parameter which expresses the connection between product yield and microorganism growth (h^{-1}).

The new temperature-considered model was obtained by supplementing the fundamental model with an additional differential equation for calculating the temperature in the bioreactor and by replacing the constant parameters μ_m and α with temperature-dependent ones, as described by the Equations (4) and (5). The derived mathematical model is expressed with the following equations:

$$\dot{x}_1(t) = \frac{\mu_m\left(1 + k_{\mu_m}(x_4(t) - \vartheta_0)\right)\left(1 - \frac{1}{P_i}x_3(t)\right)x_2(t)}{S_m + x_2(t) + \frac{1}{S_i}(x_2(t))^2}x_1(t) \tag{6}$$

$$\dot{x}_2(t) = -\frac{\mu_m\left(1 + k_{\mu_m}(x_4(t) - \vartheta_0)\right)\left(1 - \frac{1}{P_i}x_3(t)\right)x_2(t)}{S_m + x_2(t) + \frac{1}{S_i}(x_2(t))^2}x_1(t) \tag{7}$$

$$\dot{x}_3(t) = \left(\alpha\ \frac{(1+k_\alpha(x_4(t)-\vartheta_0))\mu_m\left(1+k_{\mu_m}(x_4(t)-\vartheta_0)\right)\left(1-\frac{1}{P_i}x_3(t)\right)x_2(t)}{S_m+x_2(t)+\frac{1}{S_i}(x_2(t))^2} + \beta\right)x_1(t) \tag{8}$$

$$\dot{x}_4(t) = \frac{1}{T_{\vartheta cs}}(u(t) - x_4(t)) \tag{9}$$

where additionally to the symbols in (1)–(5):

$u(t)$—indicates the reference temperature of the bioreactor's temperature control system (°C), and
$T_{\vartheta cs}$—is the time constant of the simple 1st order model of the controlled heating system (h).

The newly developed model considers the influence of temperature on the transient and steady-state of the fermentation process, which is why we named this model the temperature-considered model. The developed enhanced model (6–9) represents the transition between different models (1–3).

When developing and testing the temperature-considered model, instead of linear expressions (4) and (5), we tried different non-linear analytical expressions to describe the dependence of the model parameters on the temperature in the bioreactor. Tested functions were identified in such a way that they enable good fitting of the mathematical model's responses to the measured trajectories [21]. The selected linear functions represent a good compromise—they are easy to identify and, at the same time, allow a good description of the progress during the fermentation process.

The significant advantage of the temperature-considered model (6–9) is its ability to allow an analysis of the impact of temperature changes on the fermentation process dynamics and steady-state characteristics. The derived mathematical model (6–9) is in structure partially similar to the basic fundamental model (1–3), but in its functionality the newly derived model (6–9) is incomparable to the fundamental model (1–3). The fundamental model (1–3) is an autonomous model and allows only simulations of responses of the concentrations to the initial values of fermentation substances. On the contrary, the model (6–9) represents a non-autonomous model of non-linear differential equations, which additionally enables the calculation of the courses of concentrations of bioreactor substances in the case of temperature variations in a bioreactor. Therefore, model (6–9) is suitable for the design of fermentation control systems based on temperature changes in the bioreactor, which is not possible with model (1–3).

3.4. Parameters Identification of the Temperature-Considered Model

The parameters of the mathematical model μ_m, P_i, S_m, S_i, α, β, $k_{\mu m}$, k_α, and $T_{\vartheta cs}$ (all in Equations (1)–(9)) depend on the quality and quantity of the substances, and of the external operating parameters. In the case of unchanged external conditions, the parameters remain more or less constant during the fermentation process [21]. The parameters of the mathematical models for the fermentation processes in the laboratory or industrial bioreactors can be calculated by different optimization methods from the measured trajectories of the bioreactors' substances. In our study, the particle swarm algorithm was used to obtain the parameters of the mathematical model.

Particle swarm is a population-based algorithm [28]. Particle swarm optimization (PSO) is originally attributed to [29]. During the optimization the swarm of particles varies throughout the selected area. The optimization algorithm calculates the objective function at each step. After their calculation, the algorithm sets the new particles' velocities. The algorithm moves each particle to the best-founded location. PSO is a metaheuristic procedure that may provide a sufficiently good solution to an optimization problem, in case of few, incomplete, imperfect, or no assumptions about the problem being optimized. PSO can search very large spaces of candidate solutions. PSO does not use the gradient of the problem being optimized, which means that it does not require that the optimization problem be differentiable. However, PSO does not guarantee that an optimal solution is ever found.

Functions from MathWorks MATLAB/Optimization Toolbox library were used for faster realization of the PSO for the calculation of the model's parameters. Matlab function particleswarm.m is based on the algorithm described in [29], using modifications suggested in [30,31]. Details of the PSO algorithm in the particleswarm.m function are written in [32].

The optimization was accomplished from the measured time course of the CO_2 concentration. For the objective function requires, the error between the measured and calculated response was computed. The integral absolute error (IAE) objective function was implemented to calculate the fitting of the mathematical model with the laboratory batch bioreactor [10].

Because we approximately estimated the range of values of the model's parameters, we constrained the area where the algorithm was searching for optimal solutions. The optimization algorithm changed the parameters of the mathematical model (6–9) for so long in order to reach the minimum of the IAE function. Optimization was finished when the relative change of the objective function reached the stopping criterion suggested in the default defined value.

For the determination of the temperature-considered model, it was necessary to identify nine parameters of the mathematical model. PSO can be used to identify all parameters simultaneously, but we can also identify the model's parameters in several stages. With the help of a systematic trial procedure, we found that the best results are obtained by separating the parameters identification into two phases. In the first phase, the parameters μ_m, P_i, S_m, S_i, α, and β were identified. In the second phase, the remaining parameters $k_{\mu m}$, k_α, and $T_{\vartheta cs}$ were calculated.

3.4.1. Identification of the Parameters μ_m, P_i, S_m, S_i, α, and β

In the first stage of the optimization procedure, the parameters μ_m, P_i, S_m, S_i, α, and β were identified. These represent the fundamental part of the mathematical model, which coincides with the constant temperature environmental conditions. This models the fermentation process's transient and steady-state behavior resulting from the initial concentrations of the substances in the bioreactor.

First, the initial values of the microorganisms, substrate, and fermentation product for the fermentation process in the studied batch bioreactor were measured. The measured values are written in Table 1.

Table 1. Initial values of the fermentation process in the studied bioreactor.

Variable	Value
the initial value of the microorganisms' concentration	$x_1(0) = 2.6$ mg/L
the initial value of the substrate's concentration	$x_2(0) = 9.0$ mg/L
the initial value of the product's concentration	$x_3(0) = 0.1$ mg/L
the initial temperature of the bioreactor's contents	$x_4(0) = 22$ °C

After that, the fermentation process was carried out. Only the dissolved CO_2 concentration needed to be measured. The dissolved CO_2 concentration represents the fermentation product and is the most important substance in the bioreactor. During the fermentation process, all external conditions (temperature, stirrer's speed) were constant. The measured time course of the dissolved CO_2 concentration for constant temperature 22 °C is shown in Figure 2. The PSO with IAE objective function was applied to calculate the parameters of the fundamental part of the model, valid for the constant bioreactor's temperature. The IAE function was calculated as the integral of the absolute error between the measured course of the dissolved CO_2 concentration (Figure 2) and the calculated state-space variable $x_3(t)$ of the non-linear mathematical model (1–3). The PSO algorithm changed the values of the identified model μ_m, P_i, S_m, S_i, α, and β for so long that the objective function reached a minimum. The time course of the objective function during the PSO procedure is shown in Figure 10.

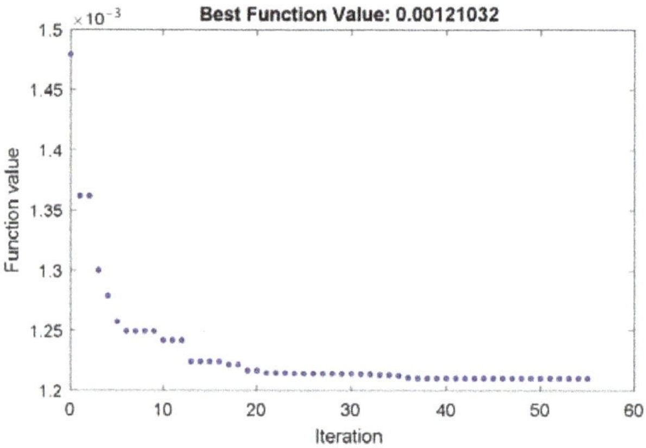

Figure 10. The time course of the IAE objective function during the PSO with the parameters μ_m, P_i, S_m, S_i, α and β.

As can be seen from Figure 10, less than 60 iterations were required to calculate the parameters μ_m, P_i, S_m, S_i, α, and β of the fundamental part of the mathematical model. All identified parameters are presented in Table 2.

Table 2. Parameters of the Mathematical Model for the Fermentation Process in the Studied Bioreactor.

Parameter	Value
the maximum microorganisms' growth rate	$\mu_m = 2.1\ h^{-1}$
the product inhibition constant	$P_i = 0.75\ g/L$
the substrate saturation constant	$S_m = 0.03\ g/L$
the substrate inhibition constant	$S_i = 1.0\ g/L$
the parameter of the product yield related to microorganisms' growth	$\alpha = 0.38\ \frac{g/L}{g/L}$
the parameter of the product yield independent of the microorganisms' growth	$\beta = 0.002\ h^{-1}$
the temperature of the bioreactor's contents during the fermentation process	$\vartheta_0 = 22\ °C$

3.4.2. Identification of the Parameters $k_{\mu m}$, k_α, and $T_{\vartheta cs}$

In the second stage of the optimization, the parameters $k_{\mu m}$, k_α, and $T_{\vartheta cs}$ were identified. These parameters are important for the supplementary part of the mathematical model, which describes the influence of the variable temperature on the fermentation process transient and steady-state behavior. The parameters $k_{\mu m}$ and k_α characterize the impact of the variable temperature on the modifying of the parameters $\mu_{m\vartheta}(t)$ and $\alpha_\vartheta(t)$. The parameter $T_{\vartheta cs}$ specifies the temperature time constant that describes the alternating of the temperature in the bioreactor as a result of the reference temperature of the bioreactor's temperature control system.

To identify these parameters, a fermentation process was carried out again. This time, we changed the reference temperature during the fermentation process. The step change of the temperature from 22 °C to 27 °C arises at $t = 3$ h. The time course of the dissolved CO_2 during the fermentation with changeable temperature is shown in Figure 3. Again, the PSO with IAE objective function was used for the identification of the model parameters. The IAE objective function was calculated as the integral of the absolute error between the measured course of the dissolved CO_2 concentration (Figure 3) and the calculated state-space variable x_3(t) of the non-linear mathematical model (3–6). The PSO algorithm changed the values of the identified model $k_{\mu m}$, k_α, and $T_{\vartheta cs}$ for so long that the objective function reached a minimum. During the optimization, the parameters μ_m, P_i, S_m, S_i, α, and β have values that were identified in the first stage of the identification procedure (Table 2). The time course of the objective function during the PSO procedure is shown in Figure 11.

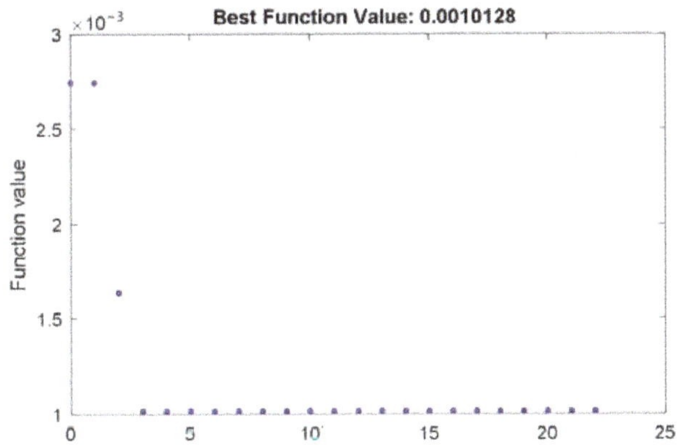

Figure 11. The time course of the IAE objective function during the PSO with the parameters $k_{\mu m}$, k_α, and $T_{\vartheta cs}$.

Figure 11 shows that less than 25 iterations were required for the calculation of the parameters $k_{\mu m}$, k_{α}, and $T_{\vartheta cs}$ of the supplemental part of the mathematical model. The identified parameters are presented in Table 3.

Table 3. Additional parameters of the augmented mathematical model for the fermentation process in the studied bioreactor.

Parameter	Value
the coefficient of the impact of the temperature changing on the maximum growth rate μ_m	$k_{\mu m} = 0.14\ (°C)^{-1}$
the coefficient of the impact of the temperature changing on the parameter α	$k_{\alpha} = 0.03\ (°C)^{-1}$
the time constant of the 1st order model of the controlled heating system	$T_{\vartheta cs} = 0.1$ h
the temperature of the bioreactor's contents at the beginning of the fermentation process	$\vartheta_0 = 22$ °C

To estimate the temperature-considered mathematical model parameters, we can also use the computationally less demanding sensitivity method [24–26]. Its advantage is especially evident in models with a large number of parameters.

3.5. Simulation Results of the Temperature-Considered Model

The simulation results of the identified temperature-considered model of the fermentation in the laboratory bioreactor are shown in Figure 12. Presented are the time responses of the microorganisms, substrate, and product in the case of constant temperature (solid lines) $T = 22$ °C, and the same variables in the case of the bioreactor's reference temperature step change from 22 °C to 27 °C in $t = 3$ h (drawn with dashed lines). It is evident that temperature changes generate substantial variations in the dynamics of all quantities of the fermentation process. The time course of the actual temperature in the bioreactor follows the reference temperature step change. The delay in the actual temperature is small—it corresponds to the short time constant of the controlled heating system $T_{\vartheta cs} = 0.1$ h.

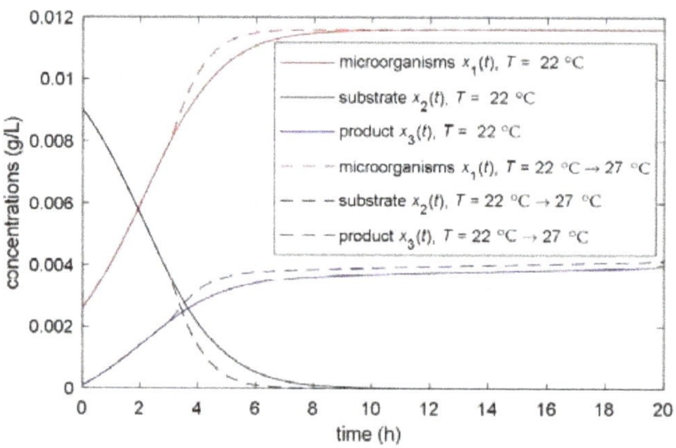

Figure 12. Time courses of the concentrations of the microorganisms, substrate, and product during the fermentation processes with constant and changeable bioreactor temperatures.

3.6. Comparison of Simulations and Experimental Results

The matching of the response of the measured CO_2 concentration in the laboratory bioreactor with the response of the CO_2 concentration calculated with the identified temperature considered model is displayed in Figures 13–15.

The results of the fermentation process with a constant temperature 22 °C are presented in Figure 13. It can be seen that in the case of a constant temperature, the time course of the produced CO_2 is continuous. The response of the identified mathematical model matches very well with the measured response of the laboratory bioreactor. Matching is seen in all three phases of the fermentation process: in the starting induction phase, in the successive exponential growth phase and in the subsequent stationary phase and the end dying phase.

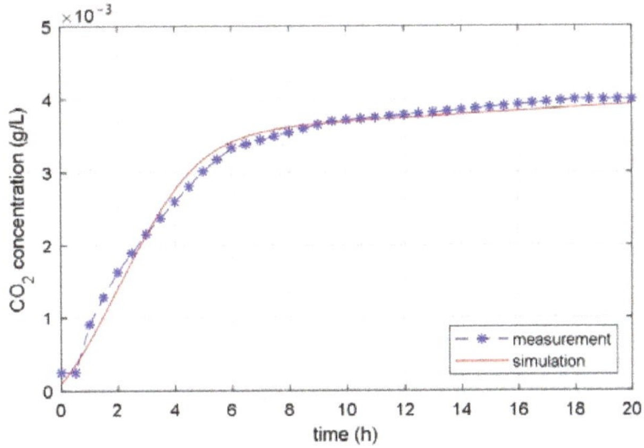

Figure 13. Measured and simulated time courses of the CO_2 concentration during the fermentation process with constant bioreactor temperature 22 °C.

Figure 14 shows the results of the fermentation with the changeable bioreactor's temperature. Note, in the experiment and in the simulations, the step increase of the reference temperature from 22 °C for 5 °C occurred at t = 3 h. The consequence of the step change in temperature in the bioreactor is visible in the different slope of the time course at the time of the change. In this case, the matching of the responses of the identified model and the laboratory measurements is also very good.

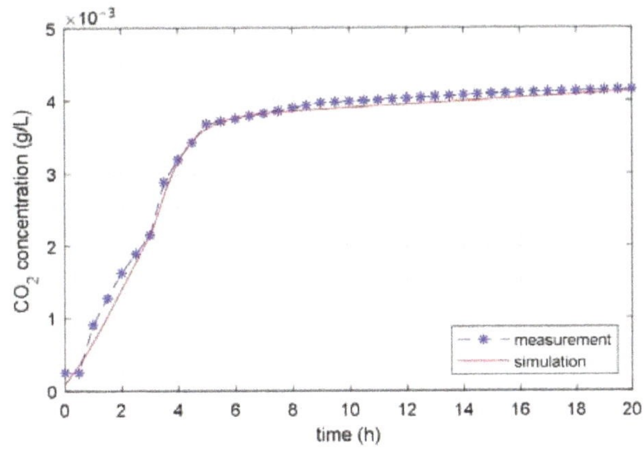

Figure 14. Measured and simulated time courses of the CO_2 concentration during the fermentation process with changeable bioreactor temperature (step change of temperature from 22 °C to 27 °C occurred at time t = 3 h).

Figure 14 shows a comparison of simulations and measurements for the input signal, which was previously used to identify model parameters. It makes sense to test the accuracy of the identified mathematical model for a new signal dataset. This is the case where the test input signal is different from the input signal used for identification. Figure 15 shows the results of the fermentation where the step increase of the reference temperature from 22 °C for 5 °C occurred at t = 6 h. Simulation and experimental results are presented. As in the previous cases, both responses match very well. The non-continuity in time course, smaller as in the previous case, is seen at t = 6 h. Figures 14 and 15 confirm the empirical findings that the effect of temperature change on CO_2 production is bigger at the beginning of the fermentation (high growing rate) and smaller in the second part of the fermentation process (low growing rate).

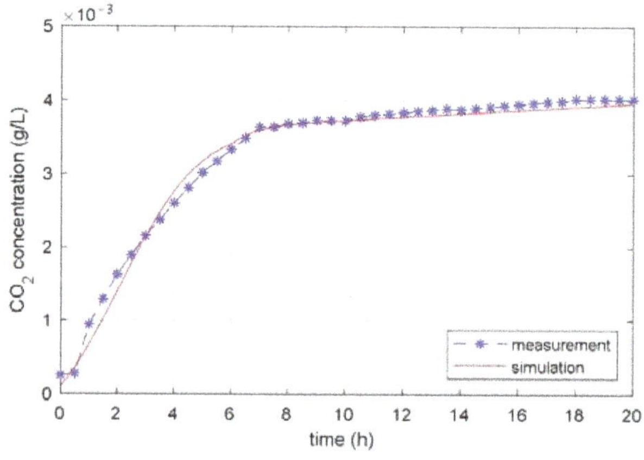

Figure 15. Measured and simulated time courses of the CO_2 concentration during the fermentation process with changeable bioreactor temperature (step change of temperature from 22 °C to 27 °C occurred at time t = 6 h).

The simulation results show that the derived model can justifiably be used for bioprocess analysis, simulations, and control system development. The identification (with optimization techniques) of the parameters of the presented model is not complicated but can take a lot of time. For identification purposes, a fermentation experiment with a constant temperature must first be performed. After that, the fermentation process must be carried out again. This time the temperature must change during fermentation.

4. Discussion and Conclusions

The study's main contribution is the development of a new dynamic model that considers the impact of temperature changes on the fermentation process dynamics and steady-state characteristics. The most important features of the presented temperature-considered model are:

- The derived model is a fourth order non-linear state-space model,
- The model's input is the reference variable of the heating/cooling system, the model's state-space variables are the concentrations of microorganisms, substrate and product, and the temperature in the bioreactor,
- The derived model is compact and suitable for the analysis of the fermentation process, for simulations, and for the implementation of the control system.
- The derived model represents a further development of the models presented in the authors' previous publications [10,21,22]. The advantage of the new model is in its simpler structure—previous models were based on a combination of a non-linear model for calculating the response to initial conditions and a linear model for calculating the response to temperature change.

- The derived model has nine parameters that depend on the biochemical substances and the determination of parameters for a specific bioreactor is possible by identification. For parameters' identification, it is necessary to carry out the fermentation process at least twice: once with a constant temperature and once with a change in temperature during the fermentation. The paper does not discuss the theoretical possibility of determining the model parameters from the biochemical data. Particle swarm optimization was used for the identification of the mathematical model.
- An important finding obtained from the derived model analysis is that the application of adaptive control theory for non-linear systems is reasonable for developing a control system that will control the fermentation process along the prescribed reference trajectory [33].
- Identification and analysis of the mathematical model were made on different fermentation processes on two different batch bioreactors. The obtained results of a multitude of experiments and calculations confirmed the presented findings.
- The article shows the use of scientific methods for the needs of engineering research. Our expectations are that academics and engineers will use the developed mathematical model for the design and synthesis of advanced control systems.

Author Contributions: Conceptualization, J.R.; Data curation, D.P.; Formal analysis, J.R. and B.P.; Investigation, J.R.; Methodology, J.R. and T.P.; Resources, A.G.; Software, J.R.; Writing-original draft, J.R.; Writing–review & editing, J.R. All authors have read and agreed to the published version of the manuscript.

Funding: This research received no external funding.

Institutional Review Board Statement: Not applicable.

Informed Consent Statement: Not applicable.

Data Availability Statement: The data presented in this study are available on request from the corresponding author.

Conflicts of Interest: The authors declare no conflict of interest.

References

1. Watabe, K.N.; Pehu, E. *Biotechnology Inteligence Unit: Plant Biotechnology and Plant Genetic Resources for Sustainability and Productivity*; R.G. Landes Company and Academic Press: Austin, TE, USA; San Diego, CA, USA, 1997.
2. Biotechnology Market Size, Share & Trends Analysis Report by Application (Health, Food & Agriculture, Natural Resources & Environment, Industrial Processing Bioinformatics), by Technology, And Segment Forecasts, 2018–2025. 2017. Available online: https://www.grandviewresearch.com/industry-analysis/biotechnology-market (accessed on 2 March 2021).
3. Global Biotechnology Market to Reach US$414.5 Billion by 2017: Transparency Market Research. Available online: https://www.businesswire.com/news/home/20150318005457/en/Global-Biotechnology-Market-Reach-US414.5-Billion-2017 (accessed on 2 March 2021).
4. Electric Motor Market by Type, Output Power, Voltage Range, Application & Speed: Global Opportunity Analysis and Industry Forecast, 2018–2025. Available online: https://www.alliedmarketresearch.com/electric-motor-market (accessed on 2 March 2021).
5. Villadsen, J. Innovative technology to meet the demands of the white biotechnology revolution of chemical production. *Chem. Eng. Sci.* **2007**, *62*, 6957–6968. [CrossRef]
6. Council, N.R. *Putting Biotechnology to Work: Bioprocess Engineering*; The National Academies Press: Washington, DC, USA, 1992.
7. Barcelos, M.C.S.; Lupki, F.B.; Campolina, G.A.; Nelson, D.L.; Molina, G. The colors of biotechnology: General overview and developments of white, green and blue areas. *FEMS Microbiol. Lett.* **2018**, 365. [CrossRef] [PubMed]
8. DaSilva, E.J. The Colours of Biotechnology: Science, Development and Humankind. *Electron. J. Biotechnol.* **2004**, *7*, 1–2.
9. Shuler, M.L.; Kargi, F. *Bioprocess Engineering: Basic Concepts*, 2nd ed.; Prentice Hall: Engelwood Cliffs, NJ, USA, 2002.
10. Ritonja, J. Implementation of Stir-Speed Adopted Controllers onto a Batch Bioreactor for Improved Fermentation. *IEEE Access* **2021**, *9*, 16783–16806. [CrossRef]
11. Cinar, A.; Parulekar, S.J.; Undey, C.; Birol, G. *Batch Fermentation: Modeling: Monitoring, and Control*; Marcel Dekker Inc.: New York, NY, USA, 2003.
12. Ziadi, M.; M'Hir, S.; Aydi, A.; Hamdi, M. Bioreactor Scale-Up and Kinetic Modeling of Lactic Acid and Biomass Production by Enterococcus faecalis SLT13 during Batch Culture on Hydrolyzed Cheese Whey. *J. Chem.* **2020**, *2020*, 1236784. [CrossRef]
13. Miller, K.V.; Block, D.E. A review of wine fermentation process modeling. *J. Food Eng.* **2020**, 273. [CrossRef]

14. Nosrati-Ghods, N.; Harrison, S.T.L.; Isafiade, A.J.; Tai, S.L. Analysis of ethanol production from xylose using Pichia stipitis in microaerobic conditions through experimental observations and kinetic modelling. *Biochem. Eng. J.* **2020**. [CrossRef]
15. Babiker, A.B.; Hoshida, H.; Ano, A. High-temperature fermentation: How can processes for ethanol production at high temperatures become superior to the traditional process using mesophilic yeast? *Appl. Microbiol. Biotechnol.* **2009**, *85*, 861–867. [CrossRef]
16. Kucharczyk, K.; Tuszynski, T. The effect of temperature on fermentation and beer volatiles at an industrial scale. *J. Inst. Brew.* **2018**, *124*, 230–235. [CrossRef]
17. Guzzon, R.; Roman, T.; Larcher, R. Impact of different temperature profiles on simultaneous yeast and bacteria fermentation. *Ann. Microbiol.* **2020**, *70*, 1–14. [CrossRef]
18. Krychowska, A.; Kordas, M.; Konopacki, M.; Grygorcewicz, B.; Musik, D.; Wójcik, K.; Jędrzejczak-Silicka, M.; Rakoczy, R. Mathematical Modeling of Hydrodynamics in Bioreactor by Means of CFD-Based Compartment Model. *Processes* **2020**, *8*, 1301. [CrossRef]
19. Dimitrova, N.; Zlateva, P. Global Stability Analysis of a Bioreactor Model for Phenol and Cresol Mixture Degradation. *Processes* **2021**, *9*, 124. [CrossRef]
20. Fitzpatrick, J.J. Insights from Mathematical Modelling into Energy Requirement and Process Design of Continuous and Batch Stirred Tank Aerobic Bioreactors. *Chem. Eng.* **2019**, *3*, 65. [CrossRef]
21. Goršek, A.; Ritonja, J.; Pečar, D. Mathematical model of CO_2 release during milk fermentation using natural kefir grains. *J. Sci. Food Agric.* **2018**, *98*, 4680–4684. [CrossRef] [PubMed]
22. Ritonja, J.; Gorsek, A.; Pecar, D. Control of Milk Fermentation in Batch Bioreactor. *Elektron. Ir Elektrotechnika* **2020**, *26*, 4–9. [CrossRef]
23. Ritonja, J.; Goršek, A.; Pečar, D. Use of a Heating System to Control the Probiotic Beverage Production in Batch Bioreactor. *Appl. Sci.* **2021**, *11*, 84. [CrossRef]
24. Kiparissides, A.; Koutinas, M.; Kontoravdi, C.; Mantalaris, A.; Pistikopoulus, E.N. 'Closing the loop' in biological systems modeling—From the *in silico* to the *in vitro*. *Automatica* **2011**, *47*, 1147–1155. [CrossRef]
25. Lee, D.; Ding, Y.; Jayaraman, A.; Kwon, J.S. Mathematical Modelling and Parameter Estimation of Intracellular Signaling Pathway: Application to LPS-induced NFκB Activation and TNFα Production in Macrophages. *Processes* **2018**, *6*, 21. [CrossRef]
26. Chu, Y.; Jayaraman, A.; Hahn, J. Parameter sensitivity analysis of IL-6 signaling pathways. *IET Syst. Biol.* **2007**, *1*, 342–352. [CrossRef] [PubMed]
27. Thatipamala, R. On-Line Monitoring, State and Parameter Estimation, Adaptive Computer Control and Dynamic Optimization of a Continuous Bioreactor. Ph.D. Thesis, University of Saskatchewan, Saskatoon, SK, Canada, 1993.
28. Price, K.V.; Storn, R.M.; Lampinen, J.A. *Differential Evolution, a Practical Approach to Global Optimization*; Springer: Berlin/Heidelberg, Germany, 2005.
29. Kennedy, J.; Eberhart, R.C. Particle Swarm Optimization. In Proceedings of the ICNN'95—International Conference on Neural Networks, Perth, WA, Australia, 27 November–1 December 1995; IEEE: Piscataway, NJ, USA, 1995; pp. 1942–1948.
30. Mezura-Montes, E.; Coello, C.A. Constraint-handling in nature-inspired numerical optimization: Past, present and future. *Swarm Evol. Comput.* **2011**, *1*, 173–194. [CrossRef]
31. Pedersen, M.E.H. *Good Parameters for Particle Swarm Optimization*; Technical Report No. HL1001; Hvass Laboratories: Luxembourg, 2010.
32. Particle Swarm Optimization Algorithm. Available online: https://mathworks.com/help/gads/particle-swarm-optimization-algorithm.html (accessed on 12 May 2021).
33. Ritonja, J.; Goršek, A.; Pečar, D. Model Reference Adaptive Control for Milk Fermentation in Batch Bioreactors. *Appl. Sci.* **2020**, *10*, 9118. [CrossRef]

foods

MDPI

Article

Use of Food Spoilage and Safety Predictor for an "A Priori" Modeling of the Growth of Lactic Acid Bacteria in Fermented Smoked Fish Products

Angela Racioppo, Daniela Campaniello, Milena Sinigaglia, Antonio Bevilacqua, Barbara Speranza * and Maria Rosaria Corbo

Department of Agriculture, Food, Natural Resources and Engineering (DAFNE), University of Foggia, 71122 Foggia, Italy; angela.racioppo@unifg.it (A.R.); daniela.campaniello@unifg.it (D.C.); milena.sinigaglia@unifg.it (M.S.); antonio.bevilacqua@unifg.it (A.B.); mariarosaria.corbo@unifg.it (M.R.C.)

* Correspondence: barbara.speranza@unifg.it

Abstract: Fermentation is one of the oldest methods to assure the safety and quality of foods, and to prolong their shelf life. However, a successful fermentation relies on the correct kinetics depending on some factors (i.e., ingredients, preservatives, temperature, inoculum of starter cultures). Predictive microbiology is a precious tool in modern food safety and quality management; based on the product characteristics and the conditions occurring in food processing, the inactivation of or increase in microbial populations could be accurately predicted as a function of the relevant intrinsic or extrinsic variables. The main aim of this study was the optimization of the formula of a smoked fermented fish product using predictive modeling tools (tertiary and secondary models) in order to define the role of each factor involved in the formulation and assure a correct course of fermentation. Product optimization was conducted through the software Food Spoilage and Safety Predictor (FSSP), by modeling the growth of lactic acid bacteria (LAB) as a function of some key parameters such as temperature, pH, salt, liquid smoke, carbon dioxide, and nitrites. The variables were combined through a fractional design of experiments (DoE) (3^{k-p}), and the outputs of the software, i.e., the maximal growth rate (μ_{max}) and the time to attain the critical threshold (t_{crit}), were modeled through a multiple regression procedure. The simulation, through FSSP and DoE, showed that liquid smoke is the most critical factor affecting fermentation, followed by temperature and salt. Concerning temperature, fermentation at 20–25 °C is advisable, although a low fermentation temperature is also possible. Other parameters are not significant.

Keywords: predictive microbiology; FSSP; DoE; smoke; fermentation; fish

Citation: Racioppo, A.; Campaniello, D.; Sinigaglia, M.; Bevilacqua, A.; Speranza, B.; Corbo, M.R. Use of Food Spoilage and Safety Predictor for an "A Priori" Modeling of the Growth of Lactic Acid Bacteria in Fermented Smoked Fish Products. *Foods* **2022**, *11*, 946. https://doi.org/10.3390/foods11070946

Academic Editors: Carlos Vilas, Míriam R. García and Jose A. Egea

Received: 28 February 2022
Accepted: 21 March 2022
Published: 25 March 2022

1. Introduction

The application of fermentation to fish goes back many thousands of years with evidence of fermented fish products in ancient Greece (*aimeteon*) and in the Roman era (*garum*). However, although the fermentation process was once only used as a preservation method, today, it is also used for health purposes. In fact, a recent trend in the design of innovative fish products is the individuation of fish formulas fermented with appropriate starter cultures (sometimes with probiotic or other functional traits) to produce foods with improved sensory attributes, high nutritive value, and health benefits [1,2].

Fermented fish products usually have unique characteristics, especially in terms of aroma, flavor, and texture: this is the result of the transformation of organic materials into simpler compounds by the activity of microorganisms or enzymes during the fermentation process. The fermented products are more digestible, at the same time preserving their nutritional properties, even better than unfermented raw fish. These foods are a source not only of proteins, amino acids, and polypeptides, but also of minerals (calcium and iron), some B group vitamins, and, above all, polyunsaturated fatty acids [3].

Unfortunately, most fermented fish products are still local and not so easily found nationwide, with Asia and Africa being the main producers; only some fish sauces and shrimp pastes are widely known in Europe [4]. Each country has its own types of fermented fish products characterized by different production processes and formulations.

Recently, these products have been gaining increasing interest among consumers due to their healthy characteristics supported by several studies in the literature [3,5–15]. Apart from the well-known health effects of fish meat, correct fermentation, led by pro-technological microorganisms, could result in the biofortification of amino acids and peptides with some physiological and beneficial functions (antioxidant, antihypertensive, antiproliferative, hypoglycemic, immune-stimulating, and anticoagulant effects) [1]. In addition, a correct and predictable process piloted by starter cultures, either autochthonous or commercial, can limit the production of undesired metabolites and produce a good product; the choice to rely the quality of fermented products on spontaneous fermentation (natural microbiota) is no longer advisable, because of the very high risk of incurring arrested fermentation and health problems. This is why recent research is increasingly focused on the selection of microorganisms able to guarantee the best performances [16–18]. The most promising microorganisms selected as starters are generally those that are isolated from the native microbiota of traditional products since they are well adapted to the environmental conditions of the considered food [19–21].

Regardless of the type of starter, the correct course of fermentation is essential to produce a safe and high-quality product. It is also crucial to define the effects of each factor involved in the formulation (ingredients, preservatives, temperature, inoculum of starter cultures) on the process of evolution. In this context, predictive microbiology stands up as a precious tool in modern food safety and quality management. Based on the product characteristics and the conditions occurring in food processing, the inactivation of or increase in microbial populations in foods, as a function of the relevant intrinsic or extrinsic variables, could be accurately predicted [22]. A new frontier goal in predictive microbiology is the use of models and databases for product optimization, as also shown by the authors of some seafood applications [23,24].

Generally, the optimization of new formulas relies on different steps and phases and could start with an "a priori" modeling as a step prior to actual challenge tests at the laboratory level, followed by scaling up and optimization. However, before planning and executing some experiments, it is advisable to understand the role of most variables involved, and to perform a screening to exclude less significant or non-significant factors. The reduction in the number of variables is important from a mathematical point of view, because the experience of authors suggests that the use of many variables could lead to confounding phenomena; in addition, a priori modeling could help researchers minimize the number of experiments required to be performed in the lab, as well as helping them to better define the conditions of assays.

Therefore, the main aim of this study was an "a priori" optimization of the formula of a smoked fermented fish product using predictive modeling tools to define the role of each factor involved in the formulation (temperature, pH, salt, liquid smoke, carbon dioxide, and nitrites) for a correct course of fermentation. The variables were combined through a fractional design of experiments (DoE) (3^{k-p}), and the outputs of the software, i.e., the maximal growth rate (μ_{max}) and the time to attain the critical threshold (t_{crit}), were modeled through a multiple regression procedure.

2. Materials and Methods

2.1. Software and Design

Product optimization was conducted through Food Spoilage and Safety Predictor (FSSP), a free software package available as an additional tool in ComBase or on the website of the developers (http://fssp.food.dtu.dk/, accessed on 6 September 2021).

The focus was the modeling of lactic acid bacteria (LAB) growth as a function of some key parameters for a fermented fish product including intrinsic factors or extrinsic factors (temperature and pH), processing parameters (salt, liquid smoke, nitrites, or carbon dioxide), and parameters connected to LAB growth (relative lag time). The choice of liquid smoke relies on the fact that the consumer appeal towards smoked fish products is increasing and the use of liquid smoke instead of the traditional method is advisable because it is more eco-friendly, while carbon dioxide is crucial to simulate fermentation and storage under anoxic conditions. Finally, nitrites are the preservatives generally used in fermented meat products to counteract *Clostridium botulinum* growth. The variables were combined through a fractional design (3^{k-p}), and each factor had three levels (minimum or "−1", mean value or "0", and maximum or "+1"), as shown in Table 1. The combination of these variables resulted in a design consisting of 243 runs, as shown in the Supplementary Materials File S1. The input conditions for modeling were as follows: initial concentration of LAB, 5 log CFU/g; maximal population density, 9 log CFU/g; critical threshold, 8.9 log CFU/g; weak acids at 0 ppm. For modeling purposes, weak acids were excluded to reduce the number of variables to 7; in a mixed design, the randomization of 7 variables set at 3 levels produces a total of 243 combinations and makes the estimation of both the individual effects of each variable and their two-way interactions possible.

Table 1. Coded levels of the variables.

	−1	0	+1
RTL (days)	0	2	4
Temperature (°C)	10	17.5	25
NaCl (%)	0	3	6
pH	5	6	7
Smoke (ppm)	0	20	40
CO_2 (%)	0	15	30
Nitrite (ppm)	0	75	150

An increase in the number of variables causes an increase in the number of the combinations of the design (up to 729 combinations with 9 variables); a secondary effect of increasing the number of variables is the possibility of a statistical artifact due to the high number of individual and interactive terms. Thus, weak acids were not used for this design, because they are added at the end, and not throughout fermentation, to prolong the shelf life of the product.

The tool works on psychrotolerant LAB (*Latilactobacillus sakei*; *Latilactobacillus curvatus*; *Lactobacillus* spp.; and other related genera from autochthonous microbiota of fish) [25].

2.2. Modeling

All 243 combinations built by the randomization of the variables were used as inputs for the software, and for each combination, the maximal growth rate (μ_{max}) and the time to attain the critical threshold (t_{crit}), as an indirect measure of the time required to attain the steady state, were evaluated; the values are presented in Supplementary Materials File S1. Then, μ_{max} and t_{crit} were modeled through a multiple regression procedure, using the option DoE (design of experiments) of the software Statistica for Windows (Version 7, Tulsa, OK, USA), to obtain a second-order model as follows:

$$y = B_0 + \sum B_i x_i + \sum B_{ii} {x_i}^2 + \sum B_{ij} x_i x_j \quad (1)$$

where "y" is the modeled dependent variable (μ_{max} and t_{crit}); B_i, B_{ii}, and B_{ij} are the coefficients of the model, associated with independent variables; the term "x_i" highlights the individual effect of the predictor; the symbols "${x_i}^2$" and "$x_i x_j$" indicate the quadratic and interactive effects. Variables showing a significance <95% (p > 0.05) were not included in the equation by the software; moreover, due to the high number of variables, interactive terms were excluded.

The effect of each independent variable on the outputs was also evaluated through the individual desirability functions, estimated as follows:

$$d = \begin{pmatrix} 0, & y \leq y_{unfav} \\ \frac{y - y_{unfav}}{y_{fav} - y_{unfav}} & y_{unfav} \leq y \leq y_{fav} \\ 1, & y \geq y_{fav} \end{pmatrix} \quad (2)$$

where y_{unfav} and y_{fav} are the most unfavorable (high value of t_{crit} and low value of μ_{max}) and the most favorable (low value of t_{crit} and high value of μ_{max}) values of the dependent variables.

3. Results and Discussion

The use of tertiary models to predict safety and food shelf life has also been regarded by the European Union as a tool for quality assurance. The tertiary model used in this work was originally designed for seafood products [26,27], but it has been recently updated for other foods (for example, cheese), and the name has been changed to FSSP. In the latest version, it contains a collection of several options/models, including RRS models (relative rate of spoilage), and models to predict spoilers (*Photobacterium phosphoreum, Morganella psychrotolerans, Morganella morganii, Shewanella putrefaciens*) and biogenic amine formation, pathogens (*Listeria monocytogenes*), and lactic acid bacteria. As observed in the study of Mejlholm et al. [28], predictions performed through the FSSP software have good precision when the complexity of the growth model matches the complexity of the foods of interest; this was also confirmed by the study of Bolívar et al. [29] during the modeling of the growth of *L. monocytogenes* in Mediterranean fish species, where the software was used to estimate μ_{max} based on the values of pH, a_w, storage temperature, and atmospheric conditions of the studied fish species. The goodness of fit of FSSP was regarded as acceptable for *Lactobacillus* spp. (now *Lactobacillus* and related genera); in addition, the tool is flexible and focuses on a high number of parameters, and for some of them (weak acids and preservatives), it is reliable a substitution of some compounds with other ones chosen by users [25].

LAB, mainly psychrotolerant strains, grow at high concentrations during the storage of seafood, resulting in the production of off-odors, gas, slime, and undesired metabolites [25,30]. However, there are various positive and pro-technological species and strains able to contribute to shelf life and healthy characteristics, because of their metabolism and health properties [21,23,30,31]. Apart from strain characterization and validation in some matrices, to the best of the authors' knowledge, there are no models able to predict the performances of LAB during fish fermentation. These models, hereby labeled as a priori functions, could be extremely useful in the first phase of product design, because they could help researchers to choose only some variables for laboratory confirmation.

It is important to include in such models both the matrix's variables and processing factors (pH, temperature, carbon dioxide), including preservatives used in the process (salt and nitrites). Another variable hereby tested was the relative lag time (RLT). This parameter is different from the classical lag time and relies on the physiological state of cells introduced into a new environment (for a starter culture from a bioreactor to food) and should be read as the "amount of work done by cells to adapt to the new environment" [32]. From a theoretical and practical point of view, RLT is at its minimum when the microorganism has a very low lag phase and grows at its optimal growth rate, and could be 0, when the microorganism has no lag phase.

The inclusion of RLT as a main variable of the model relies on the fact that, in a guided fermentation, it is crucial to inoculate well-adapted microorganisms in the matrix (low RLT), or to use them in an environment with the same characteristics of the media used to produce their biomass. This factor was described in the past as the physiological function of the Baranyi and Roberts model and was shown to strongly affect the growth curve [33].

The steps taken in this research are similar to those used when a second-level model of predictive microbiology is developed: the first step was the evaluation of the growth curves in some conditions; then, the fitting parameters of the growth curves were modeled as a function of pH, temperature, composition, etc. In this research, FSSP substitutes for the first step.

Figure 1 shows an example of the growth curves (combinations S34 and S41) predicted for LAB using FSSP. For each combination studied (S1), the outputs of the software (μ_{max}, maximal growth rate, and t_{crit}, time to attain the critical threshold of 8.9 log CFU/g) were recorded and subsequently used as input data for a multiple regression procedure.

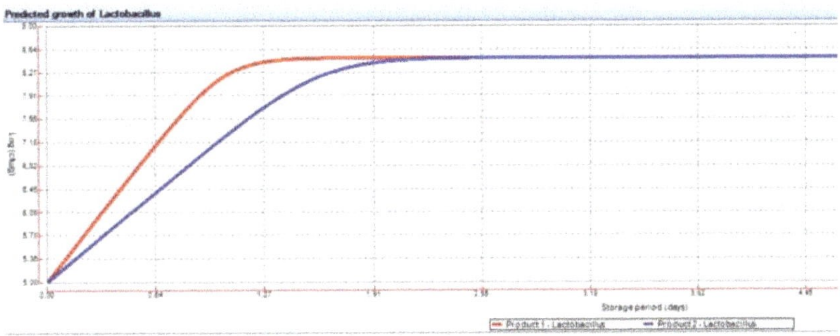

Figure 1. Example of growth curves (combinations S34 and S41) predicted for LAB using FSSP.

The growth rate is the classical parameter for secondary models, but t_{crit} was added in this research. To the authors' knowledge, this is the first time t_{crit} has been added in a modeling study of LAB cultures. In a simple way, t_{crit} could be defined as the time to attain the steady state; thus, it is an indirect measure of the performances of starter cultures and of their acidification kinetics. Moreover, the time to attain the steady state could have other implications related to the Jameson effect: a dominant group (a starter culture) could stop the growth of the other subpopulations (spoiling microorganisms) and induce a stationary phase, thus reducing the maximum level they could attain [34]. For a starter culture, the Jameson effect, and thus the time to attain the steady state, could have a strong effect on the inhibition of undesirable microorganisms and on bioprotective effects.

The first output of the statistical approach used was the table of standardized effects (Table 2), which shows the statistical weight of each studied factor; in particular, it summarizes the effects of the linear (L), quadratic (Q), and interactive terms of RLT (relative lag time), temperature, concentration of NaCl, liquid smoke, and the amount of nitrite and CO_2 at equilibrium in the headspace on the maximal growth rate (μ_{max}) and on t_{crit} of LAB in fish products. The standardized effect was evaluated as the ratio of the mathematical coefficient of each factor (from multiple regression) vs. its standard error.

Table 2. Standardized effects of linear (L), quadratic (Q), and interactive terms of RLT (relative lag time), temperature, concentration of NaCl, liquid smoke, and the amount of nitrite and CO_2 at equilibrium in the headspace on the maximal growth rate (μ_{max}) and time to attain the steady state (t_{crit}) of lactic acid bacteria in fish products. The standardized effect was evaluated as the ratio of the mathematical coefficient of each factor (from multiple regression) vs. its standard error. The degrees of freedom for the *t*-test were 204; R^2_{ad}, regression coefficient corrected for multiple regression. * Not significant.

	μ_{max}	t_{crit}
(1) RTL (L)	- *	5.030
RTL (Q)	-	-

Table 2. *Cont.*

	μ_{max}	t_{crit}
(2) T (L)	29.729	−7.516
T (Q)	-	−3.191
(3) NaCl (L)	−17.806	-
NaCl (Q)	-	-
(4) pH (L)	3.109	-
pH (Q)	2.455	-
(5) Smoke (L)	−48.844	8.146
smoke (Q)	-	−5.215
(6) CO_2 (L)	−2.502	-
CO_2 (Q)	-	-
(7) nit (L)	−2.943	-
nit (Q)	-	-
1L by 2L	-	−4.980
1L by 3L	-	-
1L by 4L	-	-
1L by 5L	-	5.023
1L by 6L	-	-
1L by 7L	-	-
2L by 3L	−10.036	-
2L by 4L	-	-
2L by 5L	−24.746	−8.288
2L by 6L	-	-
2L by 7L	-	-
3L by 4L	−2.843	-
3L by 5L	14.087	-
	μ_{max}	t_{crit}
3L by 6L	-	-
3L by 7L	-	-
4L by 5L	−4.546	-
4L by 6L	-	-
4L by 7L	-	-
5L by 6L	-	-
5L by 7L	2.378	-
6L by 7L	-	-
R^2_{ad}	0.947	0.551

For the maximal growth rate, the most significant factor was liquid smoke, followed by temperature and salt; pH, nitrites, RLT, and CO_2 did not play a significant role. For the time to attain the critical threshold (t_{crit}), the data point out the significance of liquid smoke, temperature, RLT, and salt. Although only for qualitative purposes, the main benefit of this model, compared to the boundary approach used for LAB in FSSP [25], is that it does not offer a negative overview (combinations of parameters causing growth inhibition); the ratio behind this approach is a positive method (from a mathematical point of view), that is, some of the factors and interactions are related to growth and are the parameters able to cause an increase in the dependent variables (rate and t_{crit}).

A quantitative estimation of the role of the significant variables could be gained through the use of surface response plots. Concerning the interactions of liquid smoke × temperature (Figure 2a) and liquid smoke × salt (Figure 2b), the growth rate was at its maximum at the highest value of temperature (coded level "+1", 25 °C) and with the lowest value of salt and liquid smoke; however, the model predicted the complete inhibition of LAB growth only for a few combinations of temperature and smoke (level "−1" for temperature and "0.8–1" for liquid smoke, corresponding to actual values of 10 °C and 35–40 ppm of liquid smoke). The surface response plot for the interaction of nitrite × liquid smoke (Figure 2c) underlines the non-significant effects of nitrites. The effect of liquid smoke on the growth of LAB was also confirmed by the results obtained for t_{crit}, as the highest values of this parameter were predicted for amounts of liquid smoke over 20–25 ppm (data not shown).

Surface plots are a good tool for treating data from a multiple regression, and for highlighting the individual and quadratic terms of each factor, as well as for showing the interactive effects amongst the different independent variables. However, despite these benefits, it is not possible to see the effect of each variable excluding the others, i.e., the effect of each factor is linked to the effects of other variables. Thus, to this scope, other approaches could be used, such as desirability which is a dimensionless parameter able to provide an answer to the following issue: how much is an output desired? [35] In addition, through mathematical extrapolation, desirability is an important tool to highlight critical values for each factor.

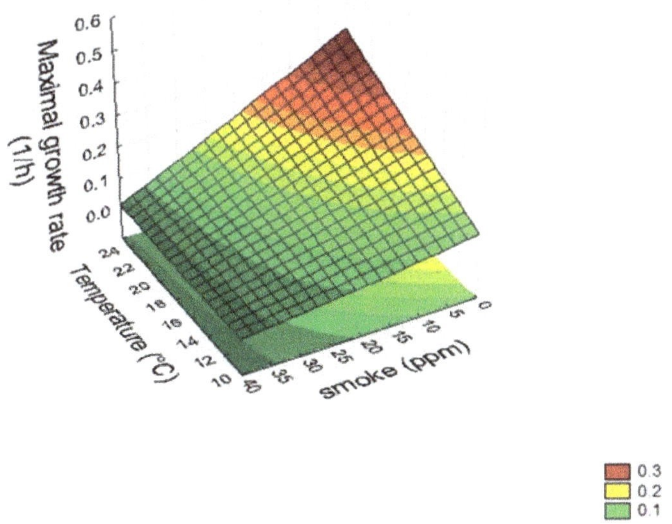

(a)

Figure 2. *Cont.*

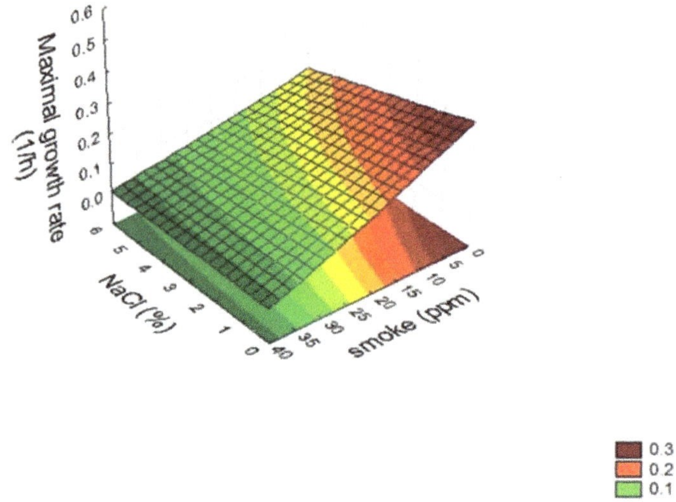

(**b**)

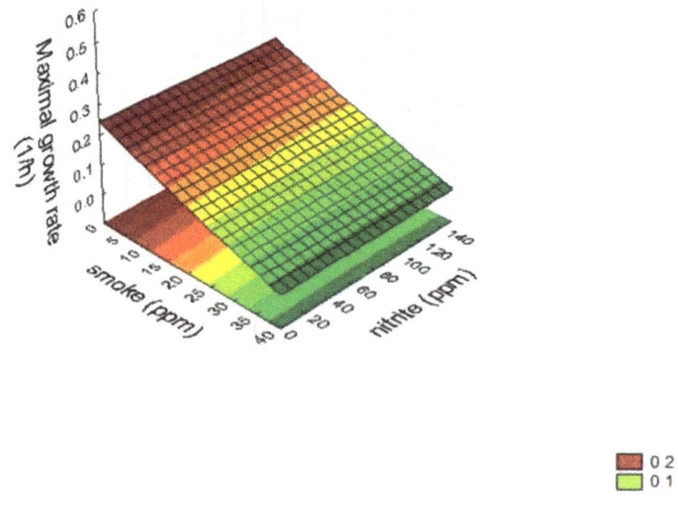

(**c**)

Figure 2. Surface response plots for the interactions liquid smoke × temperature (**a**), liquid smoke × salt (**b**), and nitrite × liquid smoke (**c**).

Figures 3 and 4 show the desirability profile for μ_{max} and t_{crit}, respectively; the profile highlights the different quantitative weights of each studied variable. Focusing on the results obtained for μ_{max} (Figure 3), only temperature, salt, and liquid smoke are able to affect fermentation, whereas other parameters such as pH, use of anoxic conditions, and nitrites are not significant. Concerning temperature, fermentation at 20–25 °C is advisable, although a low fermentation temperature is also possible (10–15 °C). This result is corroborated in numerous other studies using mathematical models to predict the growth of different microorganisms in foods at various storage temperatures [36–38], where a strong dependency was found between microbial growth and the storage temperature assayed.

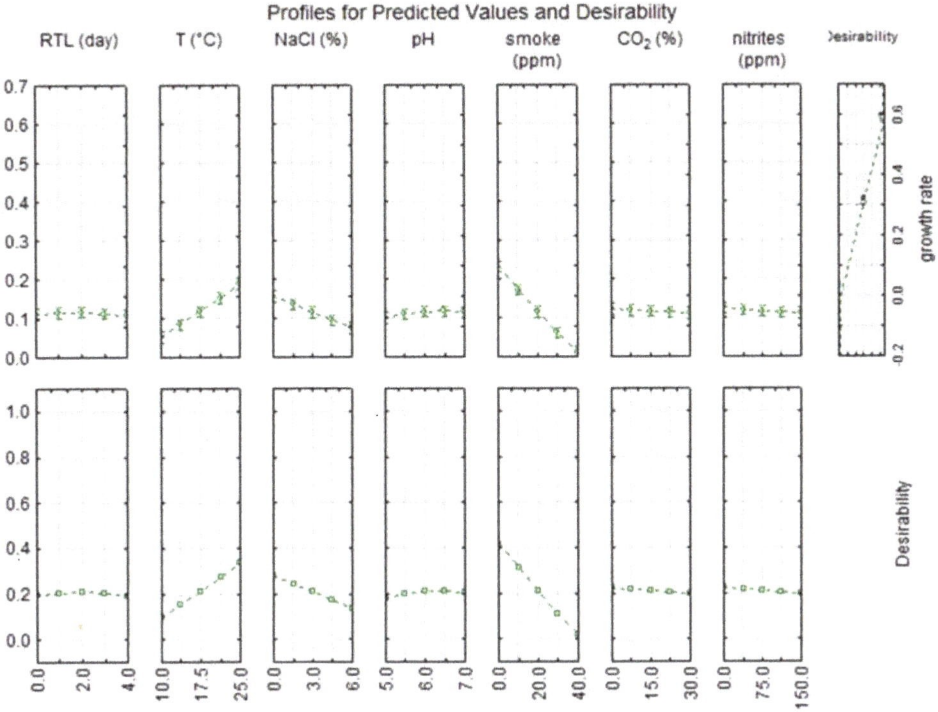

Figure 3. Desirability profile for the maximal growth rate (μ_{max}) of lactic acid bacteria in fish products.

Although a temperature of 25 °C is generally not the optimal temperature of LAB, and a further increase in desirability for higher temperatures could be hypothesized, the choice of the range was due to two main reasons: (i) higher temperatures could have masked the effect of other variables because of the strongest effect of temperatures for starter culture kinetics [39]; (ii) fish fermentation is advisable at a lower temperature to avoid massive production of undesirable compounds, such as biogenic amines [40]. Finally, 25 °C is the maximum level of temperature for LAB in FSSP.

As expected, the effect of salt was less significant than temperature (at least in the range tested in this study, 0–6%): lactic acid bacteria, in fact, are able to tolerate high salt concentrations, and this tolerance gives them an advantage over other less tolerant species, allowing a rapid start of fermentation [21]. The results (see box of NaCl in Figure 3) show that the decrease in μ_{max} begins to be significant beyond a salt concentration of 3%; however, 3–3.5% (*w*/*w*) is the salt concentration recommended by the Codex Alimentarius [41] in smoked-flavored fish products to avoid the proliferation of spoilage bacteria. Both Figures 3 and 4 point out that a smoke concentration >20 ppm (coded level 0) is not advisable because it could result in a significant delay of starter growth. This inhibitory effect is probably attributable to the bactericidal effects of smoke components (phenols, polycarboxylic acids) on most types of bacteria and/or fungi, including LAB [42]. Many other recent studies were focused on the use of liquid smoke on fish products [43–45], since consumer appeal towards these foods has been increasing recently. The traditional smoking method is under investigation since it can cause the production of harmful compounds, such as polycyclic aromatic hydrocarbons, mainly benzo(α)pyrene [46]. Consequently, the use of liquid smoke is advisable because it is more eco-friendly, requires a shorter smoking time, and ensures a longer product durability [44].

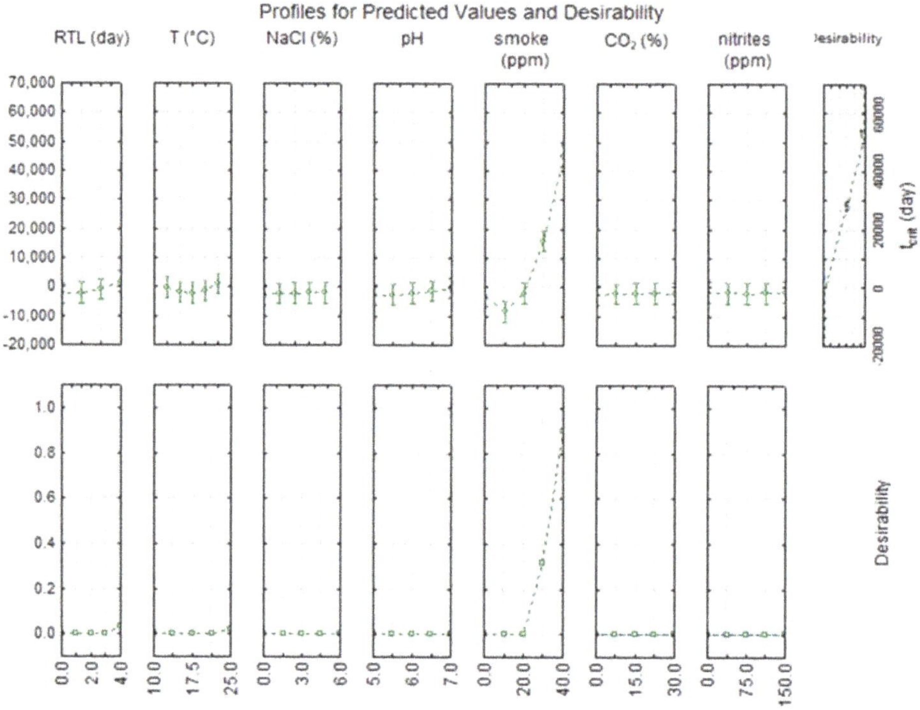

Figure 4. Desirability profile for the time to attain the steady state (t_{crit}) of lactic acid bacteria in fish products.

In this study, model building was based on a completely randomized design, which was able to estimate possible interactive and quadratic effects. This type of design is useful for screening purposes, and to reduce the number of variables; however, for robust predictive equations, other designs (for example, central composite design) are advisable, because a higher number of levels allows for good precision in the prediction, at least in the ranges tested. Thus, further investigations are required in this field, in order to assess the significance of other factors not considered in FSSP (food structure, food components, effects of natural microbiota, etc.) which could strongly and significantly affect the growth/survival of lactic acid bacteria.

4. Conclusions

The design of a fermented product is a complex process, and the definition of the formulation is a critical step above all for the performances of the starter cultures. The use of LAB is a promising way to valorize seafood products and to design safe, stable, and consumer-appealing products; however, some variables should be considered.

The use of liquid smoke and the amount of phenols are critical, because the simulation through FSSP suggested that there is a critical amount (20 ppm) after which the fermentation could experience a delay.

Concerning temperature, fermentation at 20–25 °C is advisable because of higher growth rate values, although a low fermentation temperature is also possible. The model, in fact, also predicted growth at 10–15 °C.

RLT could strongly affect the performances of a starter culture, thus suggesting that the physiological state of cells could affect their performances during fish fermentation.

Other parameters are not significant (at least those tested in this case study: pH, use of anoxic conditions, nitrites), and if they are included in the formulation, their amount should be determined through other considerations (chemistry, costs, safety for consumers, etc.).

This model could be the background for future studies devoted to the optimization of fish fermentation, as it offers some details on the effect of many factors, although it is important to consider that each a priori modeling should be followed by validation, to check if the mathematical effects have the same weight in real systems, or if other factors could play a role.

Supplementary Materials: The following supporting information can be downloaded at: https://www.mdpi.com/article/10.3390/foods11070946/s1, Supplementary Material File S1: Coded levels and combinations used for the simulation of FSSP.

Author Contributions: Conceptualization, A.B. and B.S.; methodology, M.R.C., A.B. and B.S.; software, A.B.; investigation, A.R. and D.C.; data curation, A.B.; writing—original draft preparation, A.B. and B.S.; writing—review and editing, A.B., A.R. and B.S.; supervision, M.R.C. and M.S.; funding acquisition, M.R.C. and M.S. All authors have read and agreed to the published version of the manuscript.

Funding: This research study was funded by the Puglia region through the project PO FEAMP 2014/2020—Measure 1.26 "Valorizzazione di specie ittiche affumicate mediante tecniche tradizionali e innovative" (CUP N. B71B17000990009; project leader: UNCI-Agroalimentare).

Data Availability Statement: Not applicable.

Conflicts of Interest: The authors declare no conflict of interest.

References

1. Zang, J.; Xu, Y.; Xia, W.; Regenstein, J.M. Quality, functionality, and microbiology of fermented fish: A review. *Crit. Rev. Food Sci. Nut.* **2020**, *60*, 1228–1242. [CrossRef] [PubMed]
2. Martnez-Alvarez, O.; Lopez-Caballero, M.; Gomez-Guillen, M.; Montero, P. Fermented seafood products and health. In *Fermented Foods in Health and Disease Prevention*; Frias, J., Martinez-Villaluenga, C., Penãs, E., Eds.; Elsevier: Madrid, Spain, 2016; pp. 177–202. [CrossRef]
3. Peralta, E.M.; Hatate, H.; Kawabe, D.; Kuwahara, R.; Wakamatsu, S.; Yuki, T.; Murata, H. Improving antioxidant activity and nutritional components of Philippine salt-fermented shrimp paste through prolonged fermentation. *Food Chem.* **2008**, *111*, 72–77. [CrossRef]
4. Irianto, I.H.E. *Produk Fermentasi Ikan*; Penebar Swadaya Grup: Depok, Indonesia, 2012.
5. Fujita, H.; Yamagami, T.; Ohshima, K. Effects of an ace-inhibitory agent, Katsuobushi oligopeptide, in the spontaneously hypertensive rat and in borderline and mildly hypertensive subjects1. *Nut. Res.* **2001**, *21*, 1149–1158. [CrossRef]
6. Ichimura, T.; Hu, J.; Aita, D.Q.; Maruyama, S. Angiotensin I converting enzyme inhibitory activity and insulin secretion stimulative activity of fermented fish sauce. *J. Biosci. Bioeng.* **2003**, *96*, 496–499. [CrossRef]
7. Kim, D.C.; Chae, H.J.; In, M.J. Existence of stable fibrin-clotting inhibitor in salt-fermented anchovy sauce. *J. Food. Compos. Anal.* **2004**, *17*, 113–118. [CrossRef]
8. Lee, Y.G.; Lee, K.W.; Kim, J.Y.; Kim, K.H.; Lee, H.J. Induction of apoptosis in a human lymphoma cell line by hydrophobic peptide fraction separated from anchovy sauce. *Biofactors* **2004**, *21*, 63–67. [CrossRef]
9. Je, J.Y.; Park, P.J.; Byun, H.G.; Jung, W.K.; Kim, S.K. Angiotensin I converting enzyme (ACE) inhibitory peptide derived from the sauce of fermented blue mussel, *Mytilus edulis*. *Bioresour. Technol.* **2005**, *96*, 1624–1629. [CrossRef]
10. Thongthai, C.; Gildberg, A. Asian fish sauce as a source of nutrition. In *Asian Functional Foods*; Shi, J., Sahidi, F., Ho, C.T., Eds.; Marcel Dekker/CRC Press: Boca Raton, FL, USA, 2005; pp. 215–265.
11. Duarte, J.; Vinderola, G.; Ritz, B.; Perdigon, G.; Matar, C. Immunomodulating capacity of commercial fish protein hydrolysate for diet supplementation. *Immunobiology* **2006**, *211*, 341–350. [CrossRef]
12. Itou, K.; Nagahashi, R.; Saitou, M.; Akahane, Y. Antihypertensive effect of narezushi, a fermented mackerel product, on spontaneously hypertensive rats. *Fish Sci.* **2007**, *73*, 1344. [CrossRef]

13. Binsan, W.; Benjakul, S.; Visessanguan, W.; Roytrakul, S.; Tanaka, M.; Kishimura, H. Antioxidative activity of Mungoong, an extract paste, from the cephalothorax of white shrimp (*Litopenaeus vannamei*). *Food Chem.* **2008**, *106*, 185–193. [CrossRef]
14. Faithong, N.; Benjakul, S.; Phatcharat, S.; Binsan, W. Chemical composition and antioxidative activity of Thai traditional fermented shrimp and krill products. *Food Chem.* **2010**, *119*, 133–140. [CrossRef]
15. Kleekayai, T.; Saetae, D.; Wattanachaiyingyong, O.; Tachibana, S.; Yasuda, M.; Suntornsuk, W. Characterization and in vitro biological activities of Thai traditional fermented shrimp pastes. *J. Food Sci. Technol.* **2014**, *52*, 1839–1848. [CrossRef] [PubMed]
16. Rattanachaikunsopon, P.; Phumkhachorn, P. Lactic acid bacteria: Their antimicrobial compounds and their uses in food production. *Ann. Biol. Res.* **2010**, *1*, 218–228.
17. O'Bryan, C.; Crandall, P.; Ricke, S.; Ndahetuye, J. Lactic acid bacteria (LAB) as antimicrobials in food products: Types and mechanisms of action. In *Handbook of Natural Antimicrobials for Food Safety and Quality*; Taylor, T.M., Ed.; Elsevier: Amsterdam, The Netherlands, 2015; pp. 117–129.
18. Dittoe, D.K.; Ricke, S.C.; Kiess, A.S. Organic acids and potential for modifying the avian gastrointestinal tract and reducing pathogens and disease. *Front. Vet. Sci.* **2018**, *5*, 216. [CrossRef]
19. Laranjo, M.; Potes, M.E.; Elias, M. Role of Starter Cultures on the Safety of Fermented Meat Products. *Front. Microbiol.* **2019**, *26*, 583. [CrossRef]
20. Pereira, G.V.M.; De Carvalho Neto, D.P.; Junqueira, A.C.D.O.; Karp, S.G.; Letti, L.A.; Magalhães Júnior, A.I.; Soccol, C.R. A review of selection criteria for starter culture development in the food fermentation industry. *Food Rev. Int.* **2020**, *36*, 135–167. [CrossRef]
21. Speranza, B.; Racioppo, A.; Beneduce, L.; Bevilacqua, A.; Sinigaglia, M.; Corbo, M.R. Autochthonous lactic acid bacteria with probiotic aptitudes as starter cultures for fish-based products. *Food Microbiol.* **2017**, *65*, 244–253. [CrossRef]
22. Guillier, L. Predictive microbiology models and operational readiness. *Procedia Food Sci.* **2016**, *7*, 133–136. [CrossRef]
23. Speranza, B.; Racioppo, A.; Campaniello, D.; Altieri, C.; Sinigaglia, M.; Corbo, M.R.; Bevilacqua, A. Use of Autochthonous *Lactiplantibacillus plantarum* Strains to Produce Fermented Fish Products. *Front. Microbiol.* **2020**, *11*, 615904. [CrossRef]
24. Speranza, B.; Bevilacqua, A.; Racioppo, A.; Campaniello, D.; Sinigaglia, M.; Corbo, M.R. Marinated Sea Bream Fillets Enriched with *Lactiplantibacillus plantarum* and *Bifidobacterium animalis* subsp. lactis: Brine Optimization and Product Design. *Foods* **2021**, *10*, 661. [CrossRef]
25. Mejlholm, O.; Dalgaard, P. Development and validation of an extensive growth and growth boundary model for psychrotolerant *Lactobacillus* spp. in seafood and meat products. *Int. J. Food Microbiol.* **2013**, *167*, 244–260. [CrossRef] [PubMed]
26. Dalgaard, P. Qualitative and quantitative characterization of spoilage bacteria from packed fish. *Int. J. Food Microbiol.* **1995**, *26*, 319–333. [CrossRef]
27. Gram, L.; Dalgaard, P. Fish Spoilage Bacteria Problems and Solution. *Curr. Opin. Biotechnol.* **2002**, *13*, 262–266. [CrossRef]
28. Mejlholm, O.; Gunvig, A.; Borggaard, C.; Blom-Hanssen, J.; Mellefont, L.; Ross, T.; Leroi, F.; Else, T.; Visser, D.; Dalgaard, P. Predicting growth rates and growth boundary of *Listeria monocytogenes*—An international validation study with focus on processed and ready-to-eat meat and seafood. *Int. J. Food Microbiol.* **2010**, *141*, 137–150. [CrossRef] [PubMed]
29. Bolívar, A.; Costa, J.C.C.P.; Posada-Izquierdo, G.D.; Valero, A.; Zurera, G.; Pérez-Rodríguez, F. Modelling the growth of *Listeria monocytogenes* in Mediterranean fish species from aquaculture production. *Int. J. Food Microbiol.* **2018**, *270*, 14–21. [CrossRef]
30. Borch, E.; Kant-Muermans, M.-L.; Blixt, Y. Bacterial spoilage of meat and cured meat products. *Int. J. Food Microbiol.* **1996**, *33*, 103–120. [CrossRef]
31. Zhou, Y.; Wu, S.; Peng, Y.; Jin, Y.; Xu, D.; Xu, X. Effect of lactic acid bacteria on mackerel (*Pneumatophorus japonicus*) seasoning quality and flavor during fermentation. *Food Biosci.* **2021**, *41*, 100971. [CrossRef]
32. McKellar, R.C.; Lu, X. *Modeling Microbial Responses in Foods*, 1st ed.; CRC Press: Boca Raton, FL, USA, 2003. [CrossRef]
33. Baranyi, J.; Roberts, T.A. A dynamic approach to predicting bacterial growth in food. *Int J. Food Microbiol.* **1994**, *23*, 277–294. [CrossRef]
34. Le Marc, Y.; Valík, L.; Medved'ová, A. Modelling the effect of the starter culture on the growth of *Staphylococcus aureus* in milk. *Int. J. Food Microbiol.* **2009**, *129*, 306–311. [CrossRef]
35. Bevilacqua, A.; Corbo, M.R.; Sinigaglia, M. Design of Experiments: A powerful tool in Food Microbiology. In *Current Research, Technology and Education Topics in Applied Microbiology and Microbial Biotechnology*; Microbiology Book Series; Mendez-Vilas, A., Ed.; Formatex Research Center: Badajoz, Spain, 2010; pp. 1419–1429.
36. Giménez, B.; Dalgaard, P. Modelling and predicting the simultaneous growth of *Listeria monocytogenes* and spoilage micro-organisms in cold-smoked salmon. *J. Appl. Microbiol.* **2004**, *96*, 96–109. [CrossRef]
37. Mejlholm, O.; Dalgaard, P. Modeling and predicting the growth of lactic acid bacteria in lightly preserved seafood and their inhibiting effect on *Listeria monocytogenes*. *J. Food Prot.* **2007**, *70*, 2485–2497. [CrossRef] [PubMed]
38. Vermeulen, A.; Devlieghere, F.; De Loy-Hendricks, A.; Uyttendaele, M. Critical evaluation of the EU-technical guidance on shelf-life studies for *Listeria monocytogenes* on RTE-foods: A case study for smoked salmon. *Int. J. Food Microbiol.* **2011**, *145*, 176–185. [CrossRef] [PubMed]
39. Speranza, B.; Bevilacqua, A.; Corbo, M.R.; Sinigaglia, M. A possible approach to assess acidification of meat starter cultures: A case study for some wild strains of *Lactobacillus plantarum*. *J. Sci. Food Agric.* **2017**, *97*, 2691–2698. [CrossRef] [PubMed]
40. Gardini, F.; Özogul, Y.; Suzzi, G.; Tabanelli, G.; Özogul, F. Technological factors affecting biogenic amine content in food: A review. *Front. Microbiol.* **2016**, *7*, 1218. [CrossRef]

41. Codex Alimentarius. *Standard for Smoked Fish, Smoke-Flavoured Fish and Smoke-Dried Fish*; FAO: Rome, Italy, 2013; pp. 1–7. Available online: https://www.fao.org/fao-who-codexalimentarius (accessed on 7 June 2021).
42. Belichovska, K.; Belichovska, D.; Pejkovski, Z. Smoke and Smoked Fish Production. *Meat Technol.* **2019**, *60*, 37–43. [CrossRef]
43. Utomo, B.S.B.; Singgih, W.; Tri Nugroho, W. *Asap Cair: Cara Membuat & Aplikasinya Pada Pengolahan Ikan Asap*; Penebar Swadaya Group: Depok, Indonesia, 2012.
44. Lokollo, E.; Apituley, D.A.N.; Nendissa, D.M. Pengolahan ikan cakalang (*Katsuwonus pelamis*) asap dengan menggunakan teknologi asap cair. *J. Commun. Serv.* **2012**, *1*, 165–169.
45. Rakhmayeni, D.A.; Yuniarti, T.; Sukarno, S. Application of liquid smoke from coconut shell in tandipang (*Dussumeiria acutta*) smoked fish to extend shelf life. *J. Ilm. Perikan. Kelaut.* **2020**, *12*, 315–323. [CrossRef]
46. Hokkanen, M.; Luhtasela, U.; Kostamo, P.; Ritvanen, T.; Peltonen, K.; Jestoi, M. Critical effects of smoking parameters on the levels of polycyclic aromatic hydrocarbons in traditionally smoked fish and meat products in Finland. *J. Chem.* **2018**, *2018*, 2160958. [CrossRef]

foods

MDPI

Article

Consideration of Maintenance in Wine Fermentation Modeling

Alain Rapaport [1,*], Robert David [2], Denis Dochain [3], Jérôme Harmand [4] and Thibault Nidelet [5]

1 MISTEA, Université Montpellier, INRAE, Institut Agro, 34060 Montpellier, France
2 Technord, 7500 Tournai, Belgium; r.david@technord.com
3 ICTEAM, Université Catholique de Louvain, 1348 Louvain-la-Neuve, Belgium; denis.dochain@uclouvain.be
4 LBE, Université Montpellier, INRAE, 11100 Narbonne, France; jerome.harmand@inrae.fr
5 SPO, Université Montpellier, INRAE, Institut Agro, 34060 Montpellier, France; thibault.nidelet@inrae.fr
* Correspondence: alain.rapaport@inrae.fr

Abstract: We show that a simple model with a maintenance term can satisfactorily reproduce the simulations of several existing models of wine fermentation from the literature, as well as experimental data. The maintenance describes a consumption of the nitrogen that is not entirely converted into biomass. We show also that considering a maintenance term in the model is equivalent to writing a model with a variable yield that can be estimated from data.

Keywords: wine fermentation; nitrogen; mathematical modeling; population model; maintenance; variable yield

Citation: Rapaport, A.; David, R.; Dochain, D.; Harmand, J.; Nidelet, T. Consideration of Maintenance in Wine Fermentation Modeling. *Foods* **2022**, *11*, 1682. https://doi.org/10.3390/foods11121682

Academic Editors: Carlos Vilas, Míriam R. García and Jose A. Egea

Received: 20 April 2022
Accepted: 17 May 2022
Published: 8 June 2022

Publisher's Note: MDPI stays neutral with regard to jurisdictional claims in published maps and institutional affiliations.

1. Introduction

The overall principle of wine fermentation consists of the conversion of sugar into ethanol by yeast. It has been observed for a long time that nitrogen consumed during the yeast growth also plays an important role. The fermentation can be indeed modeled by a two-step process in which the yeast first grows on nitrogen as a limiting resource and then degrades the non-limiting sugar into ethanol and carbon dioxide. However, experimental observations have shown that the consumed nitrogen was not entirely converted into biomass. Several mathematical models were proposed to take these characteristics into consideration. For instance, in [1,2], the biomass growth follows a logistic law whose carrying capacity depends on the initial quantity of nitrogen. In [3], a model that distinguishes part of nitrogen used for yeast growth from another part responsible of the synthesis of proteins (hexose transporters [4]) was developed. Both models were calibrated with different sets of experimental data and provide satisfactory fitting. However, both models present some drawbacks. The dependency of the dynamics on the initial condition of the first model makes it sensitive to the precise knowledge of the initial quantity of nitrogen (that needs to be "memorized" in the dynamical equations of the model). Moreover, it does not allow consideration of non-batch operations or continuous addition of nitrogen, such as in [5] for instance. The second model relies on the knowledge of the time-varying concentration of transporters, which is in general not easily accessible for experimental measurements, and several assumptions were necessary to estimate it from biomass measurements.

The objective of the present work is to propose a new model that reconciles both approaches in a single one.

The observation of the ratio of produced biomass over nitrogen consumption along the whole fermentation, determined on experimental database or numerical simulations of models [1,3], shows that this ratio is non-constant and depends on the initial quantities. This highlights that the conversion of nitrogen into biomass can be viewed as a variable yield process. The experimental evidence that nitrogen is not entirely converted into biomass therefore advocates for the consideration of a maintenance term in the modeling (see, for instance, [6]), without necessarily requiring a detailed representation of the internal mechanism or cells.

Indeed, different mechanisms in the internal functioning of the cells have been investigated in the literature, particularly the role of carbohydrate accumulation [7–9], which could explain that the growth dynamics of yeast in wine fermentation does not follow the classical mass-balanced models [10,11]. However, the measurements of these biochemical compounds is experimentally very difficult and is almost impossible in an industrial framework.

The rationale of the results presented here is to test if the introduction of a maintenance term (see [12–14] or [15–17]) can improve wine fermentation modeling. One of the original features of the proposed approach is to view nitrogen consumption as a global consumption for growth by considering a variable yield. This allows us to avoid to consider a specific structure to model the maintenance. Thus, the purpose of the present work is to investigate the ability of a simpler model with a maintenance term to reproduce and predict wine fermentation kinetics.

Here, we propose a new modeling approach based on a maintenance term (which gives rise to a variable yield), a feature that has not been yet considered in the wine fermentation literature, to the best of our knowledge.

It focuses mainly on the new modeling of the growth of yeast on nitrogen.

This new model was validated using both data generated by existing models (Section 4) and experimental data (Section 5).

2. The Proposed Model

We denote by N, S, E, CO_2 and X the concentrations of (total) nitrogen, sugar, ethanol, dioxide carbon and biomass, respectively. For simplicity, we derive a model under isothermal conditions.

For the first step $N \rightarrow X$ (yeast growth on nitrogen), we propose the following equations

$$\frac{dX}{dt} = \mu_N(N,X)X \tag{1}$$

$$\frac{dN}{dt} = -\frac{\mu_N(N,X)X}{Y} - m(N,X)X \tag{2}$$

where Y is the growth yield, μ the Contois growth function

$$\mu_N(N,X) = \frac{\mu_N^{max} N}{N + K_N X}$$

and m a *maintenance* function, which is positive for $N > 0$ and $X > 0$. We choose here a ratio-dependent kinetics function μ_N to reproduce the observation that the growth is slowing down under an excess of yeast, with a Contois expression as in [3]. In the literature, the maintenance m is often considered as constant [12,13], which was validated in continuous culture (chemostat). In general, continuous cultures are intended to be operated at a stationary phase, very differently to batch-operating mode. However, as already investigated in [17], maintenance terms have to depend on the level of available resources; say, R (N here). In particular, constant maintenance in a batch model would imply $\frac{dR}{dt} < 0$ when the resource is exhausted, i.e., $R = 0$, and thus R could take unrealistic negative values, as underlined in [14]. In [15,16], the maintenance is directly related to the microbial activity, which is stopped in absence of nutrients. This is why we consider a maintenance function proportional to the growth activity, with a factor that might depend on the nitrogen concentration (one may expect that it decreases when the substrate N becomes rare)

$$m(N,X) = \alpha(N)\mu_N(N,X)$$

where α is a positive function equal to zero for $N = 0$. Then, one can consider the function y defined as follows

$$y(N) := \frac{Y}{1 + \alpha(N)Y}, \quad N \geq 0$$

Formally, model (1) and (2) can be rewritten equivalently as

$$\frac{dX}{dt} = \mu_N(N, X)X \tag{3}$$

$$\frac{dN}{dt} = -\frac{\mu_N(N, X)X}{y(N)} \tag{4}$$

where the function y is playing the role of a *variable yield*. Identifying the function m or the function y is thus formally equivalent. However, we shall see in the next section that identifying the function y instead of m presents some practical advantages.

For the second step $S \to E + CO_2$, we follow the model proposed in the literature [3]

$$\frac{dE}{dt} = \frac{dCO_2}{dt} = \Big[\mu_N(N, X) + \beta \nu_E(E)\Big]\mu_S(S)X \tag{5}$$

$$\frac{dS}{dt} = -k\frac{dE}{dt} \tag{6}$$

where μ_S is a Monod function and ν_E a function inhibited by the ethanol

$$\mu_S(S) = \frac{\mu_S^{max} S}{K_S + S}, \quad \nu_E(E) = \frac{1}{1 + K_E E} \tag{7}$$

Inhibition by the consumption of sugar S by ethanol E has been reported many times in the literature [18–22]. The constant yield of production k of CO_2 and consumption of S follows a mass balance assumption, verified experimentally [23], that can be determined using thermodynamics considerations [24].

Note that this model can be extended to anisothermal conditions, considering that the maximal specific rate parameters μ_N^{max}, μ_S^{max} and affinity constants K_S, K_E are temperature dependent, as in [3].

3. Calibration of the Model

From model Equation (1), the parameters of the function μ_N can be identified independently of the yield and maintenance terms. To validate the hypothesis of ratio dependency of the function μ_N, one can first use experimental data to plot the slope of the logarithm of X versus the ratio $r = N/X$ and check if it qualitatively follows a function of the form

$$\mu(r) = \frac{\mu_N^{max} r}{K_N + r}$$

A classical least-square method can be applied to fit parameters μ_N^{max}, K_N on the data. Alternatively, one can plot the inverse of the slope of the logarithm of X versus the inverse of the ratio r to check if it qualitatively follows a linear dependency, as obtained from Equation (1)

$$\left(\frac{d \log X}{dt}\right)^{-1} = \frac{1}{\mu_N^{max}} + \frac{K_N}{\mu_N^{max}}\left(\frac{N}{X}\right)^{-1} \tag{8}$$

However, for the accurate identification of the parameters μ_N^{max}, K_N, a linear regression on Equation (8) is expected to be less reliable than a non-linear least-square optimization of the solution $X(\cdot)$ of (3), because $\left(\left(\frac{d}{dt}\log X\right)^{-1}, \left(\frac{N}{X}\right)^{-1}\right)$ data might be too far to be uniformly distributed.

Note from Equations (1) and (2) that one has

$$\lim_{t \to +\infty} N(t) = 0$$

(because the derivative of N cannot vanish when N is not exhausted). In absence of the maintenance term m, one gets $\frac{dX}{dt} + Y\frac{dN}{dt} = 0$ which implies that one should have

$$Y = \frac{X(+\infty) - X(0)}{N(0) - N(+\infty)} = \frac{X(+\infty) - X(0)}{N(0)}$$

To test the validity of the model with maintenance, one can plot from experimental data the ratio $\frac{X(+\infty)-X(0)}{N(0)}$ for different values of $N(0)$ to check that it is not constant. If this is the case, one can then look at identifying a non-constant function y. For this purpose, we write from Equations (3) and (4)

$$X(+\infty) - X(0) = -\int_0^{+\infty} y(N(t))\frac{dN}{dt}(t)\,dt$$

and as $t \mapsto N(t)$ is a monotone-decreasing function, one can make the change of variable $n = N(t)$ in this last integral to obtain

$$X(+\infty) - X(0) = \int_0^{N(0)} y(n)\,dn$$

Therefore, if one fits a differential function f such that $f(0) = 0$ that satisfies

$$X(+\infty) - X(0) = f(N(0))$$

for experimental data with different values of $N(0)$, then one simply gets $y = f'$.

Let us underline that identifying the function y in this way can be achieved independently of the knowledge of the kinetics μ_N, differently to the function m, which clearly presents some robustness advantages. Once the function μ_N is identified, the maintenance function can then be determined as

$$m(N, X) = \left(\frac{1}{y(N)} - \frac{1}{Y}\right)\mu_N(N, X)$$

where $Y = y(0)$ (to fulfill $\alpha(0) = 0$).

For model Equations (5) and (6), the coefficient k is kept from the literature, and the parameters β, μ_S^{max}, K_S, K_E are identified (with a least-square method) from experimental data of CO_2 production rate.

4. Validation of the Model on Synthetic Data

We have used synthetic data generated by models of the literature that were previously validated on experimental data [1,3] for a range of initial conditions and operating conditions.

Fitting comparisons of the proposed model with the different data sets are reported in Section 6.

4.1. Validation on Simulations of a Model with Transporter

We have considered the model with transporters developed in [3], which is more complex with two additional state variables: the concentrations of hexose transporters and the nitrogen dedicated to these transporters. Data were generated by simulating this model with the parameters given in [3] and operating conditions given in Table 1.

Table 1. Operating conditions for the simulation of the model with transporters.

$X(0)$	0.02 g·L^{-1}
$N(0)$	0.071–0.57 g·L^{-1}
$S(0)$	200 g·L^{-1}
time horizon	350 h
temperature	constant equal to 24°
others	no initial transporter no nitrogen addition

This model explicitly distinguishes two forms of nitrogen, one available for the yeast N_X and the other one N_{tr} for the transporters. To compare with the variable N of our model, we have considered the total nitrogen $N = N_X + N_{tr}$.

4.1.1. Estimation of the Contois Function

We have used a non-linear least-square method based on a Newton algorithm with a finite difference approximation of the Jacobian matrix (function `leastsq` of `scilab`). Figure 1 shows a good fitting of the Contois function μ_N on data $\left(\frac{N}{X}, \frac{\frac{dX}{dt}}{X}\right)$ of the transporter model, with parameters given in Table 2.

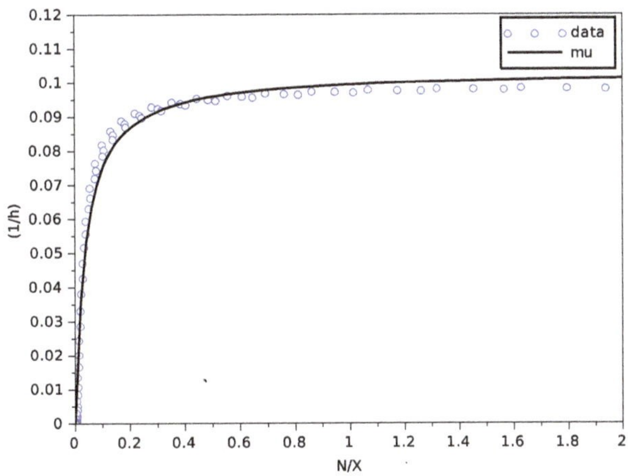

Figure 1. Result of the fitting of the Contois function on data from the model with transporters.

Table 2. Parameters of the Contois function μ_N.

μ_N^{max}	0.103 h^{-1}
K_N	0.0381 g·L^{-1}

4.1.2. Estimation of the Variable Yield Function

On Figure 2, data $X(T) - X(0)$ versus $N(0)$ from the model with transporters were plotted for $T = 350$ h (we have checked that N is quasi-null at T and that X no longer increases after T). One can see that the points are aligned. However, the line that passes through these points does not touch 0, which is not possible for a constant yield (for a constant yield, the points have to be aligned on a line that passes through 0, because when $N(0) = 0$, there is no biomass production).

Then, we fitted a C^2 function f such that $f(0) = 0$ with the following expression

$$f(N) = \begin{cases} aN + b\left(1 - \left(\frac{N_\dagger - N}{N_\dagger}\right)^3\right), & N < N_\dagger \\ aN + b, & N \geq N_\dagger \end{cases}$$

whose parameters are given in Table 3.

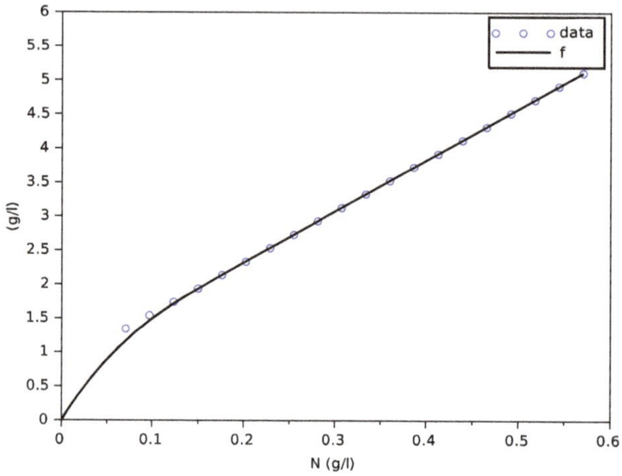

Figure 2. Result of the fitting of the function f on data from the model with transporters.

The calibration of the parameters a, b of the function f was performed with a linear regression (function `reglin` of `scilab`).

Table 3. Parameters of the variable yield function y.

a	7.55
b	0.808 g·L^{-1}
$N_\dagger$	0.176 g·L^{-1}

Then, we obtain the variable yield function y as the C^1 function

$$y(N) = f'(N) = \begin{cases} a + b\frac{3(N_\dagger - N)^2}{{N_\dagger}^3}, & N < N_\dagger \\ a, & N \geq N_\dagger \end{cases}$$

and the function α, which describes the maintenance as

$$\alpha(N) = \frac{1}{y(N)} - \frac{1}{y(0)} = \begin{cases} \frac{N_\dagger^3}{aN_\dagger^3 + 3b(N_\dagger - N)^3} - \frac{N_\dagger}{3b + aN_\dagger}, & N < N_\dagger \\ \frac{3b}{a(3b + aN_\dagger)}, & N \geq N_\dagger \end{cases}$$

which are both depicted on Figure 3.

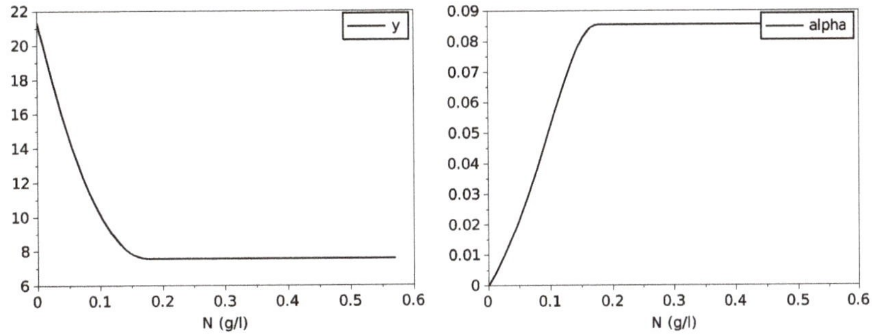

Figure 3. Graphs of the obtained variable yield function y and of the function α.

Note that the model with transporters was validated only for $N(0)$ in the interval $[0.071, 0.57]$ g·L^{-1}, and that we have no a priori information about the behavior of the yield for values of $N(0)$ smaller than 0.071 g·L^{-1}. The threshold parameter $N_{\dagger}$ was simply chosen so that the simulations of the variables X and N of the model (3) and (4) were the closest to the ones of the transporter model.

4.1.3. Estimation of the Other Parameters and Comparison of the Models

For the model of the second step $S \rightarrow E + CO_2$, the stoichiometric parameter k was taken from the literature, while the other parameters β, μ_S^{max}, K_S, K_E were estimated with a least-square optimization on the CO_2 chronicles only (the CO_2 production rate being a variable that is usually measured in experiments), starting from values in [3]. Values are given in Table 4.

Table 4. Parameters for the second step $S \rightarrow E + CO_2$ model.

k	2.17
β	2.41
μ_S^{max}	0.197 h^{-1}
K_S	21.1 g·L^{-1}
K_E	72.7 g·L^{-1}

Here, we also used a non-linear least-square method based on a Newton algorithm with a finite difference approximation of the Jacobian matrix (function `leastsq` of `scilab`). All data were re-normalized to 1 (i.e., for each variable, the figures were divided by the largest one).

Finally, we present on Figures 4–6 simulations of the new model for three largely different initial values of nitrogen from 0.170 g·L^{-1} to 0.567 g·L^{-1}. The evolution of the ethanol concentration E has not been reproduced as it is proportional to the CO_2 concentration.

These simulations show the ability of the new model to reproduce, with a single set of parameters, close simulations to the model with transporters, in terms of production of biomass and dioxide carbon, estimation of the peak of the CO_2 production rate and depletion of (total) nitrogen and sugar.

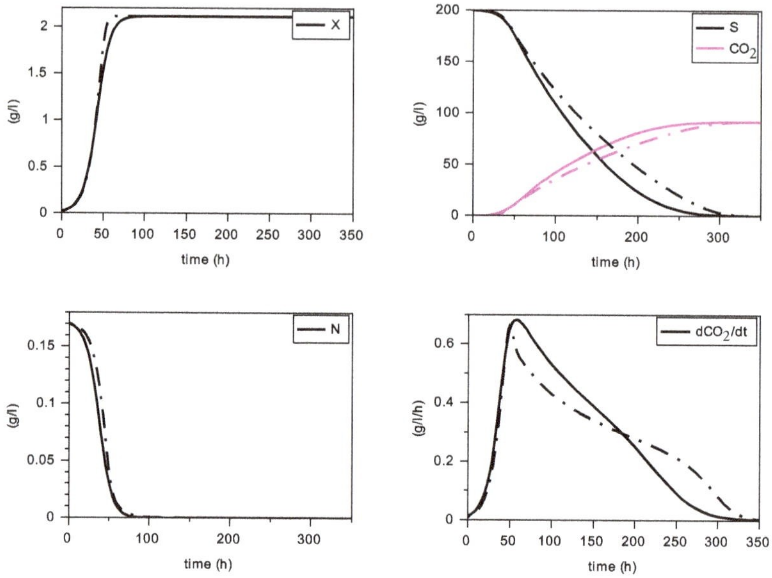

Figure 4. Comparison with the model with transporters (in dashed) for $N(0) = 0.170\ \text{g}\cdot\text{L}^{-1}$ (constant temperature of 24 °C).

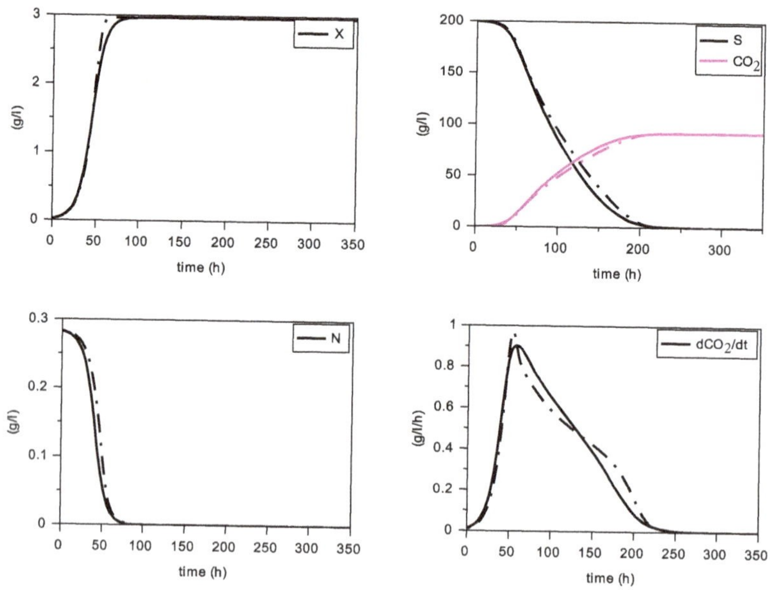

Figure 5. Comparison with the model with transporters (in dashed) for $N(0) = 0.283\ \text{g}\cdot\text{L}^{-1}$ (constant temperature of 24 °C).

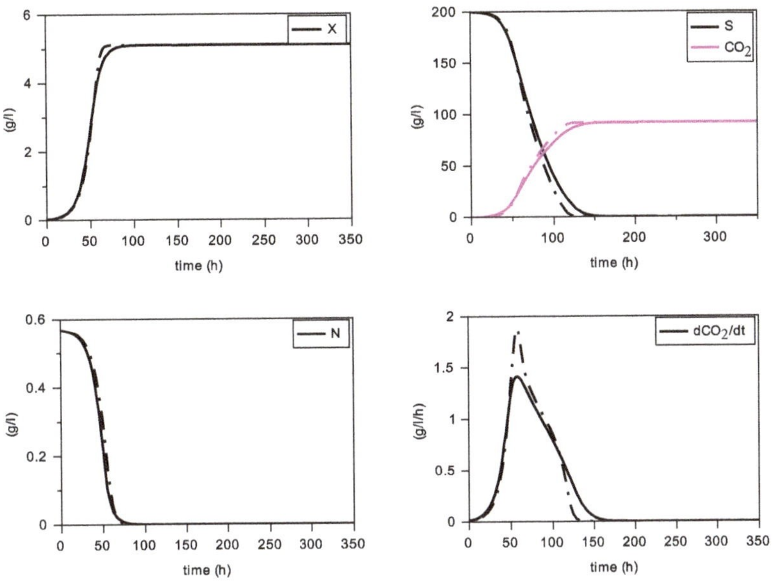

Figure 6. Comparison with the model with transporters (in dashed) for $N(0) = 0.567\ g{\cdot}L^{-1}$ (constant temperature of 24 °C).

4.2. Validation on the SOFA Model

The model proposed in [1] does not explicitly consider transporters with an additional state variable as the previous model, and instead presents a more sophisticated expression of the dynamics that depend on the initial condition, with an additional latency term at the beginning of the simulations.

Differently to the previous model, which is built as a "mass-balanced" model, this one relies on an empirical dynamics of logistic shape for the biomass growth, with some parameters that depend on the initial concentration of nitrogen $N(0)$, instead of the two-dimensional model (3) and (4).

Therefore, this is not a Markovian model. It has been validated on different operating conditions, and has been encoded into the SOFA software exploited for decision-making [2]. We launched simulations of this model for the same operating conditions than for the previous model (Table 1). Although simulations look qualitatively similar, they do not overlap, especially for the biomass chronicle. This could be explained by the fact that this model is intended to predict a number of cells and not a precise biomass (an average number of 4.15×10^9 cells for one g of biomass was used to have X expressed in $g{\cdot}L^{-1}$ as for the previous model). We proceeded to a new validation of our model on these data.

4.2.1. Estimation of the Contois Function

Figure 7 shows that the data $\left(\frac{N}{X}, \frac{\frac{dX}{dt}}{X}\right)$ do not precisely follow the graph of a function (this is most probably due to the fact that the model is not Markovian). Indeed, this happens mainly for the large value N_0 of the initial nitrogen. We believe that this could be explained by the dynamics of the biomass X of this model, which is a logistic law with a carrying capacity given by an heuristic expression that depends on N_0, and not dynamics coupled with the dynamics of N (indeed the interval of tested values of N_0 might be larger than the validity of this model). However, we have fitted the graph of a Contois function to these data with the parameters given in Table 5, which was able to satisfactorily reproduce the trajectories of the model for a large amplitude of values of N_0, as we shall see later on.

As for the previous model, we used a non-linear least-square method based on a Newton algorithm with a finite difference approximation of the Jacobian matrix (function `leastsq` of `scilab`). As one can see in Table 5, the values of μ_N^{max} and K_N are significantly larger and smaller, respectively, than in Table 2, which is consistent with the observation that this model predicts a faster convergence of the biomass to its maximal value, despite the latency term (compare Figures 4–6 with Figures 8–10).

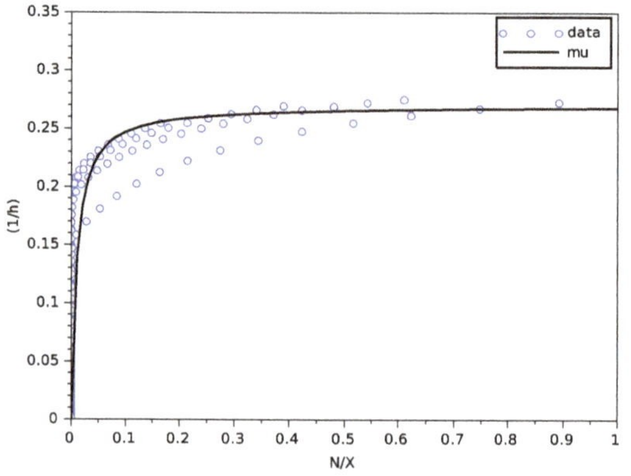

Figure 7. Result of the fitting of the Contois function on data from the SOFA model.

Table 5. Parameters of the Contois function μ_N.

μ_N^{max}	$0.270\ h^{-1}$
K_N	$0.00952\ g{\cdot}L^{-1}$

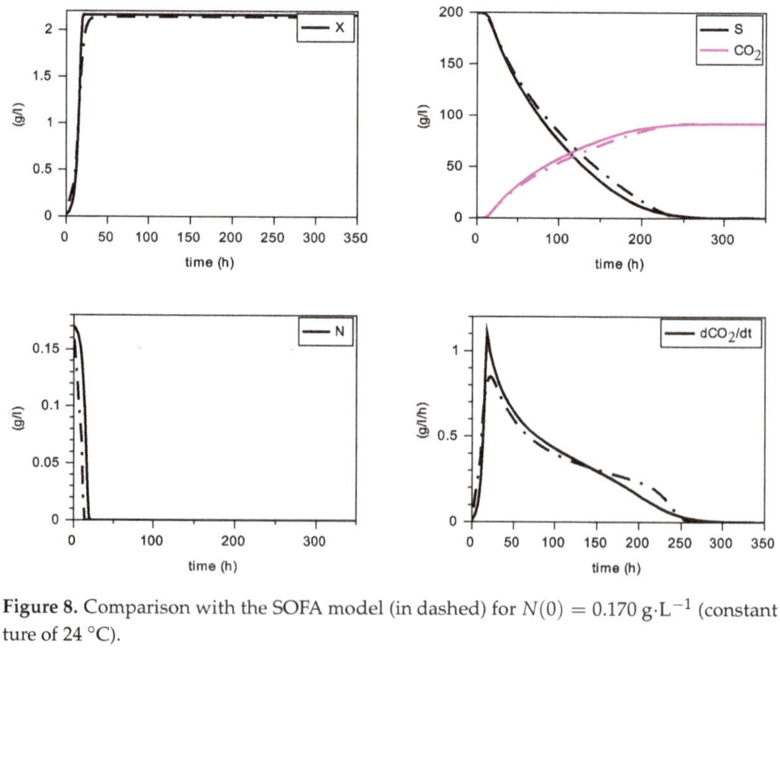

Figure 8. Comparison with the SOFA model (in dashed) for $N(0) = 0.170\ g{\cdot}L^{-1}$ (constant temperature of 24 °C).

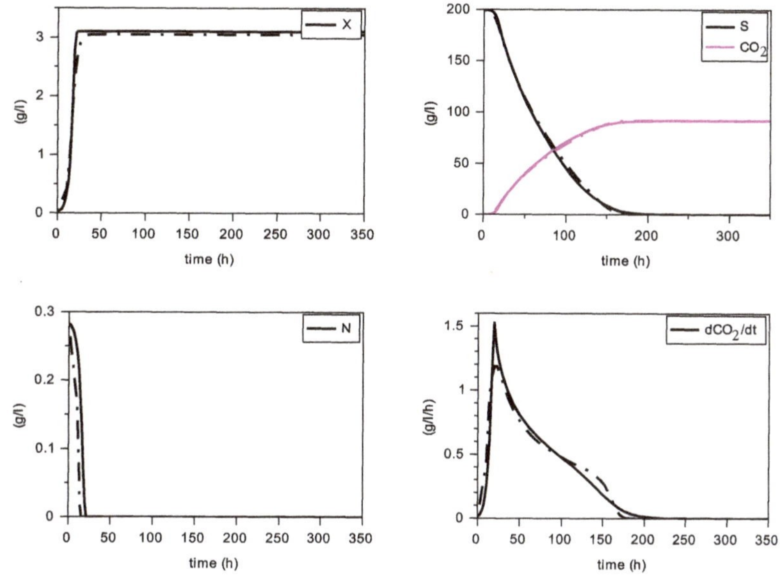

Figure 9. Comparison with the SOFA model (in dashed) for $N(0) = 0.283$ g·L^{-1} (constant temperature of 24 °C).

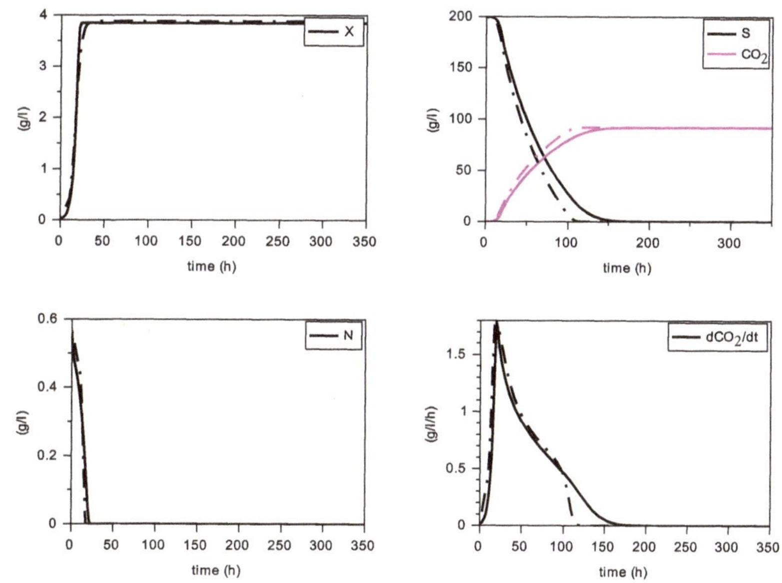

Figure 10. Comparison with the SOFA model (in dashed) for $N(0) = 0.567$ g·L^{-1} (constant temperature of 24 °C).

4.2.2. Estimation of the Variable Yield Function

Data $X(T) - X(0)$ from the simulation of the SOFA model were plotted on Figure 11 at $T = 350$ h, for different values of $N(0)$ in the interval $[0.071, 0.57]$ g·L^{-1} (here, we also checked that the fermentation was quasi-ended at T). One can see that the points follow an increasing concave curve and further increase very slowly, quite differently to the model with transporters (see Figure 2).

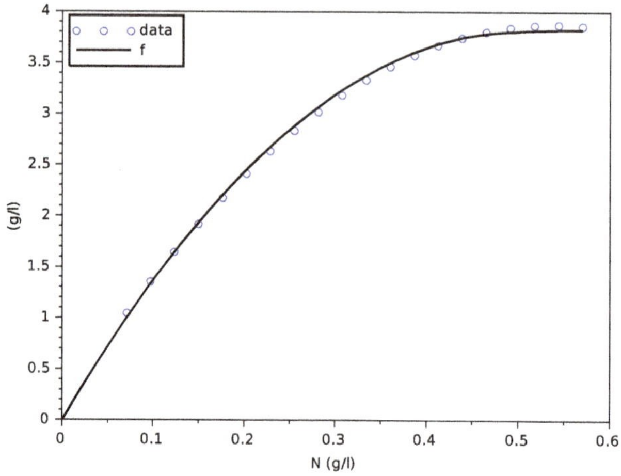

Figure 11. Result of the fitting of the function f on data from the SOFA model.

We have then fitted a C^2 function f with $f(0) = 0$ for the expression

$$f(N) = \begin{cases} bN - aN^2, & N < N_{\dagger} \\ bN - aN^2 + bN + \frac{A}{B}\left(e^{-BN_{\dagger}} - e^{-BN}\right) & N < N_{\dagger} \end{cases}$$

with

$$A = (b - 2aN_{\dagger})e^{BN_{\dagger}}, \quad B = \frac{2a}{b - 2aN_{\dagger}}$$

and parameters a, b, $N_{\dagger}$ given in Table 6.

Parameters a and b were determined with a linear regression (function `reglin` of `scilab`).

Table 6. Parameters of the variable yield function y.

a	15.1 g·L^{-1}
b	15.2
$N_{\dagger}$	0.465 g·L^{-1}

Then, we obtain the expression of the variable yield function

$$y(N) = f'(N) = \begin{cases} b - 2aN, & N < N_{\dagger} \\ Ae^{-BN}, & N \geq N_{\dagger} \end{cases}$$

as well as the function α

$$\alpha(N) = \frac{1}{y(N)} - \frac{1}{y(0)} = \begin{cases} \frac{1}{b-2aN} - \frac{1}{b}, & N < N_{\dagger} \\ \frac{e^{b(N-N_{\dagger})}}{b-2aN_{\dagger}} - \frac{1}{b}, & N \geq N_{\dagger} \end{cases}$$

whose graphs are drawn on Figure 12.

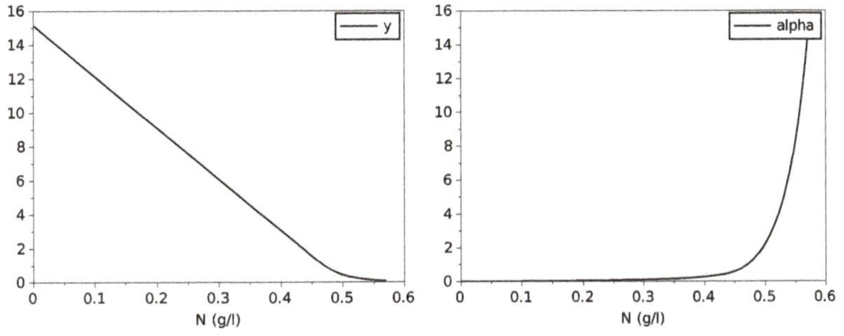

Figure 12. Graphs of the obtained variable yield function y and of the function α.

4.2.3. Estimation of the Other Parameters and Comparison of the Models

For the second step, the same stochiometric parameter k was taken for the literature, and the other parameters β, μ_S^{max}, K_S, K_E were estimated with a least-square optimization on the CO_2 chronicles only, as for data generated by the model with transporters (see Table 7).

Table 7. Parameters for the second step $S \to E + CO_2$ model.

k	2.17
β	3.22
μ_S^{max}	0.197 h^{-1}
K_S	17.6 $g \cdot L^{-1}$
K_E	36.4 $g \cdot L^{-1}$

Figures 8–10 show the comparison between the SOFA model and our calibrated model for the same initial condition than for the former comparison with the model with transporters. Here also, we see that the proposed model reproduces quite faithfully the simulations of the SOFA model, with the advantage of being a simpler Markovian model. Indeed, the difference between the model with transporters and the SOFA model can be translated into different maintenance terms (see Figures 3 and 12): for large values of nitrogen, the model with transporters behaves like a model with a maintenance proportional to the growth, while the SOFA model amounts to have a strongly increasing maintenance. Recall that the simulations for the largest value of $N(0)$ showed the most differences between these two models (for $N(0) = 0.567$ $g \cdot L^{-1}$, the model with transporters predicts a biomass production of 5.11 $g \cdot L^{-1}$, while the SOFA model predicts 3.88 $g \cdot L^{-1}$; see Figures 6 and 10). While the model with transporters was validated experimentally for $N(0)$ in the interval [0.170, 0.567] $g \cdot L^{-1}$, we believe the validation of the SOFA model for initial concentrations of nitrogen larger than 0.4 $g \cdot L^{-1}$ might need to be revisited (although our model once calibrated is able to reproduce the SOFA simulations).

5. Calibration of the Model on Real Data

We considered data from experiments conducted at SPO Lab (INRAE, Montpellier, France) in 2004, that were used to calibrate the model with transporters and the SOFA model (see [1,3]). The data consisted of a set of three experiments with the same operating conditions given in Table 1 and different initial concentrations $N(0)$ of nitrogen, exactly the same as for the simulations of Sections 4.1 and 4.2. For each experiment, one had

- Height measurement points for X.
- No measurement point for N, S or E.
- About 400 measurement points for CO_2 and dCO_2/dt.

We first calibrated a function $f(\cdot)$ to the data $(N(0), X(T) - X(0))$, with the same expression as in Section 4.2, to determine a yield function $y(\cdot)$ (see Figure 13), using a linear regression to estimate parameters a and b.

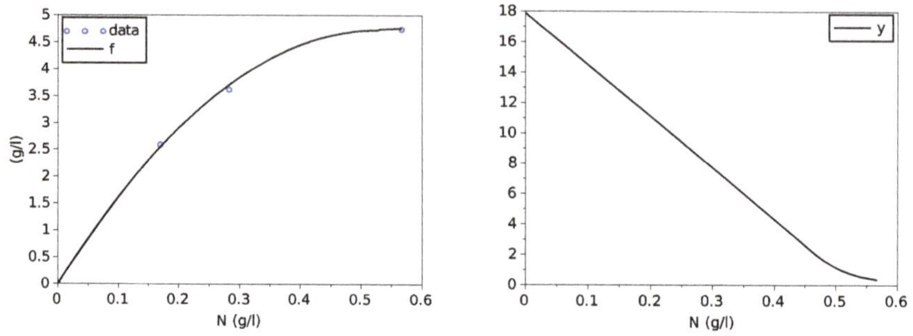

Figure 13. Results of the fitting of the function f on the experimental data (**left**) and of the corresponding variable yield function y (**right**).

As we do not have measurements of N over the time, we cannot estimate the Contois parameters independently of the CO_2 measurements, as we did with the synthetic data. All the parameters of the model were fitted simultaneously with a least-square method (values are given in Table 8), except for the sugar-conversion yield, for which we have used the value of the literature $k = 2.17$, as before.

The non-linear least-square method uses a Newton algorithm with a finite difference approximation of the Jacobian matrix (function `leastsq` of `scilab`), and the data set was re-normalized to the maximal value of 1.

Table 8. Parameters fitted on the experimental data.

μ_N^{max}	0.175 h^{-1}
K_N	0.0133 $g \cdot L^{-1}$
β	1.622
μ_S^{max}	0.393 h^{-1}
K_S	19.2 $g \cdot L^{-1}$
K_E	71.9 $g \cdot L^{-1}$

Figures 14–16 show the results of the fitting for the three experiments. One can appreciate the goodness of fit for a unique set of parameters. In particular, the production of biomass and CO_2, as well as the height and date of the peak of dCO_2/dt, are well predicted with this model.

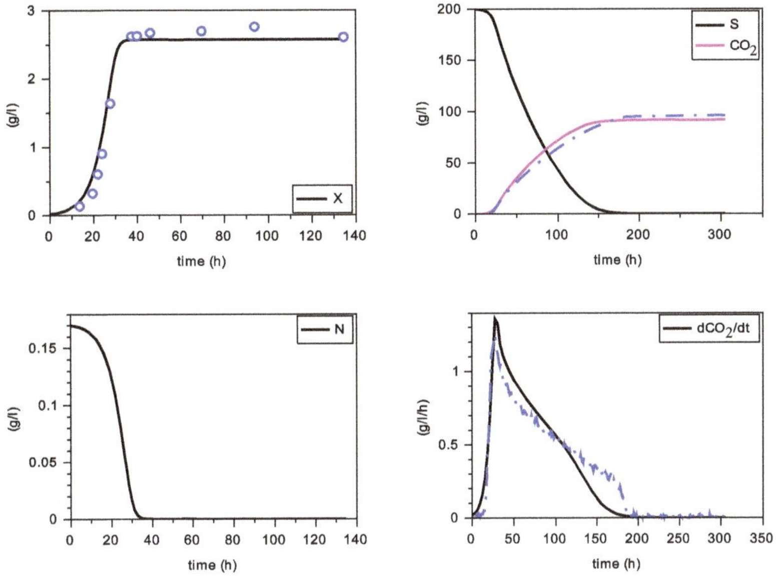

Figure 14. Simulation for $N(0) = 0.170\ \mathrm{g{\cdot}L^{-1}}$ (experimental data in blue).

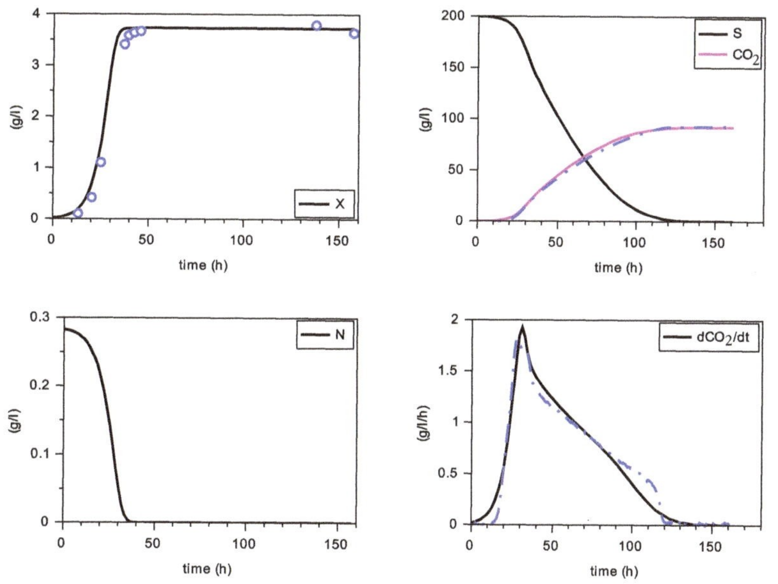

Figure 15. Simulation for $N(0) = 0.283\ \mathrm{g{\cdot}L^{-1}}$ (experimental data in blue).

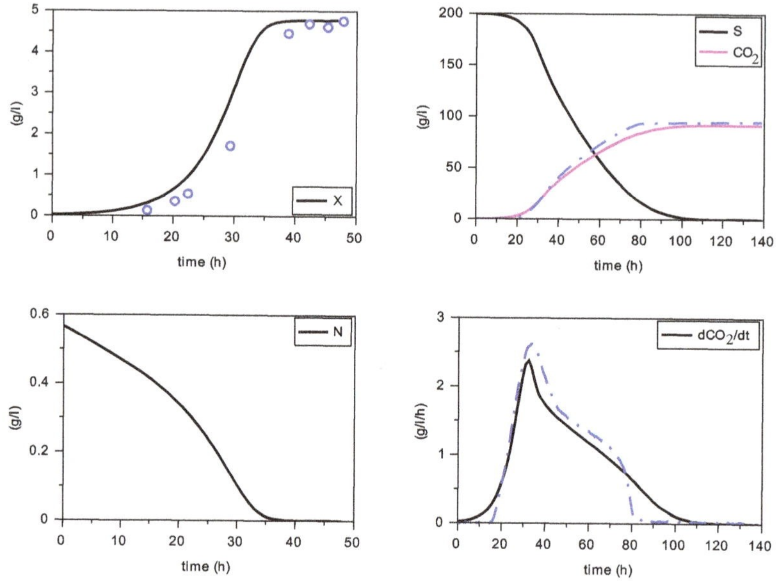

Figure 16. Simulation for $N(0) = 0.567$ g·L^{-1} (experimental data in blue).

6. Fitting Comparisons

For the calibration of the variable yield function on both synthetic and experimental data (Sections 4 and 5), we have used a linear regression (function `reglin` of `scilab`) for the determination of parameters a, b of the function f (for the model with transporter) and f' (for the SOFA model and experimental data). The residual error is given in Table 9.

Table 9. Residual standard error (RSE) for the determination of a and b.

Data	Tr Model	SOFA Model	Exp.
RSE	2.21×10^{-10}	0.199	0.225

This shows that the model with transporters behave very closely to a variable yield model. The fitting performances for the SOFA model and experimental data are more difficult to interpret, because the validity of the SOFA model for the large range of initial concentrations of nitrogen we considered is questionable, and the quantity of experimental data is quite poor compared to the synthetic data.

For the synthetic data, the calibration of the growth characteristics (parameters μ_N^{max}, K_N of the Contois function) was performed first, independently of the CO_2 data. Then, parameters for the second step (parameters k, β, μ_S^{max}, K_S, K_E for the CO_2 production) were calibrated. In both cases, a non-linear least-square method based on a Newton algorithm with a finite difference approximation of the Jacobian matrix (function `leastsq` of `scilab`) was used. Table 10 shows a good fitting quality.

Table 10. Root Mean Square Error (RMSE) for the calibration of the growth function μ and the CO_2 chronicles.

Data	Tr Model	SOFA Model	Exp.
RMSE (μ)	0.0414	0.292	-
RMSE (CO_2)	0.0543	0.0895	0.0519

We recall that for experimental data, we do not have measurement of N over time, so it was not possible to estimate the growth function independently of the CO_2 measurements. The estimation of all the parameters was made on the CO_2 measurements only. We have used the same non-linear least-square method, with data re-normalized to 1 (i.e., the figures were divided by the largest one), so that all points have equal weight in the criterion. The errors shows a good fitting of the CO_2 curves with the model with maintenance.

7. Conclusions

In this work, we demonstrated that the consideration of a maintenance term, or equivalently, a variable yield, in wine fermenting modeling can satisfactorily replace more sophisticated models with a simpler structure. Indeed, the effects of the underlying mechanisms of transporters or carbohydrate accumulation, which are difficult to capture experimentally, are somehow encoded into a maintenance term, and are translated into a variable yield between biomass and nitrogen. We showed that this variable yield, as a function of the nitrogen concentration, can be estimated from experimental data of biomass growth and nitrogen depletion, without the need to measure internal compounds. This consideration brings a flexibility to suit different kind of models or experimental data (once calibrated) with a single common structure, that could correspond to different operating conditions or hypotheses in wine fermentation. This new approach provides new perspectives of control of fermentation with nitrogen addition, based on a simple Markovian model, as well as model extensions with aromatic compounds [25] or multi-strains [26].

Author Contributions: All authors have contributed equally. All authors have read and agreed to the published version of the manuscript.

Funding: This research received no external funding.

Institutional Review Board Statement: Not applicable.

Informed Consent Statement: Not applicable.

Data Availability Statement: The data presented in this study are available on request from the corresponding author.

Conflicts of Interest: The authors declare no conflict of interest. Technord participated in the data analysis and the modeling discussion as Robert David developed in the past one of the studied model as senior researcher at Université Catholique de Louvain with Denis Dochain. Technord does not intend to use any application of Intellectual property (licensing) for these developments. The interest of this publication is purely scientific and legitimates the Data Science activity of Technord. Data were generated by INRAE, where Alain Rapaport, Jérôme Harmand and Thibault Nidelet are employed.

References

1. Malherbe, S.; Fromion, V.; Hilgert, N.; Sablayrolles, J.M. Modeling the effects of assimilable nitrogen and temperature on fermentation kinetics in enological conditions. *Biotechnol. Bioeng.* **2004**, *86*, 261–272. [CrossRef]
2. Goelzer, A.; Charnomordic, B.; Colombié, S.; Fromion, V.; Sablayrolles, J. Simulation and optimization software for alcoholic fermentation in winemaking conditions. *Food Control* **2009**, *20*, 635–642. [CrossRef]
3. David, R.; Dochain, D.; Mouret, J.R.; Vande Wouwer, A.; Sablayrolles, J.M. Nitrogen-backboned modeling of wine-making in standard and nitrogen-added fermentations. *Bioprocess Biosyst. Eng.* **2014**, *37*, 5–16. [CrossRef]
4. Elbing, K.; Larsson, C.; Bill, R.M.; Albers, E.; Snoep, J.L.; Boles, E.; Hohmann, S.; Gustafsso, L. Role of Hexose Transport in Control of Glycolytic Flux in Saccharomyces cerevisiae. *Appl. Environ. Microbiol.* **2004**, *70*, 5323–5330. [CrossRef]
5. Clement, T.; Perez, M.; Mouret, J.; Sablayrolles, J.; Camarasa, C. Use of a continuous multistage bioreactor to mimic winemaking fermentation. *Int. J. Food Microbiol.* **2011**, *150*, 42–49. [CrossRef]
6. Goma, G.; Moletta, R.; Novak, M. Comments on the "Maintenance coefficient" changes during alcohol fermentation. *Biotechnol. Lett.* **1979**, *1*, 415–420. [CrossRef]
7. Schulze, U.; Lidèn, G.; Nielsen, J.; Villadsen, J. Physiological effects of nitrogen starvation in an anaerobic batch culture of Saccharomyces cerevisiae. *Microbiology* **1996**, *142*, 2299–2310. [CrossRef]
8. Varela, C.; Pizarro, F.; Agosin, E. Biomass content governs fermentation rate in nitrogen-deficient wine musts. *Appl. Environ. Microbiol.* **2011**, *70*, 3392–3400. [CrossRef]

9. Vargas, F.; Pizarro, F.; Pérez-Correa, J.; Agosin, E. Expanding a dynamic flux balance model of yeast fermentation to genome-scale. *BMC Syst. Biol.* **2011**, *5*, 75. [CrossRef]
10. Henriques, D.; Minebois, R.; Mendoza, S.; Macías, L.; Pérez-Torrado, R.; Barrio, E.; Teusink, B.; Querol, A.; Balsa-Canto, E. A Multiphase Multiobjective Dynamic Genome-Scale Model Shows Different Redox Balancing among Yeast Species of the *Saccharomyces* Genus in Fermentation. *mSystems* **2021**, *6*, e00260-21. [CrossRef]
11. Henriques, D.; Balsa-Canto, E. The Monod Model Is Insufficient To Explain Biomass Growth in Nitrogen-Limited Yeast Fermentation. *Appl. Environ. Microbiol.* **2021**, *87*, e0108421. [CrossRef]
12. Pirt, S. The maintenance energy of bacteria in growing cultures. *Proc. R. Soc. Lond. Ser. B* **1965**, *163*, 224–231.
13. Pirt, S.J. *Principles of Microbe and Cell Cultivation*; Blackwell Science Ltd.: Hoboken, NJ, USA, 1985.
14. Rapaport, A.; Nidelet, T.; El Aida, S.; Harmand, J. About biomass overyielding of mixed cultures in batch processes. *Math. Biosci.* **2020**, *322*, 108322. [CrossRef]
15. Heijnen, J.; Roels, J. A macroscopic model describing yield and maintenance relationships in aerobic fermentation processes. *Biotechnol. Bioeng.* **1981**, *23*, 739–763. [CrossRef]
16. Beeftink, H.; van der Heijden, R.; Heijnen, J. Maintenance requirements: Energy supply from simultaneous endogenous respiration and substrate consumption. *FEMS Microbiol. Lett.* **1990**, *73*, 203–209. [CrossRef]
17. Wang, G.; Post, W.M. A theoretical reassessment of microbial maintenance and implications for microbial ecology modeling. *FEMS Microbiol. Ecol.* **2012**, *81*, 610–617. [CrossRef]
18. Leão, C.; van Uden, N. Effects of ethanol and other alkanols on the glucose transport system of Saccharomyces cerevisiae. *Biotechnol. Bioeng.* **1982**, *24*, 2601–2604. [CrossRef]
19. Viegas, C.; Sa-Correia, I.; Novais, J. Synergistic inhibition of the growth of *Saccharomyces bayanus* by ethanol and octanoic or decanoic acids. *Biotechnol. Lett.* **1985**, *7*, 611–614. [CrossRef]
20. Caro, I.; Pérez, L.; Cantero, D. Development of a kinetic model for the alcoholic fermentation of must. *Biotechnol. Bioeng.* **1991**, *38*, 742–748. [CrossRef]
21. Salmon, J.M.; Vincent, O.; Mauricio, J.C.; Bely, M.; Barre, P. Sugar transport inhibition and apparent loss of activity in Saccharomyces cerevisiae as a major limiting factor of enological fermentations. *Am. J. Enol. Vitic.* **1993**, *44*, 56–64.
22. Cramer, A.C.; Vlassides, S.; Block, D.E. Kinetic model for nitrogen-limited wine fermentations. *Biotechnol. Bioeng.* **2002**, *77*, 49–60. [CrossRef] [PubMed]
23. El Haloui, N.; Picque, D.; Corrieu, G. Alcoholic fermentation in winemaking: On-line measurement of density and carbon dioxide evolution. *J. Food Eng.* **1988**, *8*, 17–30. [CrossRef]
24. VanBriesen, J. Evaluation of methods to predict bacterial yield using thermodynamics. *Biodegradation* **2002**, *13*, 171–190. [CrossRef]
25. Mouret, J.; Farines, V.; Sablayrolles, J.; Trelea, I. Prediction of the production kinetics of the main fermentative aromas in winemaking fermentations. *Biochem. Eng. J.* **2015**, *103*, 211–218. [CrossRef]
26. Conacher, C.; Luyt, N.; Naidoo, R.; Rossouw, D.; Setati, M.; Bauer, F. The ecology of wine fermentation: A model for the study of complex microbial ecosystems. *Appl. Microbiol. Biotechnol.* **2021**, *105*, 3027–3043. [CrossRef]

foods

MDPI

Article

Parameter Estimation of Dynamic Beer Fermentation Models

Jesús Miguel Zamudio Lara [1,2], Laurent Dewasme [1], Héctor Hernández Escoto [2] and Alain Vande Wouwer [1,*]

[1] Systèmes, Estimation, Commande et Optimisation, Université de Mons, 7000 Mons, Belgium
[2] Departamento de Ingeniería Química, Universidad de Guanajuato, Guanajuato 36050, Mexico
* Correspondence: alain.vandewouwer@umons.ac.be

Abstract: In this study, two dynamic models of beer fermentation are proposed, and their parameters are estimated using experimental data collected during several batch experiments initiated with different sugar concentrations. Biomass, sugar, ethanol, and vicinal diketone concentrations are measured off-line with an analytical system while two on-line immersed probes deliver temperature, ethanol concentration, and carbon dioxide exhaust rate measurements. Before proceeding to the estimation of the unknown model parameters, a structural identifiability analysis is carried out to investigate the measurement configuration and the kinetic model structure. The model predictive capability is investigated in cross-validation, in view of opening up new perspectives for monitoring and control purposes. For instance, the dynamic model could be used as a predictor in receding-horizon observers and controllers.

Keywords: parameter estimation; mathematical modeling; beer fermentation; food industry

Citation: Zamudio Lara, J.M.; Dewasme, L.; Hernández Escoto, H.; Vande Wouwer, A. Parameter Estimation of Dynamic Beer Fermentation Models. *Foods* **2022**, *11*, 3602. https://doi.org/10.3390/foods11223602

Academic Editors: Jose A. Egea, Carlos Vilas and Míriam R. García

Received: 3 October 2022
Accepted: 7 November 2022
Published: 11 November 2022

1. Introduction

Beer is the most consumed alcoholic beverage worldwide and is produced by the fermentation of sugars in the wort by yeasts [1]. The production of beer in 2019 was 1912 million hectoliters (hl), while in 2020, the production reduced to 1820 million hl due to the COVID-19 pandemic [2]. Beer production is a complex biochemical process in which the main ingredients are water, malt (sugar source), yeast, and hops [3]; however, other products can be added, such as fruits, chocolate, and coffee grains, among others.

The fermentation stage is crucial to guarantee good quality beer since it is when all the nutrients, flavor, and odor components are produced, in addition to ethanol. At this stage, yeast is introduced in the wort (broth that is rich in sugars) from the boiling stage at the desired temperature. The main chemical reaction is the conversion of these sugars into ethanol and carbon dioxide, along with biomass growth and heat generation. At the same time, several secondary reactions occur, generating several components at lower concentrations that contribute to the flavor and aroma characteristics.

To enhance the fermentation, several factors, such as yeast pitching rate, dissolved oxygen, batch pressure, and temperature, must be taken care of by the brewers [4]. Among these factors, temperature is important as it helps accelerate the fermentation but needs to remain within controlled bonds to avoid yeast death (above 30 °C), the production of undesirable byproducts, and the growth of bacteria, damaging the final product. Therefore, rigorous control of the temperature inside the fermenter must be exercised to ensure product quality and alleviate variations between batches.

In the brewing industry, time-varying temperature profiles are established along the fermentation process in order to alleviate the above-mentioned potential issues [5]. Looking for the appropriate temperature profile is, however, not an easy task, and experimental determination can be time-consuming. Model-based optimization is, therefore, an appealing alternative. Dynamic models can be useful not only to optimize the operating conditions, but also to design state estimators reconstructing online non-measured variables or designing controllers to ensure close setpoint tracking [6].

To this end, Gee and Ramirez [7] proposed a detailed model of beer fermentation describing biomass growth and the production of flavor compounds through macroscopic reactions inferred from biological pathways. This model segregates the sugars into glucose, maltose, and maltotriose. The derivation of the corresponding sugar uptake kinetics is, therefore, the center of interest, and the related parameters are assumed to be temperature-dependent. However, this model is unlikely to be applicable for control purposes since it involves variables that require specific monitoring equipment beyond the standards of most fermenters. Andrés-Toro et al. [8], conversely, proposed to segregate the yeast (biomass) into three types: lag, active, and death, while the sugars are considered as a whole. Sugar monitoring, therefore, appears simpler, and the model also takes account of industrial operational characteristics as well as the undesired beer flavor caused by ethanol and byproduct (diacetyl and ethyl acetate) formation. The main drawback of this model is the difficulty in obtaining accurate biomass data. Trelea et al. [9] developed what can be considered as, so far, the most practical model, where only three states are considered: the dissolved carbon dioxide concentration used as an image of the growing biomass, the ethanol concentration, and the sugar concentration. The main advantage of this latter model lies in its practical control-oriented description of the fermentation process, considering variables that can be easily measured and tracked.

The objective of this work is to revisit these classical models, propose a few adaptations, and develop a thorough study of the parameter estimation problem based on a popular fermentation device, e.g., a 30-L stainless-steel Grainfather® fermenter. Two alternative mathematical models are considered, one based on yeast (biomass) and the other on carbon dioxide. The difference between these models is discussed in terms of biological interpretation, bioreactor instrumentation, and data collection (i.e., parameter estimation, model validation, and process control perspectives). As a result, models with good predictive capability are proposed together with their experimental validation.

This paper is organized as follows. The next section describes the experimental setup, while Section 3 presents a review of dynamic models of beer fermentation, together with possible model adaptations required to represent the considered case study. Section 4 develops a structural identifiability analysis based on the software tool Strike Goldd [10]. Section 5 introduces a parameter identification procedure, including parametric sensitivity analysis and model validation. The last section draws the main conclusions of this work and discusses the monitoring and control perspectives.

2. Beer Fermentation Experimental Set-Up

The pilot plant consists of a stainless-steel conical fermenter (30 L, Grainfather®), which has a built-in sensor to measure the temperature of the liquid content. This sensor is paired to a control system connected to a glycol chiller (Grainfather®) to keep the temperature regulated. Ethanol and carbon dioxide concentrations are measured online, respectively, with a tilt® hydrometer and a Plaato® airlock.

The hydrometer is introduced in the wort and keeps floating in a tilted position, measuring the specific gravity which also allows, based on some predetermined correlations, for assessing the percentage of alcohol. The sensor also has an integrated temperature sensor. The airlock consists of four components: a lid, a bubbler, a Tritan, and a smart part (containing the temperature and infrared sensors). This device measures the evenly-sized bubbles of carbon dioxide released by the wort and converts them into liters of CO_2. The data are stored and displayed in the Brewblox® interface.

Besides the two online probes, a CDR BeerLab® analyzer is used to obtain offline measurements of sugar, ethanol, and vicinal diketone (VDK) concentrations. The scheme of the full experimental setup is displayed in Figure 1.

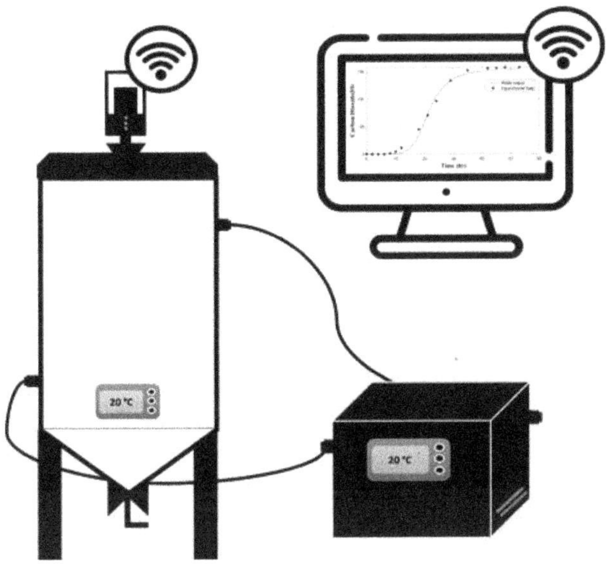

Figure 1. Stainless-steel fermenter pilot plant monitoring set-up.

In this study, ale beer fermentation was considered and carried out at a temperature ranging from 17 to 26 °C. To obtain the wort, the malt is crushed using a mill. Indeed, the grain must only be broken, but not grounded. The next step is mashing, where sugars are obtained from starch. The crushed grain is added to a boiler tank (35 L, Grainfather®) containing 19 L of water at 48 °C. The mashing step consists of four stages at different temperatures and times described in Table 1. Once mashing is finished, the grain is rinsed with water at 75 °C until 24 L of wort is obtained. Usually, the quantity of added water is 8 L. Then, the wort is boiled to sterilize the liquid. The latter step is carried out for 80 min at 100 °C. Hop is added at 40 min and 65 min. Eventually, the wort is cooled down as fast as possible to the desired fermentation temperature with the help of a counter-flow wort cooler. The cold wort is transferred in the fermentation tank, filled up to 17 L.

A set of four isothermal batch fermentations without agitation are carried out using different operating conditions described in Table 2. Each experiment is carried out once, but two replicates are taken and analyzed for each sample. The total sampling volume represents less than 10% of the initial wort volume (17 L), a condition to neglect the volume changes. Samples are taken every 2 to 3 h during the first 36 h. After this period, the process enters a stationary phase and the sampling time is therefore adapted at irregular, longer, time intervals. To analyze the samples with the CDR BeerLab®, it is necessary to achieve preprocessing, including degasification and centrifugation to eliminate everything that could interfere during the measurement.

Table 1. Operating conditions of the scheduled mash steps: temperatures and times.

Step	Temperature (°C)	Time (min)
1	48	10
2	62	30
3	71	20
4	78	5

Table 2. Operating conditions of each fermentation batch: duration, temperature, and initial sugar and biomass concentrations.

Experiment	Time (h)	Temperature (°C)	S_0 (g/L)	X_0 (g/L)
1	72	19	88	7.05×10^{-1}
2	54	19	72	5.29×10^{-1}
3	96	21	75	7.05×10^{-1}
4	72	28	81	7.05×10^{-1}

3. Mathematical Models of Beer Fermentation

Beer fermentation has been studied extensively, providing various mathematical models, and this section proposes a brief description of three of them, considered as milestones in the research field and used as a reference in the upcoming model development. These mathematical models can provide precious support to design the fermentation operating conditions.

3.1. The Model of Gee and Ramirez

This model published in 1994 includes the main components of fermentation, e.g., sugars, biomass, and ethanol, as well as amino acids, fusel alcohols, VDKs, and acetaldehydes, which impact the flavor, often in an undesirable way [7]. Esters also have an important role in the aroma and may add some pleasant character in moderate ranges, but undesired hard fruity tastes at higher levels.

The sugar uptake model reads as follows:

$$\frac{dG}{dt} = -\mu_G X, \tag{1a}$$
$$\frac{dM}{dt} = -\mu_M X, \tag{1b}$$
$$\frac{dN}{dt} = -\mu_N X, \tag{1c}$$

where G, M, and N, respectively, stand for glucose, maltose, and maltotriose. The specific growth rates are built upon classical kinetic activation (Monod law) and inhibition factors and are given by:

$$\mu_G = \frac{\mu_G G}{K_G + G}, \tag{2a}$$
$$\mu_M = \frac{\mu_M M}{K_M + M} \frac{K'_G}{K'_G + G}, \tag{2b}$$
$$\mu_N = \frac{\mu_N N}{K_N + N} \frac{K'_G}{K'_G + G} \frac{K'_M}{K'_M + M}, \tag{2c}$$

where μ_i are maximum rate constants ($i = G, M, N$), while K_i and K'_j ($j = G, M$) are, respectively, the half-saturation and inhibition constants, all assumed to depend on the temperature following an Arrhenius law of the form:

$$r = Aexp\left(\frac{B}{RT}\right), \tag{3}$$

where A and B are, respectively, the Arrhenius frequency factor and the activation energy. R is the ideal gas constant.

Biomass production is represented by the mass-balance ODE:

$$\frac{dX}{dt} = \mu_X X, \tag{4}$$

where

$$\mu_x = Y_{XG}\mu_1 + Y_{XM}\mu_2 + Y_{XN}\mu_3 \tag{5}$$

is the biomass growth rate, a function of the several sugar intake rates. Y_{Xi} $(i = G, M, N)$ are the yield coefficients of biomass with respect to the specific sugars.

The ethanol concentration is assumed to evolve proportionally to the variations of the sugar concentrations, resulting in an algebraic equation of the form:

$$E = E_0 + Y_{EG}(G_0 - G) + Y_{EM}(M_0 - M) + Y_{EN}(N_0 - N), \tag{6}$$

where Y_{Ei} $(i = G, M, N)$ are the yield coefficients of ethanol with respect to the consumed sugars.

Three main amino acids are considered, which are responsible for the formation of flavor compounds in the beer like the fusel alcohols. The amino acids are described by the following differential equations:

$$\frac{d\xi}{dt} = -Y_{\xi}\frac{\xi}{K_{\xi} + \xi}\frac{dX}{dt} = -Y_{\xi}\mu_X\frac{\xi}{K_{\xi} + \xi}X, \tag{7}$$

where Y_{ξ} and K_{ξ} are, respectively, the yield coefficients and inhibition constants with species ξ = leucine (L), isoleucine (I), and valine (V).

The impact of fusel alcohols is a plastic, solvent-like flavor. Moreover, some experiments achieved in [11] have also linked higher alcohol levels with physiological effects associated with hangovers. The four fusel alcohols represented in the model are isobutyl alcohol (IB), isoamyl alcohol (IA), 2-methyl-1-butanol (MB), and n-propanol (P).

$$\frac{dIB}{dt} = Y_{IB}\mu_V X, \tag{8a}$$

$$\frac{dIA}{dt} = Y_{IA}\mu_L X, \tag{8b}$$

$$\frac{dMB}{dt} = Y_{MB}\mu_I X, \tag{8c}$$

$$\frac{dP}{dt} = Y_P(\mu_V + \mu_I)X, \tag{8d}$$

where Y_{ζ} $(\zeta = IB, IA, MB, P)$ are yield coefficients, and μ_{ξ} are specific rates expressed as $\mu_{\xi} = -\frac{1}{X}\frac{d\xi}{dt} = Y_{\xi}\mu_X\frac{\xi}{K_{\xi}+\xi}$ $(\xi = L, I, V)$.

Esters contribute mainly to the aroma of the beer due to their high volatility. In moderate concentrations, they can confer a pleasant character to the beer. However, once in excess, the aroma becomes overly fruity, which is undesired by most consumers. Principal esters are ethyl acetate (EA), ethyl caproate (EC), and isoamyl acetate (IAc).

$$\frac{dEA}{dt} = Y_{EA}(\mu_G + \mu_M + \mu_N)X, \tag{9a}$$

$$\frac{dEC}{dt} = Y_{EC}\mu_X X, \tag{9b}$$

$$\frac{dIAc}{dt} = Y_{IAc}\mu_{IAc}X, \tag{9c}$$

where Y_{γ} are the yield coefficients $(\gamma = EA, EC, IAc)$ and μ_{IAC} is the maximum isoamyl acetate formation rate.

The common practice recommends completely removing vicinal diketones (VDKs) since they add some undesired buttery flavor notes. VDK production is assumed to be proportional to the growth rate, while their possible re-assimilation by yeast to form other by-products is proportional to their concentration. It must be noticed that acetaldehyde

(AAl) is another compound showing similar behavior to the VDKs. The mass-balance equations read:

$$\frac{dVDK}{dt} = Y_{VDK}\mu_X X - r_{VDK}VDKX, \tag{10a}$$

$$\frac{dAAl}{dt} = Y_{AAl}(\mu_G + \mu_M + \mu_N)X - r_{AAl}AAlX, \tag{10b}$$

where Y_ω defines the yield coefficients and r_ω defines first-order rate constants ($\omega = VDK, AAl$). μ_X is the biomass growth rate and μ_G, μ_M, μ_N are the sugar consumption rates.

This model therefore proposes a detailed description of the flavor and aroma of the beer but also presents the drawback of being difficult to apply in a realistic context since it requires numerous and often expensive advanced on-line monitoring devices. Indeed, in practice, biomass concentration is only measured at the beginning of a batch without further monitoring. Moreover, most of the parameters are temperature-dependent, either imposing rigorous operating conditions (i.e., only one constant temperature level) or parameter estimation of the temperature-dependent functions (which requires data collection at various temperatures).

3.2. Model of De Andrés-Toro et al.

This model, published in 1998, is more concise than the previous one as it only considers five state variables: sugars, biomass, ethanol, ethyl acetate, and diacetyl (i.e., vicinal diketones) [8]. Ethyl acetate and diacetyl are assumed to be the most influencing compounds regarding aroma and flavor. In the following, the model dynamics are described state-by-state. Biomass is segregated into three types: lagged, active, and dead. It is indeed assumed that part of the biomass goes through several states during the process, first in a lag phase when the fermentation starts, then in an active (growing) state, and eventually in an inactive (dead) state.

Lagged biomass

$$\frac{dX_L}{dt} = -\mu_L X_L, \tag{11a}$$

Active biomass

$$\frac{dX_A}{dt} = \mu_X X_A + \mu_L X_L - \mu_{DT} X_A, \tag{11b}$$

Dead biomass

$$\frac{dX_D}{dt} = \mu_{DT} X_A - \mu_{SD} X_D, \tag{11c}$$

where the lagged biomass becomes active at the specific rate μ_L, the active biomass grows at the specific rate μ_X and dies at the rate μ_{DT}, while the dead biomass settles in the bottom of the reactor at the rate μ_{SD}. μ_X and μ_{SD} are further defined as

$$\mu_X = \frac{\mu_{X_0} S}{0.5S_0 + E}, \tag{12}$$

$$\mu_{SD} = \frac{\mu_{SD_0} 0.5S_0}{0.5S_0 + E}. \tag{13}$$

μ_X represents an activation by the substrate S and an inhibition by ethanol E. The inhibition constant is assumed to be inversely proportional to half the initial substrate concentration S_0 (indeed two units of S give one unit of E in the stoichiometry of the reaction). μ_{SD} describes an inhibition by ethanol, which is directly related to CO_2, which is not a variable in this model, but whose bubbles impair the settling phenomenon. The inhibition constant is again related to half the initial substrate concentration (i.e., maximal quantity of ethanol that can be produced).

Sugar consumption follows a Monod law according to:

$$\frac{dS}{dt} = -\mu_S X_A, \tag{14}$$

with

$$\mu_S = \frac{\mu_{S_0} S}{K_S + S}, \tag{15}$$

in which μ_{S_0} is the maximum specific consumption rate and K_S is the half-saturation constant.

Ethanol production is described by

$$\frac{dE}{dt} = \mu_E X_A, \tag{16}$$

where the specific rate includes a Monod factor with respect to the substrate S and an inhibition factor related to the ethanol concentration. This factor vanishes when the ethanol reaches the maximum value $0.5S_0$:

$$\mu_E = \frac{\mu_{E_0} S}{k_E + S}\left(1 - \frac{E}{0.5S_0}\right). \tag{17}$$

Ethyl acetate is produced as a byproduct of active biomass growth:

$$\frac{dEA}{dt} = Y_{EA} \mu_X X_A, \tag{18}$$

where Y_{EA} is the yield coefficient.

Diacetyl is a component belonging to the vicinal diketones (VDKs), which is produced as the biomass grows by consuming the sugars. Afterward, diacetyl is reduced into acetoin with a reduction rate r_{VDK} activated in the presence of ethanol:

$$\frac{dVDK}{dt} = k_{VDK} S X_A - r_{VDK} VDKE. \tag{19}$$

All the parameters are assumed to be affected by temperature according to the Arrhenius law (Equation (3)).

The biomass segregation model provides an accurate and consistent description of the process but, as a drawback, requires the corresponding monitoring equipment. In the study of [8], total biomass was measured online by absorbance change detection in a photocell, while the biomass state classification was made based on pre-established assumptions.

3.3. Model of Trelea et al.

The originality of the model of Trelea et al. [9], with respect to the previous ones, is that it considers the carbon dioxide dynamics instead of the biomass dynamics. Carbon dioxide sensors are indeed easily implemented and calibrated on-line, reliable, and significantly cheaper than biomass measurement devices.

The evolution of carbon dioxide is related to yeast growth, sugar consumption, and ethanol production. CO_2 dynamics are assumed to be driven by a Monod law describing sugar activation and saturation effects, an inhibition factor taking account of the decreasing cell respiratory capacity following ethanol accumulation. The influence of the initial biomass concentration on the initial CO_2 production rate is also taken into account:

$$\frac{dCO_2}{dt} = \mu_{max} \frac{S}{K_S + S} \frac{1}{1 + K_I E^2} (CO_2 + C_0 X_0), \tag{20}$$

where K_S is the half saturation coefficient, K_I is the inhibition constant, and C_0 is a conversion factor.

Algebraic equations describe the evolution of the sugar and ethanol concentrations in relation to CO_2 as suggested in [12]:

$$S = S_0 - Y_S CO_2, \tag{21}$$

$$E = Y_E CO_2, \tag{22}$$

where Y_S and Y_E are the corresponding yield coefficients. The main advantage of this model lies in its practical and control-oriented description of the fermentation process, considering a variable that can be easily measured, such as carbon dioxide. The maximum specific growth rate, μ_{max}, is however, assumed to depend on several operating parameters, such as pressure, temperature, and initial yeast concentration, as follows:

$$\mu_{max} = a_T T_n + a_P P_n + a_X X_{0,n} + a_{TP} T_n P_n + a_{TX} P_n X_{0,n} + a_0, \tag{23}$$

where a_i are parameters to be estimated.

3.4. Proposed Mathematical Models

In this study, two mathematical models of beer fermentation are proposed, one based on biomass dynamics and the other on CO_2 dynamics. These models take inspiration from the mathematical developments of the previous sections and attempt to describe experiments performed with the beer fermenter described in Section 2.

Figure 2 shows some data collected in a batch experiment at 19 °C. It is apparent that sugar was not completely consumed at the end of the fermentation, with a residual concentration of about 12 g/L (this behavior was confirmed in repeated experiments in the same and different conditions). Two possible causes of this sluggish fermentation in the end of batch were explored in additional experiments, including biomass settling and water quality. However, gentle agitation and tests with different water sources did not influence the initial observation. Other causes such as the depletion of some components required for yeast proliferation and maintenance (such as nitrogen, sterols, fatty acids) could not be assessed. The models were, therefore, adapted to describe the experimental observations. As the published models only consider the total consumption of sugars, structural modifications were made to cope with this type of behaviour. The changes mostly impact the definition of the specific growth rate, as explained in the following sections.

3.4.1. Dynamic Model Based on Biomass

An overall alcoholic fermentation reaction can be written as follows:

$$k_S \mathrm{S} \xrightarrow{r_1} k_E \mathrm{E} + k_V \mathrm{VDK} + k_{CO_2} CO_2 + \mathrm{X} \tag{24}$$

where sugars S are consumed and converted by yeasts X into ethanol E, vicinal diketones VDK, and carbon dioxide CO_2. k_S, k_E, k_V, and k_{CO_2} represent the yield coefficients of sugar, ethanol, vicinal diketones, and carbon dioxide, respectively.

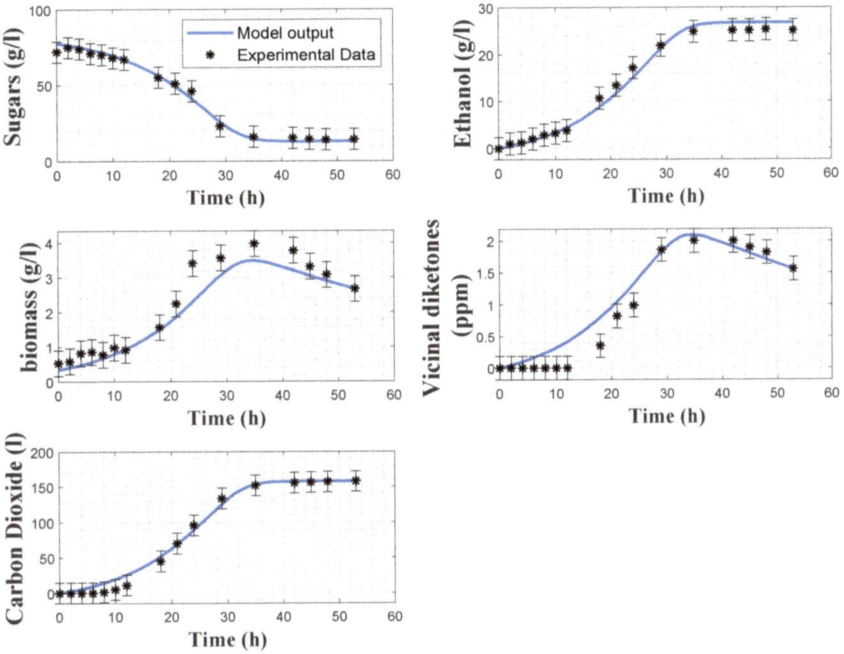

Figure 2. Direct validation of the biomass model with the data from experiment 2. Stars: experimental data. Error bars: 95% confidence intervals. Continuous blue line: model prediction.

Vicinal diketones are later reduced by yeasts, producing 2-3-butanodiol P in the following second reaction:

$$\mathrm{VDK} \xrightarrow{r_2} \mathrm{P} \tag{25}$$

From Equations (24) and (25), a set of mass-balance ODEs can be derived:

$$\frac{dX}{dt} = \mu_X X - \delta_X X, \tag{26a}$$

$$\frac{dS}{dt} = -k_S \mu_X X, \tag{26b}$$

$$\frac{dE}{dt} = k_E \mu_X X \tag{26c}$$

$$\frac{dCO_2}{dt} = -k_{CO_2} \mu_X X, \tag{26d}$$

$$\frac{dVDK}{dt} = k_V \mu_X X - r_{VDK} VDK. \tag{26e}$$

The specific growth rate is defined as:

$$\mu_X = \mu_{max}\left(1 - \frac{S_{min}}{S}\right), \qquad for \quad S \geq S_{min}, \tag{27a}$$

$$= 0, \qquad for \quad S \leq S_{min}. \tag{27b}$$

The specific growth rate usually represented by a Monod law is replaced by a Droop-like factor [13], commonly used to describe microalgae growth. This kinetic structure expresses that a minimum level of sugar S_{min} is necessary to trigger growth. Above this threshold, the Monod factor has an activation/saturation effect similar to a Monod factor.

Furthermore, a decrease in biomass is observed due to settling and is represented by the coefficient δ_X in Equation (26a).

In classical batch fermentation, the temperature is the only variable that can be manipulated during the batch to control the evolution of the fermentation. Based on the previously discussed modeling studies [7–9] and experimental observation, a temperature-dependency of the maximum specific growth rate μ_{max} is expected that can be described following several structural laws [14]. Moreover, the reduction speed of the VDKs, r_{VDK}, is also affected by temperature. In Table A1, the definitions and units of each parameter are listed.

3.4.2. Dynamic Model Based on Carbon Dioxide

The yeast concentration is difficult to monitor during the batch and, most of the time, this variable is either indirectly measured by the turbidity of the wort, or simply by sampling and cell counting using a cytometer or a dry-weight method. Hence, there is a strong motivation to monitor other, more accessible, process variables and develop dynamic models describing their evolution. Carbon dioxide may be such an indicator since it is a product of the biochemical reactions related to sugar oxidation, which provides the necessary energy to the yeast cells to grow and leads to ethanol production if sugar is in excess, activating the overflow metabolism [15,16]. In this study, the model of [9] was adapted in the following way:

$$\frac{dCO_2}{dt} = \mu_X CO_2, \tag{28a}$$

$$S = S_0 - k_S CO_2, \tag{28b}$$

$$E = E_0 + k_E CO_2, \tag{28c}$$

$$\frac{dVDK}{dt} = k_V \mu_X CO_2 - r_{VDK} VDK, \tag{28d}$$

where the specific growth rate is given by:

$$\mu_x = \mu_{max} \frac{S}{K_S + S} \left(1 - \frac{CO_2}{Cp_{max} S_0}\right). \tag{29}$$

The kinetic description contains a Monod law for the sugar activation and saturation effects, and a logistic factor to mimic the observed sigmoidal evolution of carbon dioxide. The data sets also reveal that the maximum production of carbon dioxide is correlated with the initial sugar concentration S_0, which therefore defines the maximum CO_2 level (i.e., the carrying capacity of the logistic model). The initial biomass concentration, which is assumed to be known (measured) and directly correlated to the CO_2 dynamics in [9], was not used in the current study since the initial condition of CO_2 was available. The rates μ_{max} and r_{VDK} are assumed to depend on temperature and Table A2 lists the definitions and units of some parameters.

4. Structural Identifiability and Observability of the Models

Identifiability globally refers to the possibility of identifying the model parameters from the available data. A model is structurally identifiable if all the parameters can be uniquely determined from ideal measurements of its outputs, i.e., collected in continuous time without errors or noise, and the knowledge of the dynamic equations [17]. If this property is not met, any further effort to estimate the non-identifiable parameters will be vain. However, the identifiability analysis is often omitted due to the assumed complexity of the mathematical developments required to achieve the analysis. Recently, several methodologies and toolboxes have been developed to significantly ease the task, as reviewed in [18]. Some of these software tools are DAISY [19], GENSSI [20], STRIKE-GOLDD [10], and SIAN [21].

On the other hand, practical identifiability deals with the possibility of assessing all or some of the model parameters under realistic conditions, e.g., sampled data and measurement noise. The Fisher Information Matrix (FIM) is useful to assess practical identifiability through a rank test condition. An ill-conditioned FIM can indicate poor practical parameter identifiability even if structural identifiability is met.

In this work, STRIKE-GOLDD (STRuctural Identifiability taken as Extended-Generalized Observability with Lie Derivatives and Decomposition) was used to investigate the structural identifiability of the proposed beer fermentation models. This software tool has been developed in MATLAB® and addresses identifiability based on the concept of observability. To this end, the model is extended by considering its model parameters as state variables with zero dynamics. The results obtained for both models indicate that structural identifiability is ensured only when all the state variables are measured.

Another property of interest is observability, which is a prerequisite to the design of a state observer to reconstruct nonmeasured state variables. The results of the analysis are provided in Tables 3 and 4 for the two dynamic models. For the model based on biomass, the analysis reveals that the set of three measurements $[CO_2, E, VDK]$ is necessary to guarantee observability. The set $[CO_2, E, S]$ shows partial observability as VDK cannot be reconstructed but biomass X could. The carbon dioxide model requires the measurement of VDK together with another variable (CO_2 or E or S) to fulfill the observability condition. An observer could, therefore, be designed to estimate the sugar concentration online, which is the most expensive measurement using an online or at-line hardware probe.

Table 3. Observability analysis of the biomass model for several measurement configurations.

Measured Outputs	Observable
$[CO_2, S]$	No
$[CO_2, E]$	No
[S, E]	No
$[CO_2, S, X]$	No
$[CO_2, E, VDK]$	Yes
$[CO_2, E, S]$	VDK: No X: Yes
$[CO_2, S, X, VDK]$	No

Table 4. Observability analysis of the carbon dioxide model for several measurement configurations.

Measured Outputs	Observable
$[CO_2, S]$	No
$[CO_2, E]$	No
[S, E]	No
[S, VDK]	Yes
[E, VDK]	Yes
$[CO_2, VDK]$	Yes

5. Parameter Identification Problem

Parameter identification is achieved using classical nonlinear parameter estimation techniques [22]. The procedure considers a Weighted Least-Square (WLS) criterion, i.e., a weighted sum of squared differences between model predictions and experimental measurements:

$$J(\theta) = \sum_{i=1}^{M} \Big(y(t_i) - y_{model}(t_i, \theta))^T W^{-1} (y(t_i) - y_{model}(t_i, \theta)) \Big) \tag{30}$$

where J is the value of the cost function, $y(t_i)$ is the vector of N measured variables at the measurement instant t_i ($i = 1, \ldots, M$), $y_{model}(t_i, \theta)$ is the model prediction that depends on the set of P parameters θ to be identified, and W is a normalization matrix where the diagonal elements are chosen as the squares of the maximum measurements values of each component concentration. This choice allows normalization of the prediction errors, and is particularly well-suited to a relative error model where it is assumed that the error is proportional to the maximum values of the observed variables:

$$W = \begin{bmatrix} \max\limits_{t_i}(y_1^2) & 0 & \dots & 0 \\ 0 & \max\limits_{t_i}(y_2)^2 & \dots & 0 \\ \vdots & \vdots & \ddots & \vdots \\ 0 & 0 & \dots & \max\limits_{t_i}(y_N)^2 \end{bmatrix} \quad (31)$$

The estimated parameter set is obtained by minimizing a cost function $J(\theta)$ as follows:

$$\hat{\theta} = \arg\min_{\theta} J(\theta) \quad (32)$$

To achieve the minimization of (32), a two-step procedure is implemented: (a) a multi-start strategy defines random sets of initial parameter values to cover as much as possible of the search field. The minimization of (32) is first performed using the Matlab® optimizer fminsearch (Nelder-Mead algorithm); (b) the Matlab lsqnonlin optimizer is subsequently used from the identified global minimum (i.e., the smallest local minimum identified in the search space by fminsearch) to refine the minimization and compute the Jacobian matrix containing the model parameter sensitivities, denoted y_θ. These sensitivities can be exploited to compute the Fisher Information Matrix (FIM) defined as:

$$\text{FIM} = \sum_{i=1}^{M} y_\theta^T(t_i)\hat{\Omega}^{-1} y_\theta(t_i) \quad (33)$$

where $\hat{\Omega} = \hat{\epsilon}^2 W$ is the a posteriori covariance matrix of the measurement errors, which can be evaluated using the weighting matrix W (Equation (31)) and an a posteriori estimator of the relative measurement error:

$$\hat{\epsilon}^2 = \frac{J^*}{MN - P} \quad (34)$$

where J^* is the value of the cost function at the optimum, MN represents the total number of data, and P is the number of estimated parameters θ. An estimate of the parameter estimation error covariance matrix can then be inferred from the Cramer–Rao bound as follows:

$$\hat{\Sigma} = \text{FIM}^{-1} \quad (35)$$

From the diagonal of the covariance matrix $\hat{\Sigma}$, the standard deviations for each parameter can be extracted and the corresponding coefficients of variations can be calculated as:

$$\text{CV} = \frac{\sigma}{\hat{\theta}_i} \quad (36)$$

To achieve the estimation of the parameters of the beer fermentation models, a total of 4 batch experiments are considered as shown in Table 2. Out of these 4 experiments, 3 are used for parameter estimation and model direct validation (experiments 2 to 4), while experiment 1 is used for cross-validation. An important point of the current work is the use of all the samples of experiments 2, 3, 4 to achieve the identification, including two temperature-varying parameters, i.e., the specific growth rate μ_{max} and the VDK reduction rate r_{VDK}. Previous studies have indeed demonstrated that the other parameters do not change significantly with temperature. In addition to the stoichiometric and kinetic parameters, the initial conditions are also considered unknown (and are therefore estimated) since possibly corrupted by measurement noise.

5.1. Biomass Model

This model counts 8 parameters (k_S, k_E, S_{min}, g_x, k_V, k_{CO_2}, μ_{max}, r_a) to be estimated. The dependence on temperature of the parameters μ_{max} and r_{VDK}, is formulated as follows:

$$\mu_{max} = a\ln(T) + b; \qquad r_{VDK} = c\ln(T) + d; \tag{37}$$

introducing the additional parameters a, b, c, d. Several nonlinear model structures have been considered to correlate the parameters with the temperature. It turned out that the selected logarithmic structure provides the best results.

In practice, the identification proceeds in three steps: (a) a first parameter estimation without explicit temperature dependence (i.e., μ_{max} and r_{VDK} are considered constant), (b) an estimation of the four parameters linked to the temperature dependence (the others being fixed at their previously estimated values), and (c) a global identification of all the parameters starting from the previous estimates.

Good practice recommends partitioning the data using a ratio of approximately 75/25 for parameter estimation (and subsequent direct validation), and cross-validation, respectively. Accordingly, three experimental data sets are used in direct validation and the remaining one in cross-validation. Since the parameter estimation procedure aims at capturing information on the process in a wide range of operations, it is legitimate to include experiments with different initial sugar and biomass concentrations and temperature levels. Particularly, it is important to collect informative data regarding the evolution of μ_{max} and r_{VDK} with respect to temperature in (37). Among the several possible data partitions, one possible combination appears to be: experiments 2, 3, 4 for parameter estimation (and direct validation) and experiment 1 for cross-validation. Indeed, experiments 1 and 2 are carried out at the same temperature, but experiment 2 also includes different sugar and biomass initial conditions. Table 5 reports the values of the estimated parameters and their coefficients of variations.

Table 5. Parameter estimate values and coefficients of variation (CV) for the biomass model.

Parameter	Units	Value	CV (%)
k_S	gS/gX	15.3	7
k_E	gE/gX	6.31	6
S_{min}	g/L	13.1	11
g_x	h^{-1}	1.67×10^{-2}	7
k_V	gVDK/gX	6.51×10^{-1}	10
k_{CO_2}	L	37.1	5
a	h^{-1}	4.00×10^{-1}	3
b	h^{-1}	−1.1	5
c	h^{-1}	2.50×10^{-2}	10
d	h^{-1}	-5.60×10^{-2}	9

Figures 2 and 3 show some direct validation results, i.e., the fitting of the model to the experimental data collected in experiments 2 and 4 together with the a posteriori error bars on the experimental data. The model reproduces quite well the dynamics of the several variables, even if the biomass predictions sometimes deviate from the confidence intervals of the data, and some deviations in the VDK production are also observed in the early hours. The coefficients of variations confirm the good estimation results, as the maximum relative CV is 11% for the minimum substrate quota S_{min}.

In order to assess the model predictive capacity, cross-validation is achieved using the dataset from experiment 1. In this case, only the initial conditions are estimated while the parameters are kept fixed. As shown in Figure 4, the model predicts satisfactorily the experimental data. The biomass data again has some uncertainty, which can probably be linked to several factors such as cell counting errors, biomass mixing (to counteract biomass settling and collect representative samples) and nitrogen limitation [23].

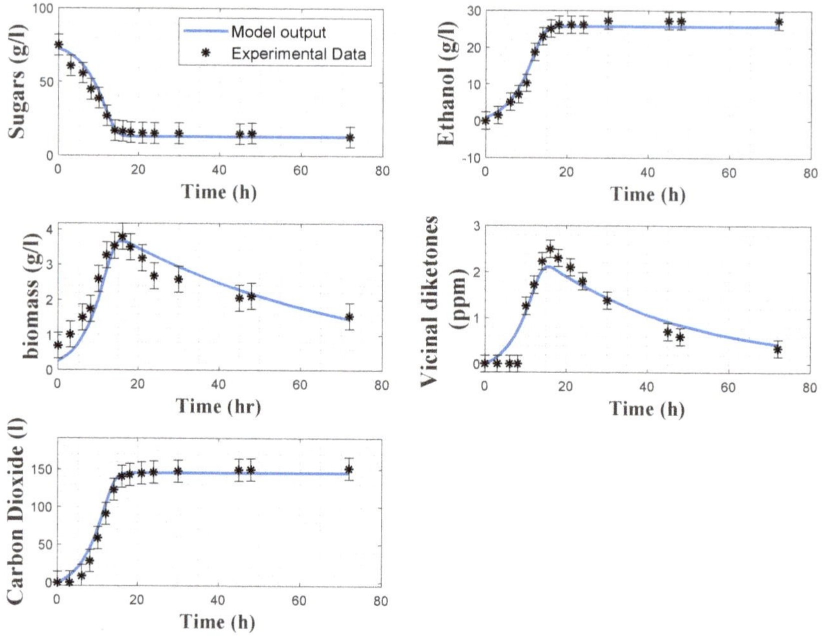

Figure 3. Direct validation of the biomass model with the data from experiment 4. Stars: experimental data. Error bars: 95% confidence intervals. Continuous blue line: model prediction.

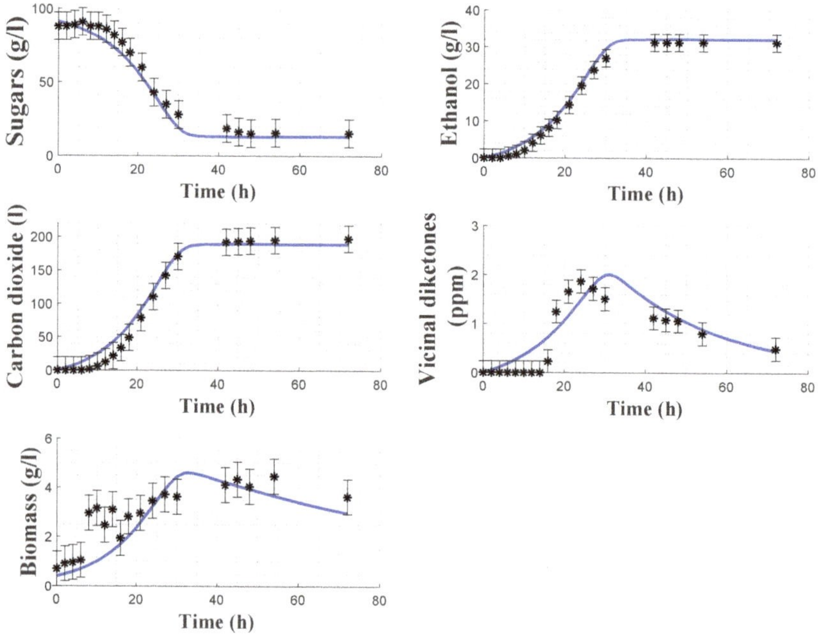

Figure 4. Cross-validation of the biomass model with the data from experiment 1. Stars: experimental data. Error bars: 95% confidence intervals. Continuous blue line: model prediction.

Vicinal diketone dynamics present a varying latency phase followed by production/consumption, both driven by biomass dynamics. The uncertainty on the latency period compromises the resulting fitting since biomass does not exhibit the same behavior in the early phase of fermentation.

5.2. Carbon Dioxide Model

A similar procedure was applied to estimate the values of the carbon dioxide model parameters. In this case, the dependence on temperature of μ_{max} and $r_V DK$ is best represented by:

$$\mu_{max} = a\ln(T) + b; \qquad r_{VDK} = cT^2 + dT + e; \tag{38}$$

Hence, the resulting model counts 10 parameters (k_S, k_E, K_S, Cp_{max}, k_V, a, b, c, d, e), and Table 6 reports the estimated values with their respective coefficients of variations. As can be noticed, the latter are smaller than the ones of the previous model, mainly due to the absence of biomass measurement and the associated uncertainty.

The identification is again decomposed into distinctive steps: (a) estimation of the parameters without temperature dependence and with an arbitrary value for K_S whose practical identifiability is poor, (b) estimation of K_S with all the other parameters fixed at their previously estimated values, (c) estimation of the parameter linked to the temperature dependence (a to e) with all the others fixed to their previous values, and (d) final re-estimation of all the parameters.

Figures 5 and 6 show the direct validation with experiments 2 and 4, as well as the a posteriori error bars on the experimental data.

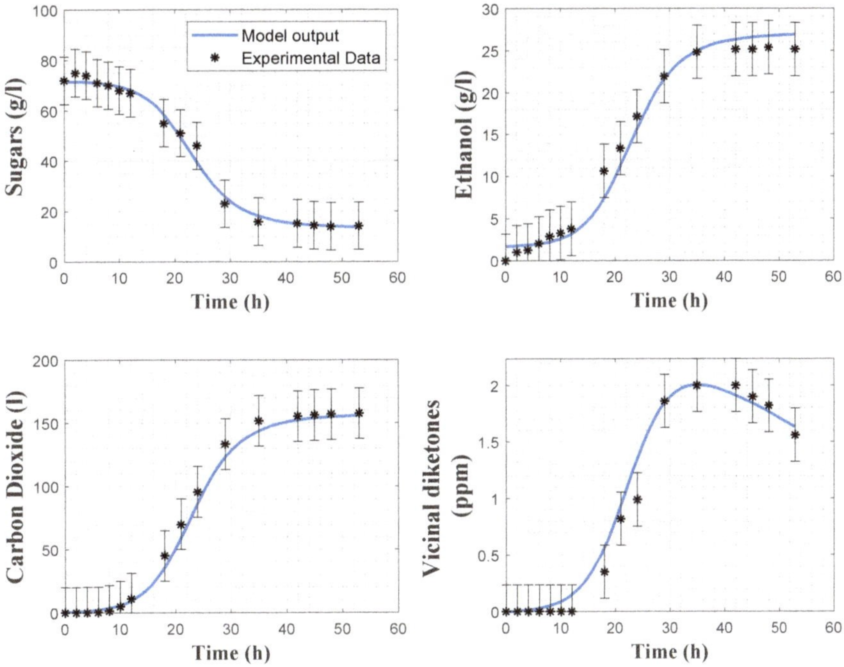

Figure 5. Direct validation of the carbon dioxide model with the data from experiment 2. Stars: experimental data. Error bars: 95% confidence intervals. Continuous blue line: model prediction.

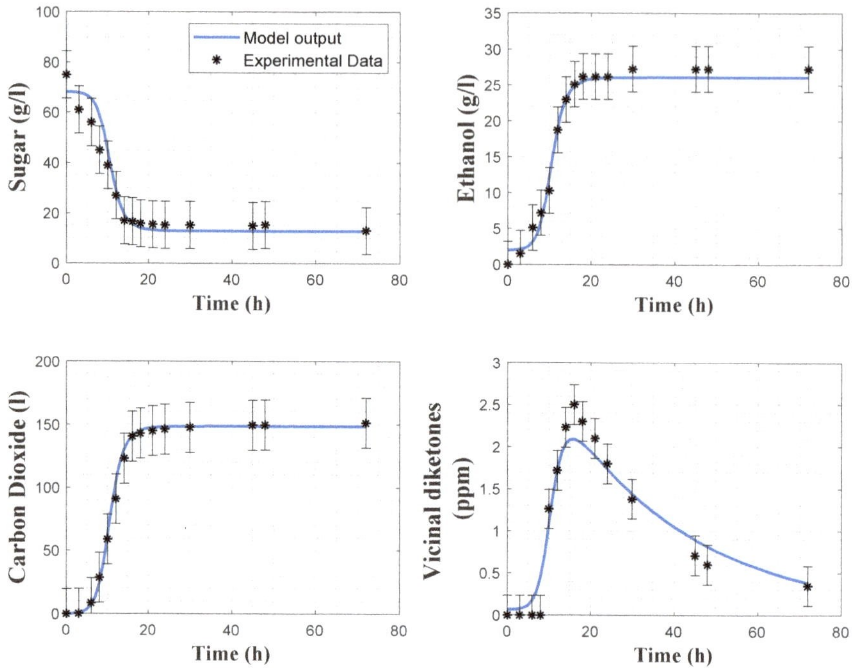

Figure 6. Direct validation of the carbon dioxide model with the data from experiment 4. Stars: experimental data. Error bars: 95% confidence intervals. Continuous blue line: model prediction.

Table 6. Parameter estimate values and coefficients of variations (CV) for the carbon dioxide model.

Parameter	Units	Value	CV (%)
k_S	gS/lCO_2	3.72×10^{-1}	4
k_E	gE/lCO_2	1.62×10^{-1}	3
K_S	g/L,	12.0	9
Cp_{max}	L/g	2.18	3
k_V	ppm	1.74×10^{-2}	4
a	h^{-1}	1.41	6
b	h^{-1}	− 3.89	8
c	$°C^{-2}\,h^{-1}$	-3×10^{-4}	12
d	$°C^{-1}\,h^{-1}$	1.6×10^{-2}	6
e	h^{-1}	-1.75×10^{-1}	5

Model cross-validation using experiment 1 is shown in Figure 7 and confirms the satisfactory predictive capacity of the model, except for some minor deviations in the evolution of the VDKs due to the presence of a time-varying latency phase.

In Table 7, the root means square errors (RMSEs) are provided for each experiment and each variable separately. The cost function residuals of the direct validations are also provided. It can be observed that, overall, the RMSE values are small for both models. Regarding the biomass model, X and VDK present larger RMSEs than the other variables, due to the observed deviations between the model prediction and the experimental data in Figures 2 and 3. Regarding the carbon dioxide model, RMSEs indicate a better fit with the experimental data. This statement is confirmed by Figures 5 and 6. Cross-validation results also support this analysis since the RMSEs of the biomass and carbon dioxide models are, respectively, 1.997 and 0.846.

Table 7. Cost function residuals and relative RMSEs for each state variable of the two models, resulting from direct validation.

Model	Cost Function Residual	Variable	Exp 2 RMSE	Exp 3 RMSE	Exp 4 RMSE	Global RMSE
Biomass	1.76	CO_2	4.86×10^{-2}	5.40×10^{-2}	6.40×10^{-2}	5.32×10^{-2}
		S	3.37×10^{-2}	8.12×10^{-2}	5.43×10^{-2}	5.80×10^{-2}
		E	4.78×10^{-2}	6.67×10^{-2}	5.15×10^{-2}	5.30×10^{-2}
		X	9.96×10^{-2}	9.63×10^{-2}	1.62×10^{-1}	1.09×10^{-1}
		VDK	1.12×10^{-1}	1.16×10^{-1}	1.51×10^{-1}	1.12×10^{-1}
Carbon dioxide	0.72	CO_2	3.45×10^{-2}	2.99×10^{-2}	4.74×10^{-2}	3.56×10^{-2}
		S	2.23×10^{-2}	7.40×10^{-2}	4.63×10^{-2}	5.36×10^{-2}
		E	5.47×10^{-2}	9.91×10^{-2}	4.11×10^{-2}	6.23×10^{-2}
		VDK	6.26×10^{-2}	8.59×10^{-2}	1.12×10^{-1}	7.50×10^{-2}

Discriminating among the proposed models is difficult since they target different variables. However, taking into account the cost function residuals, the carbon dioxide model fit better to the current operating conditions and monitoring set-up (J = 0.72) than the biomass model (J = 1.76). Furthermore, from a practical point of view, the identification of the dioxide carbon model requires a sensor configuration that is easier to set up, limiting offline analytical analysis. Conversely, the identification of the biomass model requires offline cell counting to measure yeast concentration. Moreover, considering process control, carbon dioxide online sensors are affordable, whereas biomass sensors are expensive (alternatively a biomass software sensor could be developed based on the measurements of CO_2, E, VDK). The main advantage of the biomass model lies in the provided information about the biomass metabolic state during the fermentation process, allowing a more straightforward detection of possible contamination.

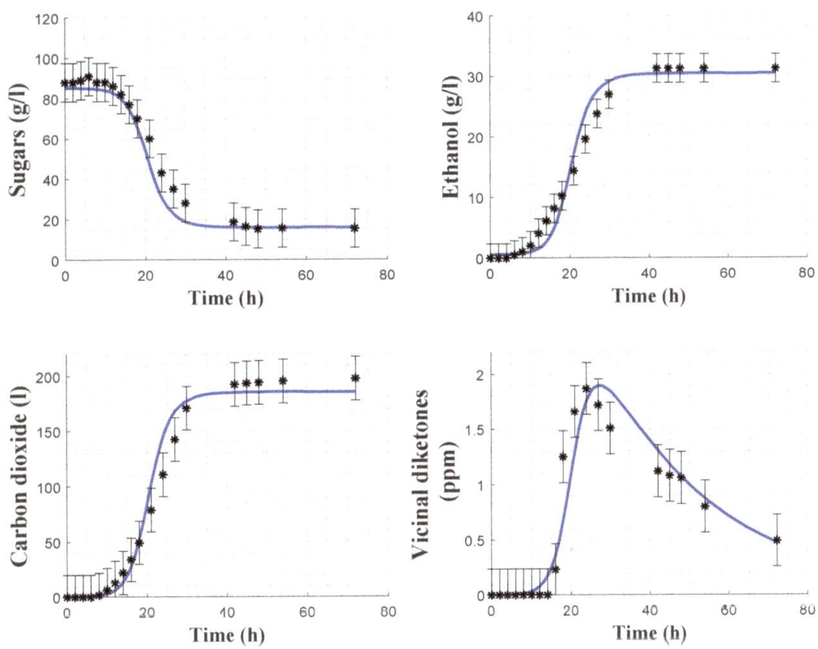

Figure 7. Cross-validation of the carbon dioxide model with the data from experiment 1. Stars: experimental data. Error bars: 95% confidence intervals. Continuous blue line: model prediction.

6. Conclusions

The demand for processes with more rigorous quality standards, as is the case in the pharmaceutical industry, has led to the development of approaches such as process analytical technologies (PAT), now being extended to the agro-food sector and, more specifically, to the brewing industry. This work is motivated by the growing importance of mathematical modeling, in the context of PAT, to design process digital twins that can support lab-scale operations. Model-based advanced monitoring and control techniques can indeed be developed in view of optimizing and improving the process. In this study, two alternative models, initially proposed in seminal works, are adapted and identified under realistic experimental conditions. One of the models is based on the description of the biomass evolution, while the other, more pragmatic, considers carbon dioxide, a more accessible variable that can be measured with cheap sensors. These models take account of the temperature influence in a simple way. A systematic identification procedure is described. Cross-validation highlights the good predictive capability of both models, which are good candidates for model-based control.

Author Contributions: Conceptualization, J.M.Z.L., L.D., and A.V.W.; methodology, J.M.Z.L., L.D., H.H.E., and A.V.W.; software, J.M.Z.L.; validation, J.M.Z.L.; formal analysis, J.M.Z.L.; investigation, J.M.Z.L., L.D., H.H.E., and A.V.W.; data curation, J.M.Z.L.; writing—original draft preparation, J.M.Z.L., L.D., and A.V.W.; writing—review and editing, L.D., H.H.E., and A.V.W.; supervision, A.V.W. and H.H.E.; project administration, H.H.E. and A.V.W.; funding acquisition, H.H.E. and A.V.W. All authors have read and agreed to the published version of the manuscript.

Funding: This research received no external funding.

Data Availability Statement: The data presented in this study are available on request from the corresponding author.

Acknowledgments: The authors are grateful to Vincent Moeyaert, Perla Melendez, and Ivette Navarro for their help in the experimental set-up. J.M.Z.L. acknowledges the support of CONACYT under grant 782850.

Conflicts of Interest: The authors declare no conflict of interest.

Appendix A

Table A1. Parameter nomenclature list for biomass model.

Parameter	Definition	Units
k_S	Sugar yield coefficient	gS/gX
k_E	Ethanol yield coefficient	gE/gX
S_{min}	Minimum sugar quota	g/L
δ_X	Settling constant	h^{-1}
k_V	VDK yield coefficient	gVDK/gX
k_{CO_2}	CO_2 yield coefficient	L
a	Temperature-dependency coefficient	h^{-1}
b	Temperature-dependency coefficient	h^{-1}
c	Temperature-dependency coefficient	h^{-1}
d	Temperature-dependency coefficient	h^{-1}

Table A2. Parameter nomenclature list for carbon dioxide model.

Parameter	Definition	Units
k_S	Sugar yield coefficient	gS/lCO_2
k_E	Ethanol yield coefficient	gE/lCO_2
K_S	Substrate limitation coefficient	g/L
Cp_{max}	Carrying capacity	L/g
k_V	VDK yield coefficient	ppm
k_{CO_2}	CO_2 yield coefficient	L
a	Temperature-dependency coefficient	h^{-1}
b	Temperature-dependency coefficient	h^{-1}
c	Temperature-dependency coefficient	$°C^{-2}\,h^{-1}$
d	Temperature-dependency coefficient	$°C^{-1}\,h^{-1}$
e	Temperature-dependency coefficient	h^{-1}

References

1. Iserentant, D. Beers: Recent Technological Innovations in Brewing. In *Fermented Beverage Production*; Lea, A.G.H., Piggott, J.R., Eds.; Springer: Boston, MA, USA, 2003; pp. 41–58. [CrossRef]
2. Barth, S.J. *BarthHass Report Hops 2020/2021*; Technical Report; BarthHaas: Nuremberg, Germany, 2021.
3. Willaert, R. The Beer Brewing Process: Wort Production and Beer Fermentation. In *Handbook of Food Products Manufacturing*; John Wiley & Sons, Ltd.: Hoboken, NJ, USA, 2007; Chapter 20, pp. 443–506. [CrossRef]
4. Guido, L.; Rodrigues, P.; Rodrigues, J.; Gonçalves, C.; Barros, A. The impact of the physiological condition of the pitching yeast on beer flavour stability: An industrial approach. *Food Chem.* **2004**, *87*, 187–193. [CrossRef]
5. Rodman, A.D.; Gerogiorgis, D.I. Multi-objective process optimisation of beer fermentation via dynamic simulation. *Food Bioprod. Process.* **2016**, *100*, 255–274. [CrossRef]
6. Shopska, V.; Denkova, R.; Lyubenova, V.; Kostov, G. 13-Kinetic Characteristics of Alcohol Fermentation in Brewing: State of Art and Control of the Fermentation Process. In *Fermented Beverages*; Grumezescu, A.M., Holban, A.M., Eds.; Woodhead Publishing: Sawston, UK, 2019; pp. 529–575. [CrossRef]
7. Gee, D.A.; Ramirez, W.F. A flavour model for beer fermentation. *J. Inst. Brew.* **1994**, *100*, 321–329. [CrossRef]
8. De Andrés-Toro, B.; Girón-Sierra, J.; López-Orozco, J.; Fernández-Conde, C.; Peinado, J.; García Ochoa, F. A kinetic model for beer production under industrial operational conditions. *Math. Comput. Simul.* **1998**, *48*, 65–74. [CrossRef]
9. Trelea, I.C.; Titica, M.; Landaud, S.; Latrille, E.; Corrieu, G.; Cheruy, A. Predictive modelling of brewing fermentation: From knowledge-based to black-box models. *Math. Comput. Simul.* **2001**, *56*, 405–424. [CrossRef]
10. Villaverde, A.F.; Barreiro, A.; Papachristodoulou, A. Structural Identifiability of Dynamic Systems Biology Models. *PLos Comput. Biol.* **2016**, *12*, 1–22. [CrossRef] [PubMed]
11. Xie, J.; Tian, X.F.; He, S.G.; Wei, Y.L.; Peng, B.; Wu, Z.Q. Evaluating the Intoxicating Degree of Liquor Products with Combinations of Fusel Alcohols, Acids, and Esters. *Molecules* **2018**, *23*, 1239. [CrossRef] [PubMed]
12. Pandiella, S.; García, L.; Díaz, M.; Daoud, I. Monitoring the carbon dioxide during beer fermentation. *MBAA Technol.* **1995**, *32*, 126–131.
13. Droop, M. Vitamin B12 and marine ecology. IV. The kinetics of uptake, growth and inhibition in Monochrysis lutheri. *J. Mar. Biol. Assoc. UK* **1968**, *48*, 689–733. [CrossRef]
14. Zwietering, M.H.; Jongenburger, I.; Rombouts, F.M.; van 't Riet, K. Modeling of the Bacterial Growth Curve. *Appl. Environ. Microbiol.* **1990**, *56*, 1875–1881. [CrossRef]
15. Crabtree, H.G. Observations on the carbohydrate metabolism of tumours. *Biochem. J.* **1929**, *23*, 536–545. [CrossRef]
16. Sonnleitner, B.; Käppeli, O. Growth of Saccharomyces cerevisiae is controlled by its limited respiratory capacity: Formulation and verification of a hypothesis. *Biotechnol. Bioeng.* **1986**, *28*, 927–937. [CrossRef] [PubMed]
17. Bellman, R.; Åström, K. On structural identifiability. *Math. Biosci.* **1970**, *7*, 329–339. [CrossRef]
18. Wieland, F.G.; Hauber, A.L.; Rosenblatt, M.; Tönsing, C.; Timmer, J. On structural and practical identifiability. *Curr. Opin. Syst. Biol.* **2021**, *25*, 60–69. [CrossRef]
19. Bellu, G.; Saccomani, M.P.; Audoly, S.; D'Angiò, L. DAISY: A new software tool to test global identifiability of biological and physiological systems. *Comput. Methods Programs Biomed.* **2007**, *88*, 52–61. [CrossRef] [PubMed]
20. Chis, O.; Banga, J.R.; Balsa-Canto, E. GenSSI: A software toolbox for structural identifiability analysis of biological models. *Bioinformatics* **2011**, *27*, 2610–2611. [CrossRef] [PubMed]

21. Hong, H.; Ovchinnikov, A.; Pogudin, G.; Yap, C. SIAN: Software for structural identifiability analysis of ODE models. *Bioinformatics* **2019**, *35*, 2873–2874. [CrossRef] [PubMed]
22. Walter, E.; Pronzato, L. *Idnetification of Parametric Models from Experimental Data*, 1st ed.; Springer: London, UK, 1997.
23. Schulze, U.; Lidén, G.; Nielsen, J.; Villadsen, J. Physiological effects of nitrogen starvation in an anaerobic batch culture of Saccharomyces cerevisiae. *Microbiology (Reading)* **1996**, *142*, 2299–2310. [CrossRef] [PubMed]

foods

MDPI

Article

Dynamics of Microbial Inactivation and Acrylamide Production in High-Temperature Heat Treatments

Jose Lucas Peñalver-Soto [1,2], Alberto Garre [3], Arantxa Aznar [1], Pablo S. Fernández [1] and Jose A. Egea [2,*]

[1] Departamento de Ingeniería Agronómica, Instituto de Biotecnología Vegetal, Universidad Politécnica de Cartagena (ETSIA), Paseo Alfonso XIII, 48, 30203 Cartagena, Spain; joselucasps@gmail.com (J.L.P.-S.); arantxa.aznar@upct.es (A.A.); pablo.fernandez@upct.es (P.S.F.)

[2] Centro de Edafología y Biología Aplicada del Segura (CEBAS-CSIC), Campus Universitario de Espinardo, 30100 Murcia, Spain

[3] Food Microbiology, Wageningen University & Research, P.O. Box 17, 6700 AA Wageningen, The Netherlands; alberto.garreperez@wur.nl

* Correspondence: jaegea@cebas.csic.es

Abstract: In food processes, optimizing processing parameters is crucial to ensure food safety, maximize food quality, and minimize the formation of potentially toxigenic compounds. This research focuses on the simultaneous impacts that severe heat treatments applied to food may have on the formation of harmful chemicals and on microbiological safety. The case studies analysed consider the appearance/synthesis of acrylamide after a sterilization heat treatment for two different foods: pureed potato and prune juice, using *Geobacillus stearothermophilus* as an indicator. It presents two contradictory situations: on the one hand, the application of a high-temperature treatment to a low acid food with *G. stearothermophilus* spores causes their inactivation, reaching food safety and stability from a microbiological point of view. On the other hand, high temperatures favour the appearance of acrylamide. In this way, the two objectives (microbiological safety and acrylamide production) are opposed. In this work, we analyse the effects of high-temperature thermal treatments (isothermal conditions between 120 and 135 °C) in food from two perspectives: microbiological safety/stability and acrylamide production. After analysing both objectives simultaneously, it is concluded that, contrary to what is expected, heat treatments at higher temperatures result in lower acrylamide production for the same level of microbial inactivation. This is due to the different dynamics and sensitivities of the processes at high temperatures. These results, as well as the presented methodology, can be a basis of analysis for decision makers to design heat treatments that ensure food safety while minimizing the amount of acrylamide (or other harmful substances) produced.

Keywords: food safety; acrylamide formation; thermal resistance; dynamic models; simulation

Citation: Peñalver-Soto, J.L.; Garre, A.; Aznar, A.; Fernández, P.S.; Egea, J.A. Dynamics of Microbial Inactivation and Acrylamide Production in High-Temperature Heat Treatments. *Foods* **2021**, *10*, 2535. https://doi.org/10.3390/foods10112535

Academic Editor: Barbara Cardazzo

Received: 10 September 2021
Accepted: 19 October 2021
Published: 21 October 2021

1. Introduction

Different conflicting objectives often arise in many food processes (e.g., quality vs. economical cost). Finding the optimal solutions which balance among all the existing objectives is not an easy task due to the complexity of the mathematical models describing such processes [1–4]. This optimization step is crucial to produce an efficient decision-making process [5]. Recent research has been devoted to optimizing food processes where two or more conflicting objectives appear. The most common are usually related to product quality and process economy [6–8], different quality parameters [9–11], economic and environmental parameters [12,13], or, as in the present study, product quality and safety [14].

One of the most important methods of food preservation in the food industry is thermal processing. Historically, the focus was on optimizing heat treatments to improve the processes related to microbial destruction, nutrients retention, cooking values and loss of quality [15]. The main function of heat treatments is to inactivate microorganisms and enzymes to achieve safe, long shelf-life food. The associated disadvantages are related to food quality (e.g., nutrients or texture preservation) due to the effect of high temperatures. Therefore, the design of thermal processes in the food safety sector must face different objectives, such as food quality or energy consumption vs. microbial inactivation. The use of high temperatures implies food degradation but also the formation of substances that can be harmful to humans [16]. One example is acrylamide, a chemical that is produced during high-temperature processes in foods that contain reducing sugars (such as fructose and glucose) and asparagine [17]. Acrylamide was found to be carcinogenic in rodents, and the International Agency for Research on Cancer has classified it as a probable carcinogen [18,19]. This has motivated food authorities to propose methodologies to minimize acrylamide content in commercial and homemade foods. The European Food Safety Authority (EFSA) has set recommendation levels for some foods [20]. The influence of high temperature on the formation of acrylamide has been previously demonstrated [21]. The higher the processing temperature, the more acrylamide is formed, thus heat treatments need to be optimized to decrease the amount of produced acrylamide. The formation of acrylamide in a process that involves the heating of food is explained by the Maillard reaction [22]. Since its formulation, this reaction has been studied from different points of view. Traditionally, the focus was put on components that affected colour, flavour, and taste, whereas more recently the focus has moved to the analysis of the formation of mutagens and carcinogens. Acrylamide is one of these chemicals in the spotlight due to its potential formation in highly consumed foods such as potato chips [23] or Asian noodles [24]. French fries, coffee, and bread have also presented high levels of acrylamide [21]. In fact, a wide range of different food products containing fructose and asparagine can contain acrylamide.

In the case of baby food, the recommendations for acrylamide levels are more restrictive than in other types of food, and a maximum allowed amount of 30 μg/kg is set [20]. Different studies [25,26] have shown that various foods exceed these limits. Specifically, potato-based products are highly susceptible to containing high levels of acrylamide. On the other hand, although the EFSA has not yet established recommendations for foods based on vegetables or fruits [27], it has been shown that products such as prune juice can contain high concentrations of acrylamide, reaching much higher values even than in potato-based foods. [28]. Therefore, this study focuses on foods such as potato puree and prune juice, which can be catalogued as baby food and may not meet EFSA's recommendation for those products.

From a quality vs. microbiological point of view, heat stability of heat-labile quality factors presents a higher z-value than those typical of bacteria. Then, high-temperature short-time (HTST) processes are less deleterious to food quality while ensuring microbial food safety and stability [29], although the impact of acrylamide formation has not been considered. In this regard, the two proposed objectives (i.e., microbial inactivation and acrylamide formation) are opposed and the problem must be analysed.

The microorganism considered here is *Geobacillus stearothermophilus*, a Gram-positive, thermophilic, and spore-forming bacterium with an optimal growth temperature around 55 °C. The spores are very heat-resistant and usually survive canning and sterilization operations. Furthermore, it has been detected in different foods such as canned vegetables, ready-to-eat meals containing meat, fruit preparations, or dehydrated ingredients [30]. Other relevant pathogenic spore formers of interest in the food industry, such as *Clostridium botulinum* or *Bacillus cereus*, have not been considered here, as their inactivation is generally not a problem within the range of temperatures considered in this study which give rise to significant amounts of formed acrylamide.

As *Geobacillus stearothermophilus* spores are used to validate heat sterilization processes [31–34], this study evaluates the inactivation of this microorganism in a heat treatment within the typical temperatures applied to the studied products (120–135 °C).

In this work, we have simulated and analysed the dynamics of the two objectives in a thermal inactivation operation. The aim is to use mathematical models to determine the conditions where the amount of formed acrylamide is minimum while ensuring microbial inactivation. To the authors' knowledge, no previous work was published facing the inactivation of microorganisms and acrylamide production.

2. Materials and Methods

2.1. Case Study

This study analyses the dynamics of two processes associated with the application of severe heat treatments in some foods: microbial inactivation and acrylamide production. In principle, higher temperatures ensure a safer operation from the microbiological point of view, but can also induce a higher acrylamide production. We have applied this analysis to two practical cases of the food industry (potato puree and prune juice) where different time/temperature treatments are simulated to assess the inactivation of a thermo-resistant microorganism and the acrylamide formation. The microorganism is first characterized (see Section 2.2) and then heat treatments scenarios are simulated for each of the considered foods to assess the dynamics of microbial inactivation and acrylamide formation and their balance.

2.2. Microbial Inactivation Model and Parameter Estimation

To characterize the behaviour of the microorganism, the Bigelow model [35] was chosen. This choice is motivated by the use of this model to characterize the microbial inactivation of *Geobacillus stearothermophilus* within the literature (e.g., [36–38]). In any case, the use of other models would not invalidate the methodology presented here, and the whole procedure would be similar. It considers a log-linear relationship between the fraction of survivors (S) and treatment time, (t), as shown in Equation (1).

$$log_{10}S = \frac{-t}{D(T)} \quad (1)$$

$$log_{10}D(T) = log_{10}D(T_{ref}) - \frac{T - T_{ref}}{z} \quad (2)$$

The influence of treatment temperature (T) in the microorganism is reflected as the D-value, which is log dependent on the temperature, as shown in Equation (2). The D-value represents the time required to reduce the microbial population by 90% at a constant temperature, and the z-value quantifies the sensitivity of the D-value to temperature changes. The reference temperature, T_{ref}, is a parameter without biological meaning but can improve parameter identifiability [39,40].

To estimate the model parameters, several D-values (n = 113) for thermal inactivation of *Geobacillus stearothermophilus* were collected from the literature (Web of Science database) as described in [41]. A temperature range between 97.5 °C and 137.5 °C was considered, and only food matrixes (especially vegetable-based) were included [38,42–50]. Non-linear regression was applied to obtain mean $log_{10}D(T_{ref})$-values and z-values with their respective standard errors.

The estimated model parameters are $log_{10}D(T_{ref}) = -0.0468 \pm 0.03789$, $z = 8.66 \pm 0.283$ (°C). The experimental data, as well as the fitted model, are presented in Figure S1, that shows a good model fitting, which confirms the suitability of the Bigelow model to characterize the microorganism in the conditions considered. The reference temperature was set to a value near the middle of the temperature range ($T_{ref} = 125$ °C) as recommended by [40]. Monte Carlo simulations were used to calculate the probability that a heat treatment (time, temperature) produces at least 6 logarithmic reductions in the microbial load (symbolized as $P(log_{10}S \geq 6)$ within the text), which is considered to be sufficient for many inactivation processes. In any case, changing this value would not change the proposed methodological approach. For this, we select 1000 pairs of values of the model parameters $(log_{10}D(T_{ref}), z)$ obtained by simulation from two independent normal distributions, where $log_{10}D(T_{ref}) \sim N(\mu = 0.0468, \sigma = 0.03789)$ and $z \sim N(\mu = 8.66, \sigma = 0.283)$. Next, the expected log reduction was calculated with Equations (1) and (2) from the heat treatment conditions (time, temperature) for each of the 1000 pairs of parameter values. Finally, the probability of achieving the target inactivation was calculated by dividing the number of cases that comply with $log_{10}S \geq 6$ by 1000. This procedure was used to address the first objective: maximize $P(log_{10}S \geq 6)$.

2.3. Acrylamide Production Objective

To quantify the acrylamide formation, the multi-response kinetics in a glucose-asparagine reaction at high temperatures (120–200 °C) proposed by [51] were used. The model is based on the reaction network shown in Figure S2.

Glucose and asparagine react to form a Schiff base. Fructose is formed by glucose isomerization, and it also reacts with asparagine to form the Schiff base. At the same time, the Schiff base is degraded into melanoidins and acrylamide, whereas acrylamide is degraded into unknown species (named *Product X*). Study [51] calculated the equilibrium constants for the temperature range (120–200 °C), which showed a logarithmic relationship with temperature. Using this observed relationship, a temperature-dependent function was fitted for each constant ($K_i(T)\ \forall i = 1, 2 \ldots, 6$). Although the range of temperatures used by Knol et al. does not coincide with the range considered for t$log_{10}D(T)$ vs. T values in this study (e.g., from 97.5 to 137.5), we consider that the same logarithmic relationship applies in our case. The estimated kinetic constant values were used in the system of ordinary differential equations that quantifies the amount of acrylamide formed for a specific time and temperature, shown in Equation (3).

$$\begin{cases} \frac{d[Glucose]}{dt} = -K_1(T)\cdot[Glucose]\cdot[Asparagine] - K_2(T)\cdot[Glucose] \\ \frac{d[Fructose]}{dt} = -K_3(T)\cdot[Fructose]\cdot[Asparagine] + K_2(T)\cdot[Glucose] \\ \frac{d[Asparagine]}{dt} = -K_1(T)\cdot[Glucose]\cdot[Asparagine] - K_3(T)\cdot[Fructose]\cdot[Asparagine] \\ \frac{d[Schiffbase]}{dt} = K_1(T)\cdot[Glucose]\cdot[Asparagine] + K_3(T)\cdot[Fructose]\cdot[Asparagine] \\ \quad -K_4(T)\cdot[Schiffbase] - K_5(T)\cdot[Schiffbase] \\ \frac{d[Acrylamide]}{dt} = K_4(T)\cdot[Schiffbase] - K_6(T)\cdot[Acrylamide] \end{cases} \tag{3}$$

One of the solutions to the system of differential equations is the acrylamide concentration formed for a certain heat treatment (time, temperature), which is the second objective in our formulation. In this second objective, apart from the heat treatment conditions, that is, the time and temperature variables, it is necessary to set the initial amounts of glucose, fructose, and asparagine. Two foods were selected: potato puree (typical pH of 5.1–6.0) and prune juice (typical pH of 4.0–5.0). For each of the two selected foods, initial concentrations are shown in Table 1. Details on these calculations are provided in Supplementary Material. From the existing potato varieties, the *red potato* was chosen, as it produces the highest amount of acrylamide [52].

Table 1. Initial concentrations of glucose, fructose, and asparagine in pureed potato and prune juice.

Concentration (mM)	Pureed Potato	Prune Juice
Glucose	10.740	213.426
Fructose	7.190	8.731
Asparagine	36.551	213.421

3. Results and Discussion

This section analyses the results of the simulation of isothermal heat treatments with temperatures between 120 and 135 °C for two food matrices, pureed potatoes and prune juice. On the one hand, to quantify inactivation, a Monte Carlo simulation approach is used, as explained in Section 2.2, which provides the probability that the heat treatment (time, temperature) produces at least six logarithmic reductions ($P(log_{10}S \geq 6)$). Therefore, all results for inactivation in both foods are related to that probability.

On the other hand, acrylamide production is reported in the units of measurement $\left(\frac{\mu g}{kg}\right)$ recommended by EFSA to make comparisons [20].

3.1. Single-Objective Analysis

Regarding the first objective (i.e., microbial inactivation), $P(log_{10}S \geq 6)$ as a function of temperature and time of heat treatment is presented in Figure 1. As expected, the areas with the highest probability of having more than six log-reductions in the microbial count are those related to higher temperatures and longer treatment times. It is notable that, in the ranges of time and temperature considered (Figure 1), $P(log_{10}S \geq 6)$ is very sensitive to small changes in temperature or in processing time. This sensitivity is higher as the temperature increases.

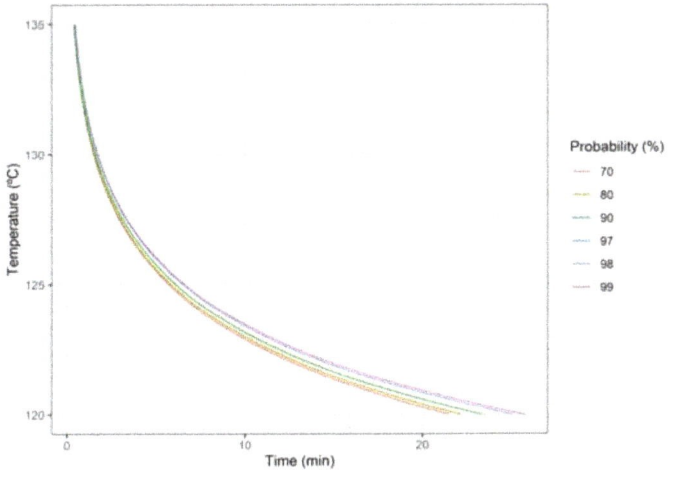

Figure 1. Probability of six or more logarithmic reductions $P(log_{10}S \geq 6)$ for *Geobacillus stearothermophilus* as a function of heat treatment (time, temperature).

On the other hand, the impact of the heat treatment on acrylamide formation is represented in Figure 2A for pureed potato and Figure 2B for prune juice. The area showing the highest acrylamide formation is defined by the highest treatment temperature. As expected, higher temperature and/or longer duration of the heat treatment had a positive correlation with acrylamide formation, although significant differences were found between the foods tested. The main difference was the amount of acrylamide that could be produced, which was higher for prune juice at all the time/temperature combinations. For example, for the most severe treatments (upper right corner of Figure 2A,B) the concentration was around two times higher in the case of prune juice. Comparing the isolines of both objectives, larger changes in temperature or processing time are needed to produce significant changes in acrylamide production (Figure 2) than in $P(log_{10}S \geq 6)$ (Figure 1). In other words, acrylamide formation is less sensitive to temperature and time than inactivation.

In any case, this mono-objective analysis confirms that both objectives counter each other, thus a balance in the operating parameters must be achieved. As an expected conclusion of this analysis, the increase in temperature or processing time in thermal treatments favours the inactivation of the microorganism and disfavours (i.e., increases) acrylamide formation. However, the sensitivity of both responses to changes in temperature or time is significantly different.

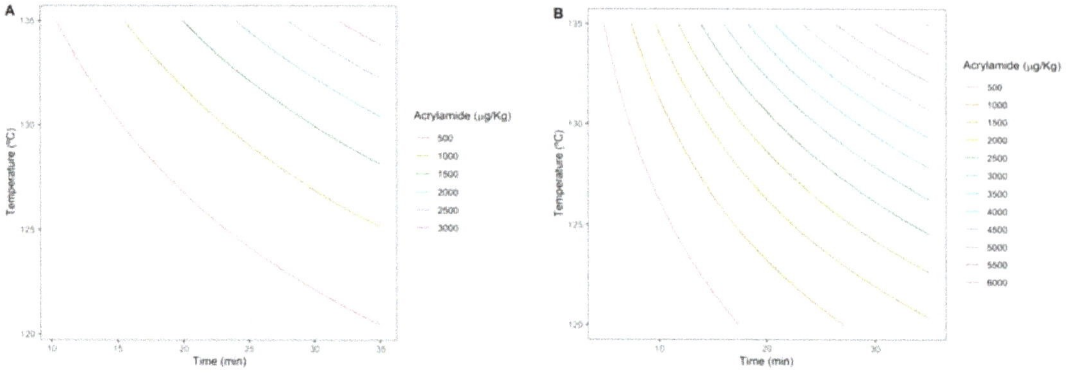

Figure 2. Amount of acrylamide formed as a function of heat treatment (time, temperature) for the potato puree (**A**) and prune juice (**B**).

3.2. *Dynamics of Geobacillus Inactivation and Acrylamide Formation*

As previously discussed, both objectives conflict. Therefore, to analyse the influence of the temperature on both objectives simultaneously, the inactivation rates and the amount of formed acrylamide for different treatment temperatures are evaluated. Simulated inactivation curves for the ranges of temperature and time considered are represented in Figure 3 following the Bigelow model (i.e., Equations (1) and (2)). The increase in temperature produces an increase in the slope of the inactivation curve and therefore a faster inactivation. On the other hand, acrylamide formation is explained by the Maillard reaction (Equation (3)) and its dynamics are represented in Figure 4. Figure 4A shows the formation rates for pureed potato. It is observed that the higher the temperature, the higher the formation. This is confirmed by Figure 4B, which corresponds to the amount of acrylamide formed in prunes juice.

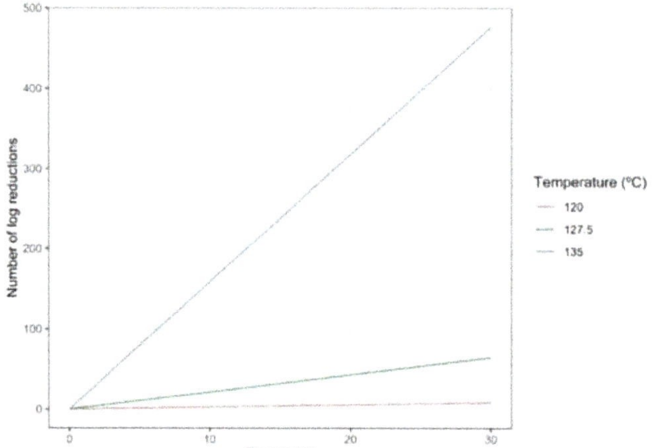

Figure 3. *Geobacillus stearothermophilus* inactivation dynamics according to the Bigelow model.

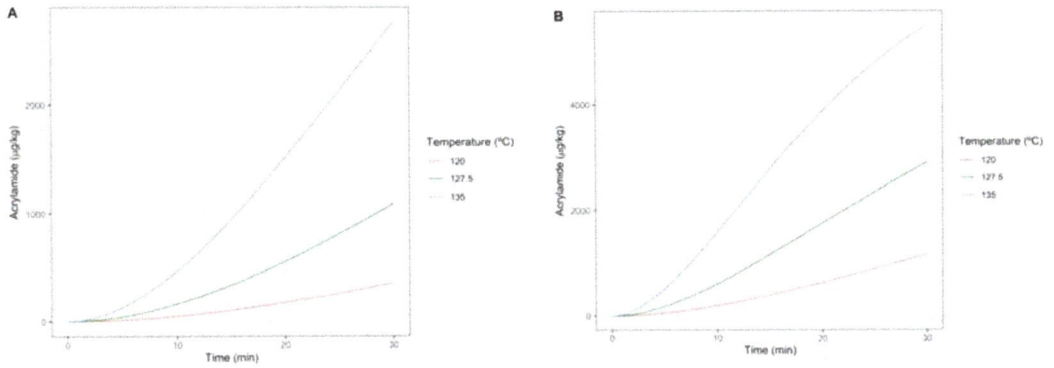

Figure 4. Acrylamide formation dynamics according to the Maillard reaction: potato puree (**A**) and prune juice (**B**).

Figure 5A,B represent the inactivation curves for five different temperatures between 120 and 135 °C and the acrylamide formed for each temperature when $P(log_{10}S \geq 6) = 0.95$ for pureed potato and prune juice, respectively. Inactivation curves are represented as coloured lines and the left y-axis measures the number of logarithmic reductions. As discussed before, higher temperatures result in a higher slope of the inactivation rate. Bars placed at the times required for each treatment temperature show that processes at higher temperatures need less treatment time to achieve at least six log-reductions with a 95% probability. On the other hand, the height of each bar represents the acrylamide formed for each treatment (time–temperature).

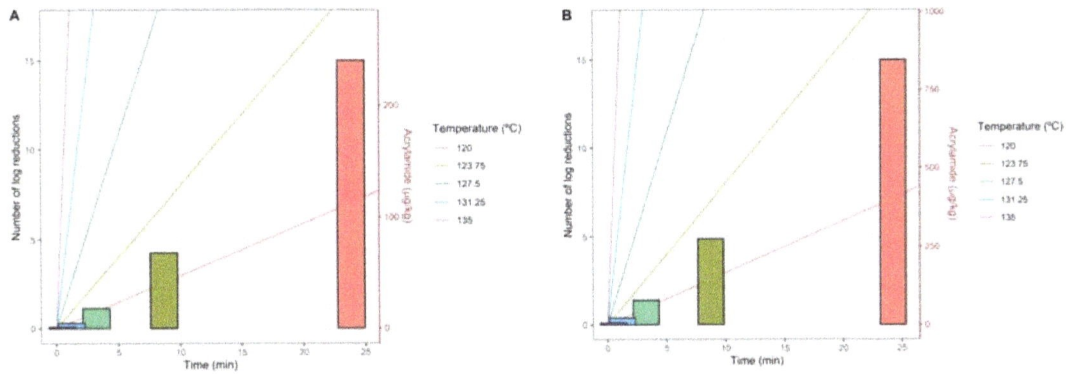

Figure 5. Inactivation rates represented as lines and acrylamide formed represented as bars for different temperatures when $P(log_{10}S \geq 6) = 0.95$, potato puree (**A**) and prune juice (**B**).

It is observed that both for the potato puree (Figure 5A) and the prune juice (Figure 5B), the lower the temperature, the greater the amount of acrylamide is formed due to the higher treatment time needed to achieve $(log_{10}S \geq 6) = 0.95$. As the temperature increases, the treatment time required to inactivate the spores decreases, and therefore the amount of acrylamide formed is also lower. We must recall that both objectives (i.e., microbial inactivation and acrylamide formation) are determined by the combination of temperature and time. The processing time for each temperature is determined by the defined level of inactivation (i.e., $(log_{10}S \geq 6) = 0.95$). This time is different for each temperature and the balance between dynamics explains the unexpected differences in the calculated acrylamide amount.

Table 2 quantifies the time and acrylamide formed for each temperature and three values of $P(log_{10}S \geq 6)$. For a probability of inactivation of 95%, if the temperature is increased by 12.5% (from 120 to 135 °C), the treatment time is reduced by 98% (from 23.92 to 0.46 min). Higher acrylamide amounts were observed for prune juice, as the amount of reducing sugars is higher than in pureed potato, although the qualitative appearance of the dynamics of the objectives is the same for both foods. The previously mentioned 12.5% increase in temperature (that would imply a 98% reduction in the treatment time) would result in a 99.5% and 99.3% reduction in acrylamide for pureed potato and prune juice, respectively. Therefore, as kinetics behave differently for temperature changes, an additional analysis to find the optimal trade-off between both objectives was performed, based on a multi-objective approach.

Table 2. Acrylamide formed and duration for different heat treatments and inactivation probabilities.

Temperature (°C)	90%			95%			99%		
	Time (min)	Acrylamide (μg/kg)		Time (min)	Acrylamide (μg/kg)		Time (min)	Acrylamide (μg/kg)	
		Pureed Potato	Prune Juice		Pureed Potato	Prune Juice		Pureed Potato	Prune Juice
120.00	23.15	226.52	789.23	23.92	241.29	829.20	25.66	275.45	913.46
123.75	8.43	64.94	257.49	8.71	68.14	269.85	9.28	77.56	300.61
127.50	3.11	17.43	73.78	3.22	18.15	78.79	3.47	21.06	87.71
131.25	1.16	4.50	19.70	1.21	4.74	21.38	1.31	5.55	23.79
135.00	0.44	1.14	5.18	0.46	1.21	5.61	0.49	1.40	6.23

3.3. Multi-Objective Approach

The dependence of both objectives on time and temperature calls for a multi-objective optimization approach where the aim would be to find the pairs (temperature, time) that provide the Pareto front of optimal solutions (e.g., set of temperature/time solutions for which no objective can be improved without sacrificing the other one). However, the application of such an approach led to a set of Pareto solutions which consisted of the maximum temperature tested (135 °C) and different processing times (data not shown). This behaviour can be explained by the different dynamics of the objectives analysed above: an increase in temperature drastically reduces the processing time needed to achieve $P(log_{10}S \geq 6) = 0.95$, which at the same time reduces the amount of acrylamide formed due to the slower dynamics of that process. For that reason, the non-dominated solutions consist of pairs of the maximum temperature tested and different processing times.

While this is a perfectly valid mathematical result, it is not so useful from the engineering point of view if we want to check the effects of other temperatures over the objectives, which can be critical in the case of considering additional objectives. For this reason, we eliminated the temperature as a decision variable and fixed it to some discrete values within the tested range, analysing the evolution of both objectives for each of the discrete temperatures over time. Figure 6 shows the acrylamide formed (y-axis) in a heat treatment that inactivates the microorganism for different values of $P(log_{10}S \geq 6)$, represented in the x-axis.

The highest temperature tested (135 °C, orange dots) caused the lowest amount of formed acrylamide at any level of probability as treatment time was short. On the other hand, for the lowest temperature (associated with long treatment times, 120 °C, purple dots) the highest amount of acrylamide is formed. This reaffirms the results obtained in the previous section and also allows us to broaden the perspective of the problem, as it can be observed that increasing the probability of reaching the target microbial inactivation promotes the acrylamide formation, although with a very low sensitivity, except in the area of probability close to 100%, where the acrylamide formation tends to rise more significantly. The differences between foods (Figure 6A,B) are only manifested in the acrylamide levels, but the qualitative behaviour is similar.

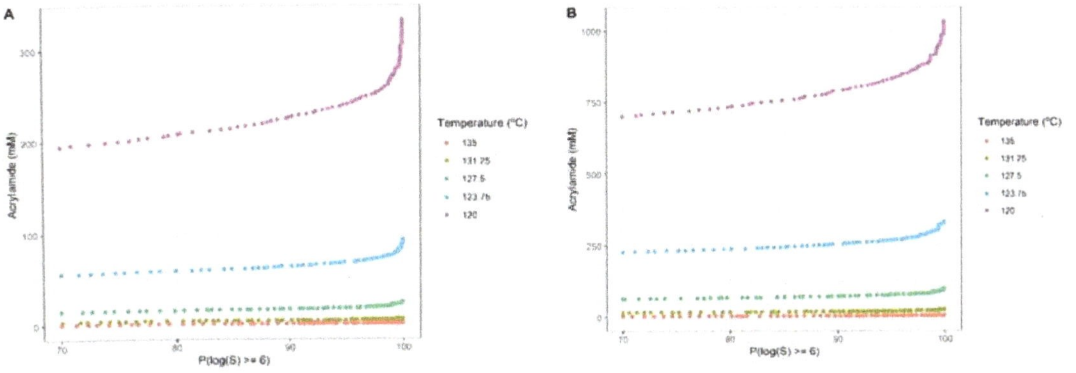

Figure 6. Acrylamide formed in a heat treatment that inactivates the microorganism for different values of $P(log_{10}S \geq 6)$: potato puree (**A**) and prune juice (**B**).

For the lowest temperature (120 °C) the duration of the treatment exceeds 23 min for $P(log_{10}S \geq 6) = 90\%$ (Table 2). On the other hand, increasing the temperature from 120 °C to 135 °C (+12.5%), the time needed for a 90% probability is around half a minute and acrylamide could be decreased by up to 99.5% and 99.3% for pureed potato and prune juice, respectively.

To increase $P(log_{10}S \geq 6)$ from 90% to 99%, the processing time must increase between approximately 10 and 13% depending on the temperature considered. Due to this variation, the amount of acrylamide increases by about 15–23% depending on the chosen temperature, considering both products again.

To assist the decision-making process, Figure 7A,B collect the necessary information to design a heat treatment that seeks to maximize food safety by inactivating a microorganism and minimizing acrylamide formation. These figures show the amount of acrylamide formed (y-axis) as a function of the temperature of the heat treatment (x-axis), as well as $P(log_{10}S \geq 6)$ (colour of the lines). The treatment time is determined by both the chosen temperature and the probability level, as explained below. The horizontal black line represents the maximum amount recommended in baby food 30 (μg/kg) [20], that was considered as a worst-case scenario. Figure 7A refers to pureed potato, whereas Figure 7B refers to prune juice. These figures provide a global vision of how the heat treatment directly determines the amount of formed acrylamide. The area that is above the horizontal line is an undesirable area, as the amount of acrylamide exceeds the recommendation. Ideally, we should remain in the lower region to ensure that the product has a low level of acrylamide while ensuring the inactivation of the microorganism given a defined probability value for $P(log_{10}S \geq 6)$. In both cases, it is observed that, when the temperature increases, acrylamide falls for this temperature range, as discussed above.

In Figure 7A, for pureed potato, the horizontal line divides the desirable region at around 126–127 °C for all the considered probabilities. Therefore, the heat treatments should be above that temperature. At 126–127 °C we would lie within the recommended limit of acrylamide (25–30 μg/kg) whereas an increase of 8–9 °C would produce around 1 μg/kg (Table 2). For the case of prune juice, Figure 7B, the same behaviour is observed. However, the amount of acrylamide formed is higher, as the initial concentrations of fructose and asparagine are also higher. In this case, the recommended limit for all the considered probabilities would lie within 129–130 °C. This information can be useful to consider the use of these foods in, e.g., baby-food products, to avoid exceeding the recommendation.

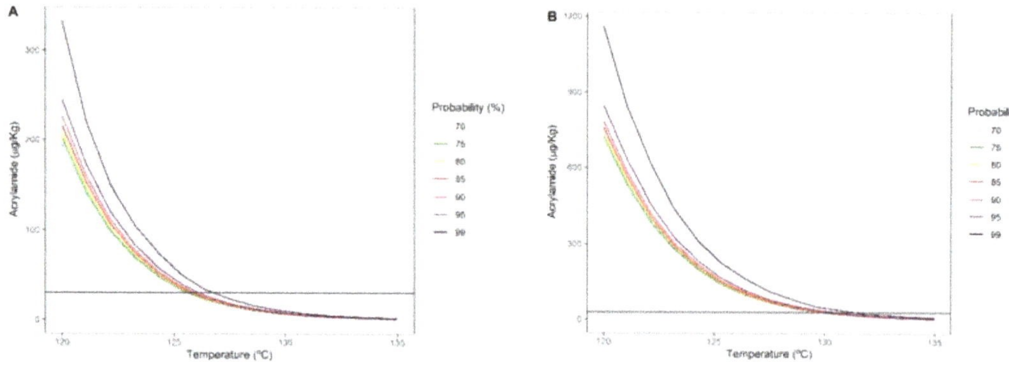

Figure 7. Acrylamide formed as a function of time for each temperature and $P(log_{10}S \geq 6)$: potato puree (**A**) and prune juice (**B**).

Decision makers can use plots such as Figure 7 to decide which levels of acrylamide are likely to be present in the final product depending on $P(log_{10}S \geq 6)$ and temperature, which could affect other properties of the food not considered here. Another approach would be to select the desired acrylamide level and $P(log_{10}S \geq 6)$. In this way, the treatment temperature is determined. Figure 7 shows that, for high temperatures, as the duration of the treatments are very short regardless of the inactivation probability chosen, the acrylamide formation is low in every case. However, for lower temperatures the sensitivity is higher: small changes in temperature (linked to higher exposure times) result in significant changes in acrylamide production.

To complement these design steps, the duration of treatment should be calculated. For that purpose, Figure 8 shows the time (x-axis) for each temperature (y-axis) for three selected probability values of $P(log_{10}S \geq 6)$. Therefore, in the example, if the required temperature is 130 °C for a level of 85%, in the middle plot a treatment duration of 2.5 min is obtained. This value can be more easily retrieved by simulating both models (acrylamide production and microbial inactivation).

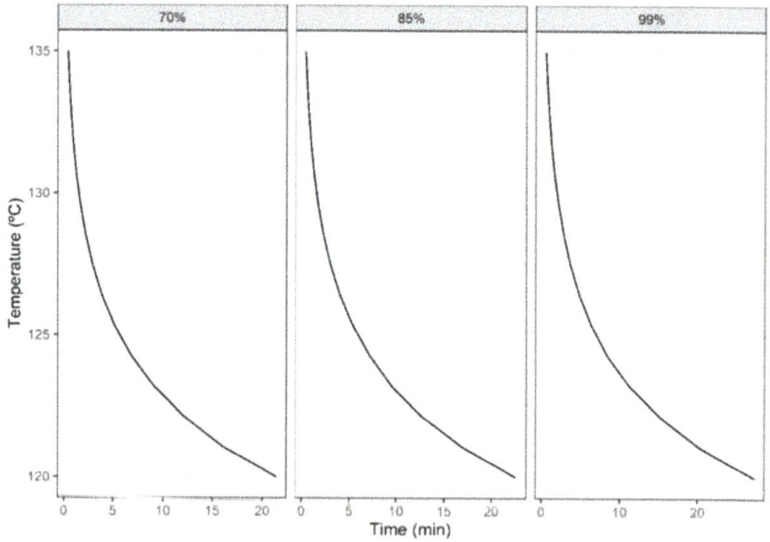

Figure 8. Treatment time required for each temperature and probability.

4. Conclusions

We have analysed the balance between microbial inactivation and acrylamide formation in the case where a thermoresistant microorganism can be present in food in which acrylamide can be formed due to high temperature thermal treatments. As a case study, we have chosen the inactivation of *Geobacillus stearothermophilus* in two particular foods (i.e., pureed potato and prune juice) that we have characterized with the Bigelow model. The acrylamide formation has been modelled with the Maillard equation. The analysis of the dynamics of both processes reveals that, to ensure a certain level of microbial inactivation, heat treatments at higher temperatures lead to decreased acrylamide formation, similar to the behaviour of quality components. This is due to the processes' different sensitivities to temperature. While microbial inactivation is very sensitive (i.e., the times to produce a level of inactivation with a certain probability dramatically decreases with temperature), acrylamide formation is not. The methodology presented here can be used by decision makers to design heat treatments when food safety objectives are faced, maximizing inactivation and minimizing the amount of acrylamide (or other target substances). As the selected foods can be used as ingredients in baby foods, the obtained outcomes were compared with the EFSA's recommendations about maximum acrylamide concentrations in baby food. Even at the highest temperatures, where the least amount of acrylamide is formed due to the short processing times to ensure the microbial inactivation, the expected acrylamide amounts are very close to the maximum EFSA's recommendation, therefore it should be taken into account when using them in infant formulations.

The methodology presented here can be a basis to re-design processes where food safety and acrylamide formation are important issues. It can be used to make decisions when there are unexpected process variables deviations (e.g., lower treatment temperatures than expected). It can also be extended to other processes and microorganisms, and in future work we plan to include other conflicting objectives depending on temperature and time, such as quality or cost. This future work will also be addressed to experimentally validate the conclusions obtained here to refine the model fitting to other conditions and to check the effects of different food matrixes.

Supplementary Materials: The following are available online at https://www.mdpi.com/article/10.3390/foods10112535/s1, Figure S1: Experimental data and fitted model for the considered $log_{10}D/T$ values, Figure S2: Kinetic model for acrylamide formation from glucose, fructose and asparagine proposed by Knol et al. 2005. Supp_file1: Calculation of initial concentrations of glucose, fructose and asparagine for pureed potato and prune juice.

Author Contributions: Conceptualization: J.A.E. and P.S.F.; methodology, J.A.E., A.G. and J.L.P.-S.; software, J.L.P.-S. and A.G.; validation, J.A.E., A.G., A.A. and P.S.F.; formal analysis, J.A.E., A.G., A.A. and P.S.F.; investigation, J.L.P.-S.; resources, J.A.E. and P.S.F.; data curation, J.L.P.-S. and A.G.; writing—original draft preparation, J.L.P.-S.; writing—review and editing, J.A.E., A.G., A.A., P.S.F. and J.L.P.-S.; visualization, J.L.P.-S.; supervision, J.A.E., A.G. and P.S.F.; project administration, J.A.E. and P.S.F.; funding acquisition, J.A.E. and P.S.F. All authors have read and agreed to the published version of the manuscript.

Funding: The financial support of this research work was provided by the Ministry of Science, Innovation and Universities of the Spanish Government and European Regional Development Fund (ERDF) through project AGL2017-86840-C2-1-R. J.L.P.-S. is grateful to the JAE-INTRO program from CSIC (Grant no JAEINT19_EX_0797). A.G. was supported by a postdoctoral grant from the Fundación Séneca (20900/PD/18).

Institutional Review Board Statement: Not applicable.

Informed Consent Statement: Not applicable.

Conflicts of Interest: The authors declare no conflict of interests.

References

1. Perrot, N.; Trelea, I.; Baudrit, C.; Trystram, G.; Bourgine, P. Modelling and analysis of complex food systems: State of the art and new trends. *Trends Food Sci. Technol.* **2011**, *22*, 304–314. [CrossRef]
2. Trystram, G. Modelling of Food and Food Processes. *J. Food Eng.* **2012**, *110*, 269–277.
3. Banga, J.R.; Balsa-Canto, E.; Moles, C.G.; Alonso, A. Improving food processing using modern optimization methods. *Trends Food Sci. Technol.* **2003**, *14*, 131–144. [CrossRef]
4. Erdoğdu, F. *Optimization in Food Engineering*, 1st ed.; CRC Press: Boca Raton, FL, USA, 2008; ISBN 9781420061420.
5. Madoumier, M.; Trystram, G.; Sébastian, P.; Collignan, A. Towards a holistic approach for multi-objective optimization of food processes: A critical review. *Trends Food Sci. Technol.* **2019**, *86*, 1–15. [CrossRef]
6. Gergely, S.; Békássy-Molnár, E.; Vatai, G. The use of multiobjective optimization to improve wine filtration. *J. Food Eng.* **2003**, *58*, 311–316. [CrossRef]
7. Sendín, J.O.H.; Alonso, A.; Banga, J.R. Efficient and robust multi-objective optimization of food processing: A novel approach with application to thermal sterilization. *J. Food Eng.* **2010**, *98*, 317–324. [CrossRef]
8. Kiranoudis, C.; Markatos, N. Pareto design of conveyor-belt dryers. *J. Food Eng.* **2000**, *46*, 145–155. [CrossRef]
9. Holdsworth; Simpson, R. Multiobjective Optimization in Thermal Food Processing. In *Thermal Processing of Packaged Foods*; Springer International Publishing: Cham, Switzerland, 2016; pp. 415–424. ISBN1 978-3-319-24902-5. ISBN2 978-3-319-24904-9.
10. Abakarov, A.; Sushkov, Y.; Mascheroni, R.H. A multi-criteria optimization and decision-making approach for improvement of food engineering processes. *Int. J. Food Stud.* **2013**, *2*, 1–21. [CrossRef]
11. Winiczenko, R.; Górnicki, K.; Kaleta, A.; Martynenko, A.; Janaszek-Mańkowska, M.; Trajer, J. Multi-objective optimization of convective drying of apple cubes. *Comput. Electron. Agric.* **2018**, *145*, 341–348. [CrossRef]
12. Krüger, C.; Castellani, F.; Geldermann, J.; Schöbel, A. Peat and pots: An application of robust multiobjective optimization to a mixing problem in agriculture. *Comput. Electron. Agric.* **2018**, *154*, 265–275. [CrossRef]
13. Bortolini, M.; Faccio, M.; Ferrari, E.; Gamberi, M.; Pilati, F. Fresh food sustainable distribution: Cost, delivery time and carbon footprint three-objective optimization. *J. Food Eng.* **2016**, *174*, 56–67. [CrossRef]
14. Vilas, C.; Mauricio-Iglesias, M.; García, M.R. Model-based design of smart active packaging systems with antimicrobial activity. *Food Packag. Shelf Life* **2020**, *24*, 100446. [CrossRef]
15. Holdsworth, S. Optimisation of thermal processing—A review. *J. Food Eng.* **1985**, *4*, 89–116. [CrossRef]
16. Mogol, B.A.; Gökmen, V. Thermal process contaminants: Acrylamide, chloropropanols and furan. *Curr. Opin. Food Sci.* **2016**, *7*, 86–92. [CrossRef]
17. Mottram, D.S.; Wedzicha, B.L.; Dodson, A.T. Acrylamide is formed in the Maillard reaction. *Nature* **2002**, *419*, 448–449. [CrossRef] [PubMed]
18. Johnson, K.A.; Gorzinski, S.J.; Bodner, K.M.; Campbell, R.A.; Wolf, C.H.; Friedman, M.A.; Mast, R.W. Chronic toxicity and oncogenicity study on acrylamide incorporated in the drinking water of Fischer 344 rats. *Toxicol. Appl. Pharmacol.* **1986**, *85*, 154–168. [CrossRef]
19. Rice, J.M. The carcinogenicity of acrylamide. *Mutat. Res. Toxicol. Environ. Mutagen.* **2005**, *580*, 3–20. [CrossRef]
20. European Food Safety Authority. Update on acrylamide levels in food from monitoring years 2007 to 2010. *EFSA J.* **2012**, *10*, 2938. [CrossRef]
21. Tareke, E.; Rydberg, P.; Karlsson, P.; Eriksson, S.; Törnqvist, M. Analysis of Acrylamide, a Carcinogen Formed in Heated Foodstuffs. *J. Agric. Food Chem.* **2002**, *50*, 4998–5006. [CrossRef]
22. Maillard, L.C. Action Des Acides Aminés Sur Les Sucres: Formation Des Mélanoïdines Par Voie Méthodique. *Comptes. R. Acad. Sci.* **1912**, *154*, 66–68.
23. Caixeta, A.T.; Moreira, R.; Castell-Perez, M.E. IMPINGEMENT DRYING OF POTATO CHIPS. *J. Food Process. Eng.* **2002**, *25*, 63–90. [CrossRef]
24. Markowski, M.; Cenkowski, S.; Hatcher, D.W.; Dexter, J.E.; Edwards, N.M. The effect of superheated—Steam dehydration kinetics on textural properties of asian noodles. *Trans. ASAE* **2003**, *46*, 389–395. [CrossRef]
25. Elias, A.; Roasto, M.; Reinik, M.; Nelis, K.; Nurk, E.; Elias, T. Acrylamide in commercial foods and intake by infants in Estonia. *Food Addit. Contam. Part A* **2017**, *34*, 1875–1884. [CrossRef]
26. Mousavi Khaneghah, A.; Fakhri, Y.; Nematollahi, A.; Seilani, F.; Vasseghian, Y. The Concentration of Acrylamide in Different Food Products: A Global Systematic Review, Meta-Analysis, and Meta-Regression. *Food Rev. Int.* **2020**, 1–19. [CrossRef]
27. Elmore, J.S.; Xu, F.; Maveddat, A.; Kapetanou, R.; Qi, H.; Oruna-Concha, M.J. Acrylamide content of vegetable chips. In *Food-Borne Toxicants: Formation, Analysis, and Toxicology*; ACS Symposium Series; American Chemical Society: Washington, DC, USA, 2019; Volume 1306, pp. 15–26, ISBN 9780841233874.
28. Becalski, A.; Brady, B.; Feng, S.; Gauthier, B.; Zhao, T. Formation of acrylamide at temperatures lower than 100 °C: The case of prunes and a model study. *Food Addit. Contam. Part A* **2011**, *28*, 726–730. [CrossRef] [PubMed]
29. Stumbo, C.R. *Thermobacteriology in Food Processing*; Elsevier: Amsterdam, The Netherlands, 1973; ISBN 978-0-12-675352-3.
30. Postollec, F.; Mathot, A.-G.; Bernard, M.; Divanac'H, M.-L.; Pavan, S.; Sohier, D. Tracking spore-forming bacteria in food: From natural biodiversity to selection by processes. *Int. J. Food Microbiol.* **2012**, *158*, 1–8. [CrossRef] [PubMed]
31. Brown, K. Spore resistance and ultra heat treatment processes. *J. Appl. Bacteriol.* **1994**, *76*, 67S–80S. [CrossRef] [PubMed]

32. Spicher, G.; Peters, J.; Borchers, U. Microbiological efficacy of superheated steam. I. Communication: Results with spores of Bacillus subtilis and Bacillus stearothermophilus and with spore earth. *Zentralblatt Hyg. Umweltmed. Int. J. Hyg. Environ. Med.* **1999**, *201*, 541–553.
33. Carlberg, D.M. *2005 Cleanroom Microbiology for the Non-Microbiologist*, 2nd ed.; CRC Press: Boca Raton, FL, USA, 2005, ISBN 978-0-8493-1996-9.
34. Fernández, P.; Ocio, M.; Rodrigo, F.; Martínez, A. Mathematical model for the combined effect of temperature and pH on the thermal resistance of Bacillus stearothermophilus and Clostridium sporogenes spores. *Int. J. Food Microbiol.* **1996**, *32*, 225–233. [CrossRef]
35. Bigelow, W. The Logarithmic Nature of Thermal Death Time Curves. *J. Infect. Dis.* **1921**, *29*, 528–536. [CrossRef]
36. Mann, A.; Kiefer, M.; Leuenberger, H. Thermal Sterilization of Heat-sensitive Products Using High-temperature Short-time Sterilization. *J. Pharm. Sci.* **2001**, *90*, 275–287. [CrossRef]
37. Leguerinel, I.; Couvert, O.; Mafart, P. Modelling the influence of the incubation temperature upon the estimated heat resistance of heated bacillus spores. *Lett. Appl. Microbiol.* **2006**, *43*, 17–21. [CrossRef] [PubMed]
38. André, S.; Leguerinel, I.; Palop, A.; Desriac, N.; Planchon, S.; Mafart, P. Convergence of Bigelow and Arrhenius models over a wide range of heating temperatures. *Int. J. Food Microbiol.* **2018**, *291*, 173–180. [CrossRef] [PubMed]
39. Poschet, F.; Geeraerd, A.; Van Loey, A.; Hendrickx, M.; Van Impe, J. Assessing the optimal experiment setup for first order kinetic studies by Monte Carlo analysis. *Food Control.* **2005**, *16*, 873–882. [CrossRef]
40. Peñalver-Soto, J.L.; Garre, A.; Esnoz, A.; Fernández, P.S.; Egea, J.A. Guidelines for the design of (optimal) isothermal inactivation experiments. *Food Res. Int.* **2019**, *126*, 108714. [CrossRef] [PubMed]
41. Van Asselt, E.D.; Zwietering, M.H. A systematic approach to determine global thermal inactivation parameters for various food pathogens. *Int. J. Food Microbiol.* **2006**, *107*, 73–82. [CrossRef]
42. Gibbiel, A.Y.; Al, A.T.H.A.-E. Measurement of Heat Resistance Parameters for Spores Isolated from Canned Products. *J. Appl. Bacteriol.* **1973**, *36*, 321–327. [CrossRef]
43. Hassan, H.F.; Ramaswamy, H.S. Heat resistance of g. stearothermophilus and c. sporogenes in carrot and meat alginate purees. *J. Food Process. Preserv.* **2011**, *35*, 376–385. [CrossRef]
44. Fernandez, P.S.; Ocio, M.J.; Sanchez, T.; Martinez, A. Thermal Resistance of Bacillus stearothermophilus Spores Heated in Acidified Mushroom Extract. *J. Food Prot.* **1994**, *57*, 37–41. [CrossRef]
45. Fernandez, P.S.; Gómez, F.J.; Ocio, M.J.; Rodrigo, M.; Sánchez, T.; Martínez, A. D Values of Bacillus stearothermophilus Spores as a Function of pH and Recovery Medium Acidulant. *J. Food Prot.* **1995**, *58*, 628–632. [CrossRef]
46. Ocio, M.; Fernández, P.; Rodrigo, F.; Martínez, A. Heat resistance of Bacillus stearothermophilus spores in alginate-mushroom puree mixture. *Int. J. Food Microbiol.* **1996**, *29*, 391–395. [CrossRef]
47. López, M. Effect of pH heating medium on the thermal resistance of Bacillus stearothermophilus spores. *Int. J. Food Microbiol.* **1996**, *28*, 405–410. [CrossRef]
48. Molin, N.; Snygg, B.G. Effect of Lipid Materials on Heat Resistance of Bacterial Spores. *Appl. Microbiol.* **1967**, *15*, 1422–1426. [CrossRef]
49. Yildiz, F.; Westhoff, D. Sporulation and thermal resistance of Bacillus stearothermophilus spores in milk. *Food Microbiol.* **1989**, *6*, 245–250. [CrossRef]
50. Etoa, F.-X.; Michiels, L. Heat-induced resistance of Bacillus stearothermophilus spores. *Lett. Appl. Microbiol.* **1988**, *6*, 43–45. [CrossRef]
51. Knol, J.J.; van Loon, W.A.M.; Linssen, J.P.H.; Ruck, A.-L.; van Boekel, M.A.J.S.; Voragen, A.G.J. Toward a Kinetic Model for Acrylamide Formation in a Glucose–Asparagine Reaction System. *J. Agric. Food Chem.* **2005**, *53*, 6133–6139. [CrossRef] [PubMed]
52. Vivanti, V.; Finotti, E.; Friedman, M. Level of Acrylamide Precursors Asparagine, Fructose, Glucose, and Sucrose in Potatoes Sold at Retail in Italy and in the United States. *J. Food Sci.* **2006**, *71*, C81–C85. [CrossRef]

foods

MDPI

Article

Multiobjective Optimization of a Frying Process Balancing Acrylamide Formation and Quality: Solution Analysis and Uncertainty Propagation

Jose Lucas Peñalver-Soto [1,2], María Muñoz-Guillermo [3], Alberto Garre [1], Asunción Iguaz [1], Pablo S. Fernández [1] and Jose A. Egea [2,*]

1 Departamento de Ingeniería de Alimentos y del Equipamiento Agrícola, Instituto de Biotecnología Vegetal, Universidad Politécnica de Cartagena (ETSIA), Paseo Alfonso XIII, 48, 30203 Cartagena, Spain
2 Departamento de Mejora Vegetal, Centro de Edafología y Biología Aplicada del Segura (CEBAS-CSIC), Campus Universitario de Espinardo S/N, 30100 Murcia, Spain
3 Departamento de Matemática Aplicada y Estadística, Universidad Politécnica de Cartagena, Antiguo Hospital de Marina (ETSII), Av. Dr. Fleming S/N, 30202 Cartagena, Spain
* Correspondence: jaegea@cebas.csic.es

Abstract: In this study, we performed multi-objective model-based optimization of a potato-frying process balancing between acrylamide production and a quality parameter (yellowness). Solution analysis revealed that, for most of the Pareto solutions, acrylamide levels exceeded the EFSA recommendation. Almost equivalent optimal solutions were found for moderate processing conditions (low temperatures and/or processing times) and the propagation of the uncertainty of the acrylamide production model parameters led to Pareto fronts with notable differences from the one obtained using the nominal parameters, especially in the ranges of high values of acrylamide production and yellowness. These results can help to identify processing conditions to achieve the desired acrylamide/yellowness balance and design more robust processes allowing for the enhancement of flexibility when equivalent optimal solutions can be retrieved.

Keywords: multi-objective optimization; model-based optimization; equivalent solutions; uncertainty; Monte Carlo; frying operation; acrylamide; quality

Citation: Peñalver-Soto, J.L.; Muñoz-Guillermo, M.; Garre, A.; Iguaz, A.; Fernández, P.S.; Egea, J.A. Multiobjective Optimization of a Frying Process Balancing Acrylamide Formation and Quality: Solution Analysis and Uncertainty Propagation. *Foods* **2022**, *11*, 3689. https://doi.org/10.3390/foods11223689

Academic Editor: Cristina Delgado-Andrade

Received: 23 September 2022
Accepted: 12 November 2022
Published: 17 November 2022

1. Introduction

1.1. Optimization in Food Engineering

Food engineering has become an increasingly important field, as evidenced by the growth of mathematical models devoted to understanding and improving food-processing operations [1]. One important application of mathematical modelling in food engineering is the optimization of food-processing operations.

Optimization is the process of finding the best possible solution to a problem. This usually involves finding the best compromise among several conflicting demands. To optimize a process, one must find the set of decision variables which, for example, maximize profitability while meeting a set of constraints. Several model-based optimization methods can be used to improve food processing. These methods are more rigorous than other empirical approaches and are thus more likely to find the best possible solution [2].

In general, optimization can be applied effectively to food processing if the changes during the process can be predicted mathematically. Heat, mass, and momentum transfers (as well as kinetics) are major mechanisms in food processing, and mathematical models describing these phenomena are essential for further mathematical-based optimization procedures [3].

Optimal operating conditions in the food industry are usually sought to ensure maximum profits and product quality, subject to constraints arising from food-safety issues and often environmental regulations. However, the dynamic, nonlinear and highly constrained nature of food-processing models can make the optimization of these processes a daunting task [4].

Achieving optimization in food processing requires some way of describing the potential alternatives and of choosing the best alternative. In the design, construction and maintenance of any engineering system, different technological and managerial decisions are required to be given at different stages of the process to either minimize the effort required or maximize the benefit desired. The formal description of any optimization problem has three parts [3]:

1. A set of variables that the optimization method can control and use to specify the alternatives (e.g., applying different process-temperature profiles during thermal processing to achieve better processing for a given objective function).
2. A set of requirements (e.g., the differential equations, boundary conditions, and integral equations specifying the constraints that the system and the variables are subjected to) that the optimization method must achieve or satisfy.
3. A measure of performance to compare one alternative to another (the objective function). The objective function, which may be continuous or in some cases discrete, is the function to be optimized (maximized or minimized). This can be accomplished by using either a mathematical model or by fitting an equation through the experimental data.

1.2. *Applications of Multi-Objective Optimization in Food Engineering*

For most industrial processes in food, simultaneous optimization of multiple objectives (e.g., product quality, operating costs, and safety) is the more realistic and desirable approach, but since these criteria are often opposing, the optimal solution is not unique. The multi-objective optimization (MOO) approach is used to find the best set of solutions for a problem with multiple objectives. In food engineering, MOO is used to optimize processes where conflicting objectives such as e.g., process economy, quality parameters or environmental indexes appear. These solutions are known as nondominated or Pareto optimal solutions [5]. Each of these solutions has no prior advantage over other Pareto optimal solutions so the objective of multi-objective optimization is to generate as many solutions as possible to evaluate and prioritize optimal trade-offs among the different objectives [6].

Multi-objective approaches have been used to solve optimization problems in the food engineering industry. For instance, Vilas et al. sought to maximize food quality and safety by developing smart active packaging systems that optimize food-packaging design and prediction of the expected shelf life along the food chain [7]. Abakarov used this technique with experimental data obtained on osmotic dehydration of carrot cubes in a sodium chloride solution to improve the assessment of criteria weights and produce fairer and more consistent products [6]. Holdsworth and Simpson obtained a set of Pareto-optimal solutions for processing time, quality retention, and texture loss under specific criteria of the processing temperature [8]. Krüger et al. proposed a multi-objective optimization to choose a pot and a growth substrate mixture such that environmental emissions and costs are simultaneously minimized [9] and Gergely et al. used this approach to improve wine filtration [10]. Sendín et al. used it to maximize the retention of several nutrients and quality factors and minimize the total process time [11]. Kiranoudis and Markatos considered the multi-objective approach to design the process of a conveyor-belt dryer using not only structural and operational process variables but the quality of treated potatoes [12]. In the same line, Olmos et al. used this approach to optimize the drying time maximizing the product quality [13] and Winiczenko et al. studied the effect of drying temperature and air velocity on apple quality parameters, such as color difference, volume ratio and water absorption capacity in convective drying [14]. In the field of sustainable distribution of foods, Bortolini et al. optimized the cost, delivery time and carbon footprint with a multi-objective approach [15]. However, no work was found where acrylamide production and food quality parameters were considered simultaneously in a multi-objective optimization approach. Peñalver-Soto et al. analysed the dynamics of acrylamide production and microbiological inactivation in certain foods by performing simulations instead of a formal optimization formulation [16].

1.3. Uncertainty in MOO

Uncertainty propagation has been extensively studied in the fields of physics and mathematics [17–21] and in particular, in the field of food intake, where it has been analyzed from different approaches [22–27]. In general, the propagation of uncertainty refers to the estimation of the variability in a given quantity. This variability can be due to several factors, such as measurement error, sampling variability, or natural variability [22]. In food engineering, the variability of a particular property or characteristic of a food product can have a significant impact on the quality and safety of the product.

In this study, the propagation of uncertainty to the MOO approach was assessed by propagating the uncertainty of model parameters to the solutions shown in the Pareto front. Parameter uncertainty can affect the shape of the Pareto optimal front in multi-objective optimization, and this can have important implications for decision-making [17].

Different methods can be used to propagate uncertainty in food engineering. Each method has its strengths and weaknesses, and the choice often depends on the type of data being studied. Some of the most common methods of uncertainty propagation include Monte-Carlo simulation [28], linear approximation [29], the sigma point method [30], and polynomial chaos expansion [31].

Specifically, this article applies the Monte-Carlo method which is a powerful tool for studying the propagation of uncertainty [32]. The method is used to calculate the probability of different outcomes by randomly selecting values from a probability distribution. This approach can be used to calculate the expected value of a function or to estimate the uncertainty in a measurement. The Monte-Carlo method can be used to study the propagation of uncertainty in food engineering. Garre et al. used this methodology in microbial inactivation of foods to select optimal experiment designs [33]. In this work, we analyzed the effects of parameter uncertainty in mathematical models describing food processes over the robustness of the Pareto set of solutions in multi-objective optimization using as a case study a frying process of potato chips where a quality parameter (yellowness) and the production of acrylamide, a potential carcinogen [34,35] were defined as opposed objectives.

2. Materials and Methods

2.1. Case Study

This study analyzed the potato-frying process to optimize food safety and culinary quality. Specifically, the impact of the heat treatment on the amount of acrylamide produced and the yellowness and moisture content were studied. The whole study was based on mathematical model simulations. We used the Maillard model (Equations (1)–(5)) to simulate the acrylamide formation using the fitted model provided by Knol et al. [36] and the models proposed by Krokida et al. [37,38] (Equations (6)–(11)) to simulate the yellowness and the moisture content. All models (see Section 2.2) were previously calibrated and validated by their original authors as reported in the respective bibliographic references The outputs of such models, which depend upon temperature and time, were used to simulate the experiments and perform the multi-objective optimization as well as the uncertainty propagation analysis. The simulated heat treatments considered were all in the range of values for which the mathematical models were validated according to their authors [36–38].

The selected potato corresponded to the Agria variety, which is used in products sold in supermarket chains and has been one of the most studied varieties in frying conditions in the literature [39–41]. The high temperatures of the considered heat treatment (frying process) brought the two considered objectives into conflict since an increase in yellowness implies an increase in acrylamide production, thus a multi-objective approach was used. Slices of 15 mm thickness were considered in this work. Likewise, no air drying or osmotic pre-treatment were considered [42]. The experiments carried out by Krokida et al. to calibrate the yellowness model used a commercial deep fat fryer with temperature control of ±1 °C that was filled with 2 l of oil and the potatoes-to-oil ratio was kept at 1:50 w/v. The concentration of hydrogenated cotton seed oil in total (refined plus hydrogenated) oil was considered as 50% [37]. Nevertheless, these authors stated that the type of oil did not have any influence over color parameters. They used a Hunterlab SAV colorimeter and reported the results in the CIE Lab color scale (non-dimensional) in their experiments [37,38]. Food safety was determined by low levels of acrylamide. EFSA [43] determines 50 µg/kg as the maximum level. On the other hand, culinary quality was determined by the maximization of yellowness and the setting of moisture content between 2 and 4% as recommended by Segnini et al. [44], as an indicator or predictor of texture.

If any of the described parameters took different values, a different potato variety was considered or additional quality variables (e.g., textural ones) were incorporated, and the methodology remained the same. Here we intend to illustrate how to design a frying process using modelling tools and, particularly, multi-objective optimization, as well as providing a global picture of the balances between objectives in the whole design variables domain, as shown in the visual scheme in Figure 1.

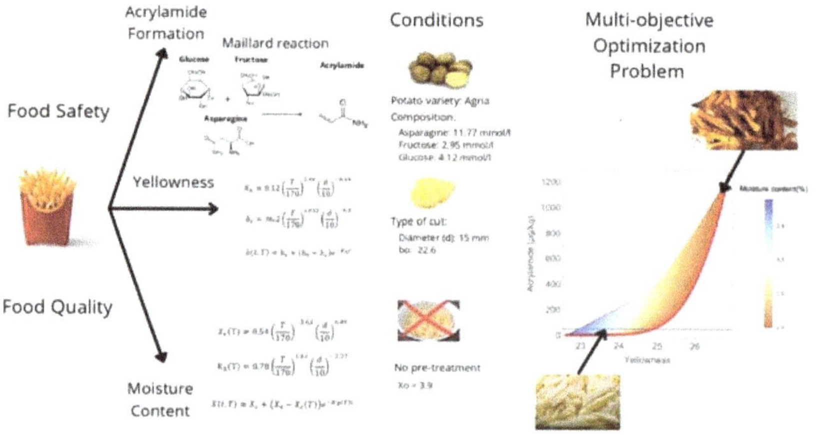

Figure 1. Outline of the case study.

2.2. Mathematical Models

2.2.1. Acrylamide Production

Because of its different applications in industry as a reactive molecule to synthesize polyacrylamide, acrylamide has been a focus of great interest [45,46]. Safe levels of exposure to acrylamide in human beings have been analyzed and studied. For a detailed review see [45] in which, among others, data on toxicology are included. EFSA in its latest report on the assessment of the genotoxicity of acrylamide [47], considered the possible modes of action of acrylamide carcinogenicity, including genotoxic and non-genotoxic effects. The paper concludes that there is substantial evidence for acrylamide genotoxicity mediated by metabolite formation, in addition to a possible contribution of non-genotoxic effects to acrylamide carcinogenicity. This is particularly interesting in food processes in which the heat treatment produces levels of acrylamide so large that they need to be controlled.

In this framework, different models have been considered. To quantify the acrylamide formation we used multi-response kinetics in a fructose–asparagine reaction at high temperatures (120–200 °C) proposed by Knol et al. [36]. The model is based on the reaction network shown in Equations (1) to (5).

Fructose and asparagine are degraded into glucose, acid acetic, Schiff base and unknown species (X_1). At the same time, the Schiff base is degraded into melanoidins and acrylamide. Knol et al. fitted the equilibrium constants for the temperature range (120–200 °C) and showed a logarithmic relationship with temperature [36]. Therefore, the equilibrium constants for each temperature are calculated ($K_i(T)$ for each $i = 1, 2, \ldots, 6$). The system of ordinary differential equations (ODEs) that relates the amount of acrylamide formed for a specific time t (minutes) and temperature T(°C) is defined in Equation (5).

$$\frac{d[Glucose]}{dt} = -K_1(T)\cdot[Glucose]\cdot[Asparagine] - K_2(T)\cdot[Glucose] \quad (1)$$

$$\frac{d[Fructose]}{dt} = -K_3(T)\cdot[Fructose]\cdot[Asparagine] + K_2(T)\cdot[Glucose] \quad (2)$$

$$\frac{d[Asparagine]}{dt} = -K_1(T)\cdot[Glucose]\cdot[Asparagine] - K_3(T)\cdot[Fructose]\cdot[Asparagine] \quad (3)$$

$$\frac{d[Schiff\ base]}{dt} = K_1(T)\cdot[Glucose]\cdot[Asparagine] + K_3(T)\cdot[Fructose]\cdot[Asparagine] - K_4(T)\cdot[Schiff\ base] - K_5(T)\cdot[Schiff\ base] \quad (4)$$

$$\frac{d[Acrylamide]}{dt} = K_4(T)\cdot[Schiff\ base] - K_6(T)\cdot[Acrylamide] \quad (5)$$

One of the outputs of the ODEs is the acrylamide concentration formed for a heat treatment (time, temperature), which was the first optimization objective in our formulation. To solve this set of ODEs, apart from the heat treatment conditions, that is, the time and temperature variables, it is necessary to set the initial amounts of fructose, glucose, and asparagine. For the "Agria" potato, which was the variety studied, the compositions were 11.77 mmol/L of asparagine, 2.95 mmol/L of fructose and 4.12 mmol/L of glucose. Details on these calculations are provided in the Supplementary Materials [48,49].

2.2.2. Yellowness

The second considered objective was a quality parameter related to the color of fried potato: yellowness. Pedreschi considered it as one of the quality parameters of interest for fried potatoes [50]. Color-related parameters are of great importance for the product to be attractive to the consumer [37–46,50,51]. In particular, different studies revealed that high values of yellowness are preferred by consumers [52]. The problem is that at the same time, an increase in temperature also implies an increase in another color-related parameter, redness, which is not a desirable quality in the final product [37]. Roughly speaking a good level of yellowness is "the goal" but, at the same time, redness must be minimized. Nevertheless, Knol et al. [36] and Pedreschi et al. [53] indicated that acrylamide concentration shows a good linear correlation with the redness of potato chips. Therefore redness can be indirectly controlled by the level of acrylamide (recall that the first objective considered in this work was to minimize the concentration of acrylamide). Here we used the model for the yellowness, namely b, proposed by Krokida et al. [37], described in Equation (6)–(8), where d is the thickness of the slice (mm) and T is the temperature (°C).

$$K_b = 0.12\left(\frac{T}{170}\right)^{2.49}\left(\frac{d}{10}\right)^{-0.44} \quad (6)$$

$$b_e = 36.2\left(\frac{T}{170}\right)^{1.012}\left(\frac{d}{10}\right)^{-0.2} \quad (7)$$

$$b(t, T) = b_e + (b_0 - b_e)e^{-K_b t} \quad (8)$$

Here we set the values of $d = 15$ mm and $b_0 = 22.6$ (corresponding to no pretreatment processes [37]). As observed in Equations (6)–(8), b, like our first-considered objective (acrylamide production), depends on both the time and temperature.

2.2.3. Moisture Content

The moisture content of the fried product is an important quality parameter. Its control is necessary to achieve the desired taste, texture, and color of the product. The moisture content indicates the water loss from the potato strips during frying. It decreases significantly when the potato is fried. The temperature of the oil has a negative effect on the moisture content of fried potatoes. The higher the temperature of the frying oil, the lower the moisture content for the same frying time. Moisture content is also related to one of the quality aspects most valued by consumers, the degree of crispiness of fried potatoes. There is a direct relationship between these variables: the higher the moisture content, the lower the crispness. Therefore, it is of utmost importance that the moisture content value is maintained between 2% and 4% as recommended by Segnini et al. [44]. Following the model proposed by Krokida et al. [38], Equations (9)–(11) define the moisture content as a function of treatment time and temperature.

$$X_e(\mathrm{T}) = 0.54\left(\frac{T}{170}\right)^{-3.63}\left(\frac{d}{10}\right)^{0.89} \tag{9}$$

$$\mathrm{K_X}(\mathrm{T}) = 0.78\left(\frac{T}{170}\right)^{1.61}\left(\frac{d}{10}\right)^{-2.27} \tag{10}$$

$$X(t,T) = X_e + (X_0 - X_e(T))e^{-K_X(T)t} \tag{11}$$

where $X_0 = 3.9$ and $d = 15$ mm, corresponding to no pretreatment, as indicated by Krokida et al. [38].

In our multi-objective optimization problem, we formulated the moisture content as a constraint whose value at the end of the frying process must lie between 2% and 4%, as recommended by Segnini et al. [44].

2.3. Multi-Objective Problem (MOP)

Multi-objective optimization aims at finding the best possible solutions to a set of conflicting objectives, Equations (12)–(16) define the mathematical formulation applied to our case study.

$$\min_{u(t)} F(x(t),\, u(t)) \tag{12}$$

$$\text{subject to :}$$

$$\frac{dx}{dt} = \Psi(x(t),\, u(t),\, t) \tag{13}$$

$$x(t_0) = x_0 \tag{14}$$

$$g(x(t),\, u(t)) \le 0 \tag{15}$$

$$u^L \le u(t) \le u^S \tag{16}$$

where the vector of objective functions, Equation (12), contains all the objectives considered in the problem. In our case, the objectives were already defined as f_1 = acrylamide production (Equation (5)) and f_2 = −yellowness (Equation (8)), (note that the negative sign indicates that this objective is maximized). x is the vector of state variables (e.g., chemical species concentrations) and u is the vector of control variables (temperature and processing time in our case). Equation (13) represents the system dynamics (dynamic mathematical models that define acrylamide production and yellowness). Equation (14) represents the values of the stated variables at the beginning of the process ($t = 0$). Equation (15) represents inequality constraints, which can be considered at the end of the process or at intermediate times (moisture content in our MOP). Finally, Equation (16) corresponds to the lower and upper boundaries for the control variables (e.g., the minimum and maximum temperature and processing time). In our problem, those boundaries were defined as [0.1, 10] min for time and [120, 200] °C for temperature.

There are several methods for solving the problem. In the first approach, we used a systematic complete search using the nominal values for the model's parameters to obtain not only the Pareto front but also the whole feasible region. The procedure is described as follows: A set D is defined as $D := \{(t, T) \in [0.1,\ 10] \times [120, 200] : 2 \leq X(t, T) \leq 4\}$, where t is the processing time, T is the temperature, and X is the moisture content. For every value b^* of yellowness within the interval [22.6, 26.9], which corresponds to the minimum and maximum yellowness values in the ranges of times and temperatures considered, we calculate the level curve $S_{b^*} := \{(t, T) \in [0.1,\ 10] \times [120, 200] : b(t, T) = b^*\} \cap D$. The numerical calculation of S_{b^*} provides the feasible region and the values that minimize the acrylamide for every S_{b_i} provide the Pareto front.

The described complete search procedure is computationally intensive since it evaluates all the solutions in the feasible regions. As stated above, this was only applied using the nominal parameter values. For assessing the uncertainty propagation, where 1000 Pareto fronts were calculated by simulating different values for the model parameters (see Section 2.4), the heuristic algorithm NSGA-II [54] was applied. Given the characteristics of the models considered in this study (nonlinear and dynamic), this type of algorithm is a suitable option to achieve good solutions (normally the optimal ones) in relatively short computational times [2]. Further, we checked that, for the nominal parameter values, the obtained Pareto front by NSGA-II coincided with that obtained with the complete search procedure. This study used the R package "nsga2" to perform the optimization of the proposed problem.

2.4. Uncertainty Propagation

Uncertainty propagation was applied to the estimation of the equilibrium constants of the differential equations describing acrylamide production (Equations (1)–(5)). The model considers up to six equilibrium constants for which confidence intervals are given in [36]. The Monte-Carlo method was used to simulate 1000 sets of positive parameter values following a normal distribution for each equilibrium constant The MOP was solved and a Pareto front for each combination of the simulated kinetic constants was obtained. Thus, 1000 Pareto fronts were obtained. These provide an idea of the uncertainty propagation of the equilibrium constants and their impact on the Pareto front.

3. Results

3.1. Multi-Objective Solutions

The multi-objective approach using the nominal values for the kinetic parameters provided in [36] led to a set of optimal (non-dominated) solutions (Pareto front) shown in Figure 2 together with the feasible space. The vertical axis represents the amount of acrylamide produced and the horizontal axis represents the yellowness. The Pareto front is represented as a thick red line. On the other hand, the colors of the feasible region represent the moisture content and the blue horizontal line represents the recommended limit for acrylamide [43].

The first relationship was, as expected, that the higher the yellowness, the lower the moisture content and the higher the amount of acrylamide. All of this was positively correlated with the treatment severity (i.e., higher temperatures and/or treatment times led to an increase in the above-mentioned variables). On the other hand, given the problem boundaries and constraints, the yellowness was limited to values between 22 and 27, while the acrylamide did not exceed 1300 μg/kg which is 26 times higher than the EFSA recommendation.

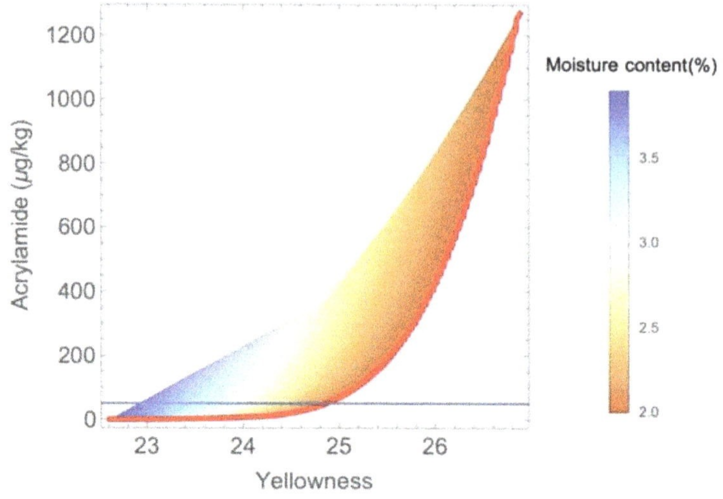

Figure 2. Feasible set of solutions and Pareto front.

A temperature–time representation of the Pareto front is shown in Figure 3. The points in red represent operational points that do not comply with the EFSA recommendation in terms of acrylamide amount, and they correspond to the highest temperatures. It can be seen that from approximately 155 °C, the dots form a curve that tends to be vertically asymptotic. This curve coincides with the conditions that keep the moisture content constraint active with a value of 2%.

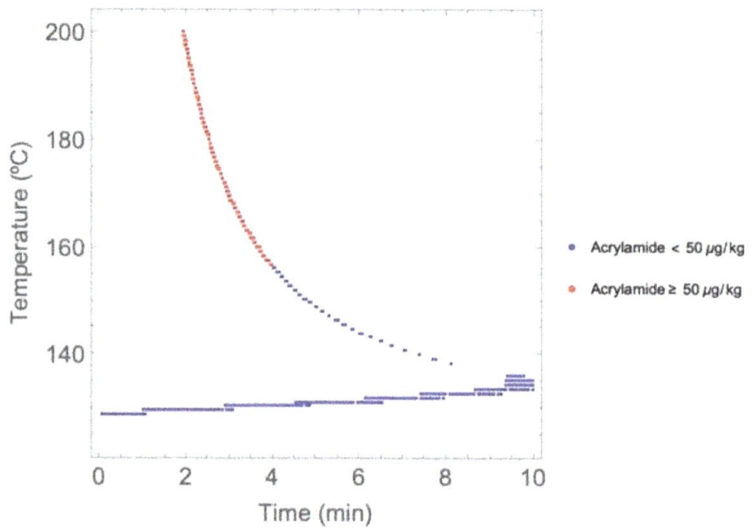

Figure 3. Pareto front temperature–time solutions.

As shown in Figures 2 and 3, most of the solutions from the Pareto front led to high levels of acrylamide, exceeding the recommended levels by up to 26 times in some cases. Working points around 200 °C with a duration of approximately 2 min generated between 1200 and 1300 μg/kg of acrylamide when the recommended upper limit is 50 μg/kg.

During the optimization process, the existence of multiple quasi-equivalent solutions in the Pareto front was found for the ranges of low acrylamide production and low yellowness (e.g., low temperatures and/or processing times). The existence of these multiple solutions was caused by the flatness of the objective functions in areas of low temperatures and processing times (Figures S1 and S2). Table 1 illustrates some of these Pareto equivalent solutions. Equivalent solutions were defined as having the same values of acrylamide and yellowness with a tolerance of 0.01 but differences in temperature and time of at least 1 °C and 0.2 min, respectively. Figure 4 shows, in the temperature–time domain, the sets of equivalent solutions found for selected points of the Pareto front. Quasi-equivalent solutions are represented with the same color in Figure 4.

Table 1. Selection of Pareto fronts and quasi-equivalent solutions.

Solutions	Time (min)	Temperature (°C)	Acrylamide (μg/kg)	Yellowness	Moisture Content (%)
#1	1.97	120.48	0.028	22.67	3.540
#1′	0.16	150.00	0.034	22.68	3.796
#2	1.99	124.74	0.069	22.76	3.425
#2′	0.61	138.00	0.076	22.76	3.614
#3	2.19	127.88	0.154	22.85	3.297
#3′	1.28	134.00	0.163	22.85	3.411
#4	2.05	132.91	0.345	22.96	3.199
#4′	3.61	127.00	0.348	22.96	3.060
#5	4.63	130.42	1.086	23.23	2.759
#5′	4.01	132.00	1.094	23.23	2.796
#6	5.03	132.39	1.821	23.39	2.617
#6′	6.29	130.00	1.823	23.39	2.571
#7	8.89	134.28	7.60	24.04	2.13
#7′	10.00	133.00	7.59	24.04	2.14

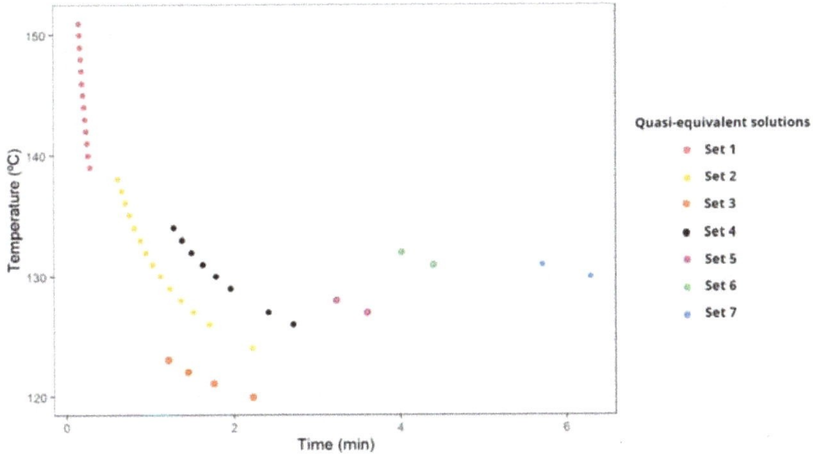

Figure 4. Selected sets of quasi-equivalent solutions.

Table 1 and Figure 4 show that the lower the time and temperature, the higher number of equivalent solutions. As time or temperature increases, the number of equivalent solutions decreases and the curve defined by them becomes more horizontal (i.e., temperature differences are relatively lower than the difference in processing time in these cases). It is of note that, mathematically speaking, no equivalent solutions for the Pareto front can be found in this problem but, due to the flatness of the objective functions in certain temperature–time ranges, a set of very similar (called equivalent here) solutions can be found, allowing processes to be flexible to achieve certain results. No equivalent solutions according to the definition above were found for temperature-time conditions where acrylamide values were above 9 μg/kg.

3.2. Uncertainty Propagation

In this section, we analyze the uncertainty propagation from the kinetic parameters of the Maillard equation characterized by Knol et al. [36] to the Pareto front of the multi-objective optimization problem. The result of the uncertainty propagation of the k's of the Maillard model, Equations (1)–(5), with the Monte-Carlo method yields the set of Pareto fronts shown in Figure 5.

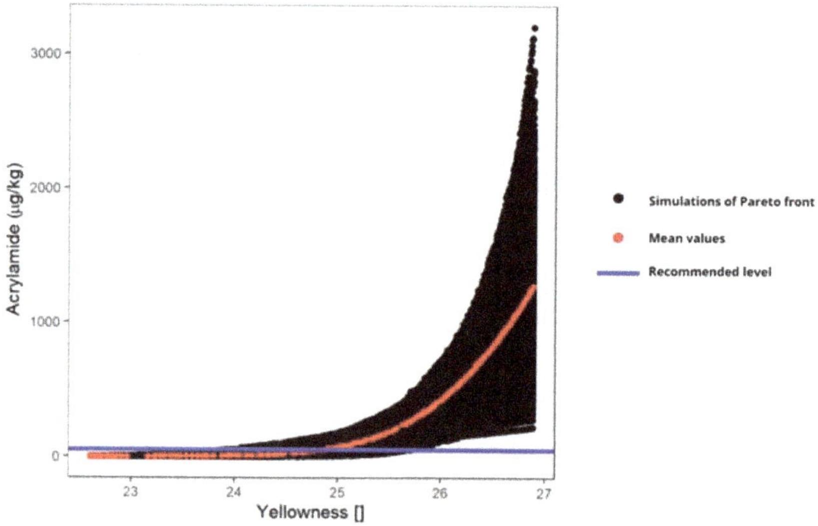

Figure 5. Set of 1000 Pareto fronts resulting from Monte-Carlo simulation of the kinetic parameters for the Maillard reaction.

Figure 5 shows, on the vertical axis, the amount of acrylamide produced and on the horizontal axis, the yellowness. The black dots are the Pareto fronts of the 1000 simulations where the red line represents the Pareto front with the mean values of the Maillard's kinetic parameters (shown in Figure 2) and the blue line represents the quantity of recommended acrylamide. The Pareto front resulting from the average kinetic values is located approximately in the middle zone of the solutions, so the assumed normal distributions for the kinetic parameters translate into a symmetric distribution of the Pareto solutions for each yellowness value. On the other hand, the uncertainty increases as the yellowness (i.e., temperature and/or time) increases. The combination of these two means that around 95% of the points considering all the 1000 Pareto fronts are outside the recommendation in terms of acrylamide production.

4. Discussion

This paper addresses the problem of food safety combined with product quality. It uses a multi-objective approach, which has been widely used in the literature [6–15]. Other studies such as that of Mestdagh [55] have studied balances between acrylamide and color but not from the quantitative and multi-objective optimization point of view addressed in this paper.

The MOP's solutions (Figure 2) show that most frying processes (considering the conditions established in Section 2.2) do not comply with EFSA recommendations. The maximum acrylamide amount recommended by EFSA could be formulated as an additional constraint (which would lead to a different Pareto front) or, alternatively, we could try to select those points corresponding to temperatures not higher than 155 °C, approximately (blue points in Figure 3). Therefore, to ensure lower acrylamide values than the maximum ones recommended by EFSA, it is recommended to use frying temperatures below 160 °C with frying times not exceeding 4 min. The next implication is that, under these conditions, the yellowness only reaches values of 22–25, so the recommended amount of acrylamide greatly limits the visual quality of the final product regarding yellowness.

These results are influenced by all the premises taken such as the potato variety and the type of thermal process, among others. Therefore, any change in these assumptions may influence the results, although the procedure and analysis is useful for studying this type of problem. For example, Johnson determined that, given their composition, not only can the potato variety modify the balance between the variables, but the way they are grown can also have an influence [56].

On the other hand, these solutions are not unique since at the practical level equivalent solutions appear. Therefore, two frying processes with different conditions (time or temperature) can produce the same amount of acrylamide while maintaining equivalent quality (yellowness and moisture content). This existence of equivalent solutions in the Pareto front was recently observed by Ortiz-Martínez et al. [57] in the multi-objective optimization of a wastewater process. In any case, an a priori analysis of the objective functions and their dynamics can help to anticipate whether multiple solutions for the Pareto front can appear [16].

The study of the propagation of the uncertainty associated with the parameters complements a study that provides an additional tool to take into account other possible scenarios. The presented analysis shows that, when performing multi-objective optimization for design purposes, the model parameters variability and their propagation must be taken into account to find sets of design options (i.e., Pareto fronts) that account for every possible scenario. In this particular application, it was also shown that the propagated variability is not the same in every part of the objectives space, being lower with soft operating conditions (low values of yellowness and acrylamide production in Figure 2 than with severe operating conditions (high values of the objectives and higher process-temperature or time).

For a proper analysis of the optimal solutions, it must be taken into account that this modelling exercise considered that the cooking temperature is uniform throughout the potato and that the temperature of the oil is equal to the temperature of the potato. Obviously, this does not actually happen [58,59], so these theoretical times could be increased without affecting the limit of acrylamide produced. However, we would still conclude that most operating conditions within the ranges of temperatures and times usually considered in real processes exceed the recommended acrylamide amount. Finally, recall that this analysis used the mean values of the estimated kinetic parameters of the acrylamide production [36] but, considering the uncertainty of such parameters, other scenarios may occur. To account for this, we used uncertainty propagation tools to take into account other possible scenarios.

5. Conclusions

The multi-objective optimization of a potato-frying process balancing between acrylamide formation and a quality parameter (yellowness) was addressed in this study. The results show that most of the optimal solutions (the Pareto front) considering the usual temperature and processing time ranges provide higher acrylamide amounts than the limit recommended by EFSA (50 µg/kg). In addition, multiple solutions for some areas of the Pareto front (namely, those providing low values of acrylamide) have been identified. The existence of these multiple solutions can be anticipated by a previous analysis of the objective function and their sensitivities to changes in the decision variables (temperature and processing time) in different areas of the search space. In our case, both objective functions showed flat areas in the ranges of low temperatures and processing times, which allows the existence of multiple optimal solutions. These multiple solutions are not equivalent from a mathematical point of view but they are from a practical point of view, allowing us to slightly change the processing conditions to obtain the same results for the objective function within a given tolerance.

The uncertainty of the kinetic parameters for acrylamide production has been propagated to the Pareto front using a Monte-Carlo simulation, showing that the uncertainty with respect to the Pareto front using the nominal values increases as the values of the objective functions do. This uncertainty must be taken into account when designing the frying process to make the design more robust and avoid undesirable solutions (e.g., too high acrylamide values).

We recommend, if possible, performing these analyses when performing model-based multi-objective optimization to design food processes. This type of methodology is of course not specific to food processes but, given their nature, where multiple objectives must be optimized simultaneously, it should be applied to them. Other objectives could be included (e.g., other quality parameters or economic or environmental factors) or other types of food/processes where acrylamide production may be an issue can be considered by applying the methodologies presented here. These methodologies can help in making optimal decisions where there are unexpected conditions deviations or in re-designing the processes.

Supplementary Materials: The following supporting information can be downloaded at: https://www.mdpi.com/article/10.3390/foods11223689/s1. Figure S1: Surface plot of the acrylamide production (Objective function 1) with respect to processing time and temperature; Figure S2: Surface plot of the yellowness (Objective function 2) with respect to processing time and temperature.

Author Contributions: Conceptualization: J.A.E., A.G. and P.S.F.; methodology, J.A.E., M.M.-G. and J.L.P.-S.; software, J.L.P.-S. and M.M.-G.; validation, J.A.E., M.M.-G., A.G., A.I. and P.S.F.; formal analysis, J.A.E., M.M.-G. and A.G.; investigation, J.L.P.-S. and A.I.; resources, J.A.E., M.M.-G. and P.S.F.; data curation, J.L.P.-S.; writing—original draft preparation, J.L.P.-S.; writing—review and editing, J.A.E., M.M.-G., A.G., A.I., P.S.F. and J.L.P.-S.; visualization, J.L.P.-S., J.A.E. and M.M.-G.; supervision, J.A.E., A.G. and P.S.F.; project administration, J.A.E. and P.S.F.; funding acquisition, J.A.E. and P.S.F. All authors have read and agreed to the published version of the manuscript.

Funding: This work has been supported by the Spanish Research Agency—AEI (MCIN/AEI/10.13039/501100011033) through projects PID2020-116318RB-C32 and MTM2017-84079-P (co-funded by "ERDF A way of making Europe", by the European Union). JLPS is grateful to the JAE-INTRO program from CSIC (Grant no JAEINT_19_00775).

Institutional Review Board Statement: Not applicable.

Informed Consent Statement: Not applicable.

Data Availability Statement: The datasets generated for this study are available on request to the corresponding author.

Conflicts of Interest: The authors declare no conflict of interests.

References

1. Ibarz, A.; Barbosa-Canovas, G.V. *Unit Operations in Food Engineering*; CRC Press: Boca Raton, FL, USA, 2002. [CrossRef]
2. Banga, J.R.; Balsa-Canto, E.; Moles, C.G.; Alonso, A. Improving food processing using modern optimization methods. *Trends Food Sci. Technol.* **2003**, *14*, 131–144. [CrossRef]
3. Erdoğdu, F. *Optimization in Food Engineering*; CRC Press: Boca Raton, FL, USA, 2008; pp. 1–762. [CrossRef]
4. García, M.-S.G.; Balsa-Canto, E.; Alonso, A.; Banga, J.R. Computing optimal operating policies for the food industry. *J. Food Eng.* **2006**, *74*, 13–23. [CrossRef]
5. Steuer, R.E. *Multiple Criteria Optimization: Theory, Computation, and Application*; Wiley: New York, NY, USA, 1986.
6. Abakarov, A.; Sushkov, Y.; Almonacid, S.; Simpson, R. Multiobjective Optimization Approach: Thermal Food Processing. *J. Food Sci.* **2009**, *74*, E471–E487. [CrossRef] [PubMed]
7. Vilas, C.; Mauricio-Iglesias, M.; García, M.R. Model-based design of smart active packaging systems with antimicrobial activity. *Food Packag. Shelf Life* **2020**, *24*, 100446. [CrossRef]
8. Holdsworth, S.D.; Simpson, R. Multiobjective Optimization in Thermal Food Processing. In *Thermal Processing of Packaged Foods*; Springer International Publishing: Cham, Switzerland, 2015; pp. 415–424. [CrossRef]

9. Krüger, C.; Castellani, F.; Geldermann, J.; Schöbel, A. Peat and pots: An application of robust multiobjective optimization to a mixing problem in agriculture. *Comput. Electron. Agric.* **2018**, *154*, 265–275. [CrossRef]
10. Gergely, S.; Békássy-Molnár, E.; Vatai, G. The use of multiobjective optimization to improve wine filtration. *J. Food Eng.* **2003**, *58*, 311–316. [CrossRef]
11. Sendín, J.O.H.; Alonso, A.; Banga, J.R. Efficient and robust multi-objective optimization of food processing: A novel approach with application to thermal sterilization. *J. Food Eng.* **2010**, *98*, 317–324. [CrossRef]
12. Kiranoudis, C.; Markatos, N. Pareto design of conveyor-belt dryers. *J. Food Eng.* **2000**, *46*, 145–155. [CrossRef]
13. Olmos, A.; Trelea, I.C.; Courtois, F.; Bonazzi, C.; Trystram, G. Dynamic optimal control of batch rice drying process. *Dry. Technol.* **2002**, *20*, 1319–1345. [CrossRef]
14. Winiczenko, R.; Górnicki, K.; Kaleta, A.; Martynenko, A.; Janaszek-Mańkowska, M.; Trajer, J. Multi-objective optimization of convective drying of apple cubes. *Comput. Electron. Agric.* **2018**, *145*, 341–348. [CrossRef]
15. Bortolini, M.; Faccio, M.; Ferrari, E.; Gamberi, M.; Pilati, F. Fresh food sustainable distribution: Cost, delivery time and carbon footprint three-objective optimization. *J. Food Eng.* **2016**, *174*, 56–67. [CrossRef]
16. Peñalver-Soto, J.L.; Garre, A.; Aznar, A.; Fernández, P.S.; Egea, J.A. Dynamics of Microbial Inactivation and Acrylamide Production in High-Temperature Heat Treatments. *Foods* **2021**, *10*, 2535. [CrossRef]
17. Costa, L.; Santo, I.E.; Oliveira, P.; Simos, T.E.; Psihoyios, G.; Tsitouras, C.; Anastassi, Z. Uncertainty on Multi-objective Optimization Problems. *AIP Conf. Proc.* **2011**, *1389*, 775–778. [CrossRef]
18. Herrador, M.; Asuero, A.G.; González, A.G. Estimation of the uncertainty of indirect measurements from the propagation of distributions by using the Monte-Carlo method: An overview. *Chemom. Intell. Lab. Syst.* **2005**, *79*, 115–122. [CrossRef]
19. Knio, O.M.; Le Maître, O.P. Uncertainty propagation in CFD using polynomial chaos decomposition. *Fluid Dyn. Res.* **2006**, *38*, 616–640. [CrossRef]
20. Luo, Y.-Z.; Yang, Z. A review of uncertainty propagation in orbital mechanics. *Prog. Aerosp. Sci.* **2017**, *89*, 23–39. [CrossRef]
21. Zhang, J. Modern Monte Carlo methods for efficient uncertainty quantification and propagation: A survey. *WIREs Comput. Stat.* **2020**, *13*, e1539. [CrossRef]
22. Akkermans, S.; Nimmegeers, P.; Van Impe, J.F. A tutorial on uncertainty propagation techniques for predictive microbiology models: A critical analysis of state-of-the-art techniques. *Int. J. Food Microbiol.* **2018**, *282*, 1–8. [CrossRef]
23. Feyissa, A.H.; Gernaey, K.V.; Adler-Nissen, J. Uncertainty and sensitivity analysis: Mathematical model of coupled heat and mass transfer for a contact baking process. *J. Food Eng.* **2012**, *109*, 281–290. [CrossRef]
24. Guillard, V.; Guillaume, C.; Destercke, S. Parameter uncertainties and error propagation in modified atmosphere packaging modelling. *Postharvest Biol. Technol.* **2012**, *67*, 154–166. [CrossRef]
25. Nicolai, B.; Verboven, P.; Scheerlinck, N.; De Baerdemaeker, J. Numerical analysis of the propagation of random parameter fluctuations in time and space during thermal food processes. *J. Food Eng.* **1998**, *38*, 259–278. [CrossRef]
26. Srivastava, S.; Chaudhuri, A.; Srivastava, R.K. Propagation of risks and their impact on performance in fresh food retail. *Int. J. Logist. Manag.* **2015**, *26*, 568–602. [CrossRef]
27. Turgut, S.S.; Feyissa, A.H.; Küçüköner, E.; Karacabey, E. Uncertainty and sensitivity analysis by Monte Carlo simulation: Recovery of trans-resveratrol from grape cane by pressurised low polarity water system. *J. Food Eng.* **2020**, *292*, 110366. [CrossRef]
28. Poschet, F.; Geeraerd, A.H.; Scheerlinck, N.; Nicolaï, B.M.; van Impe, J.F. Monte Carlo analysis as a tool to incorporate variation on experimental data in predictive microbiology. *Food Microbiol.* **2003**, *20*, 285–295. [CrossRef]
29. Van Impe, J.; Bernaerts, K.; Geeraerd, A.; Poschet, F.; Versyck, K. Modelling and prediction in an uncertain environment. *Food Process Model.* **2001**, *59*, 156–179. [CrossRef]
30. Julier, S.; Uhlmann, J.K. *A General Method for Approximating Nonlinear Transformations of Probability Distributions*; Institute of Electrical and Electronics Engineers: New York, NY, USA, 1996.
31. Wiener, N. The Homogeneous Chaos. *Am. J. Math.* **1938**, *60*, 897. [CrossRef]
32. Metropolis, N.; Ulam, S. The Monte Carlo Method. *J. Am. Stat. Assoc.* **1949**, *44*, 335–341. [CrossRef]
33. Garre, A.; Peñalver-Soto, J.L.; Esnoz, A.; Iguaz, A.; Fernandez, P.S.; Egea, J.A. On the use of in-silico simulations to support experimental design: A case study in microbial inactivation of foods. *PLoS ONE* **2019**, *14*, e0220683. [CrossRef]
34. Johnson, K.A.; Gorzinski, S.J.; Bodner, K.M.; Campbell, R.A.; Wolf, C.H.; Friedman, M.A.; Mast, R.W. Chronic toxicity and oncogenicity study on acrylamide incorporated in the drinking water of Fischer 344 rats. *Toxicol. Appl. Pharmacol.* **1986**, *85*, 154–168. [CrossRef]
35. Rice, J.M. The carcinogenicity of acrylamide. *Mutat. Res. Toxicol. Environ. Mutagen.* **2005**, *580*, 3–20. [CrossRef]
36. Knol, J.J.; van Loon, W.A.M.; Linssen, J.P.H.; Ruck, A.-L.; van Boekel, M.A.J.S.; Voragen, A.G.J. Toward a Kinetic Model for Acrylamide Formation in a Glucose–Asparagine Reaction System. *J. Agric. Food Chem.* **2005**, *53*, 6133–6139. [CrossRef] [PubMed]
37. Krokida, M.; Oreopoulou, V.; Maroulis, Z.; Marinos-Kouris, D. Colour changes during deep fat frying. *J. Food Eng.* **2001**, *48*, 219–225. [CrossRef]
38. Krokida, M.K.; Oreopoulou, V.; Maroulis, Z.B.; Marinos-Kouris, D. Deep fat frying of potato strips—Quality issues. *Dry. Technol.* **2001**, *19*, 879–935. [CrossRef]
39. Abduh, S.B.M.; Leong, S.Y.; Zhao, C.; Baldwin, S.; Burritt, D.J.; Agyei, D.; Oey, I. Kinetics of Colour Development during Frying of Potato Pre-Treated with Pulsed Electric Fields and Blanching: Effect of Cultivar. *Foods* **2021**, *10*, 2307. [CrossRef] [PubMed]

40. Kedia, P.; Kausley, S.B.; Rai, B. Development of kinetic models for prediction of reducing sugar content in potatoes using literature data on multiple potato varieties. *LWT* **2021**, *155*, 112986. [CrossRef]
41. Yang, Y.; Achaerandio, I.; Pujolà, M. Influence of the frying process and potato cultivar on acrylamide formation in French fries. *Food Control* **2016**, *62*, 216–223. [CrossRef]
42. Krokida, M.; Oreopoulou, V.; Maroulis, Z.; Marinos-Kouris, D. Effect of osmotic dehydration pretreatment on quality of french fries. *J. Food Eng.* **2001**, *49*, 339–345. [CrossRef]
43. European Food Safety Authority. Update on acrylamide levels in food from monitoring years 2007 to 2010. *EFSA J.* **2012**, *10*, 2938. [CrossRef]
44. Segnini, S.; Dejmek, P.; Oste, R. Reproducible Texture Analysis of Potato Chips. *J. Food Sci.* **1999**, *64*, 309–312. [CrossRef]
45. Friedman, M. Chemistry, Biochemistry, and Safety of Acrylamide. A Review. *J. Agric. Food Chem.* **2003**, *51*, 4504–4526. [CrossRef]
46. Friedman, M.; Mottram, D. *Chemistry and Safety of Acrylamide in Food*; Springer Science & Business Media: Berlin/Heidelberg, Germany, 2006; Volume 561.
47. Benford, D.; Bignami, M.; Chipman, J.K.; Bordajandi, L.R. Assessment of the genotoxicity of acrylamide. *EFSA J.* **2022**, *20*, e07293. [CrossRef]
48. Golmohammadi, A.; Afkari-Sayyah, A.H. Long-Term Storage Effects on the Physical Properties of the Potato. *Int. J. Food Prop.* **2012**, *16*, 104–113. [CrossRef]
49. Vivanti, V.; Finotti, E.; Friedman, M. Level of Acrylamide Precursors Asparagine, Fructose, Glucose, and Sucrose in Potatoes Sold at Retail in Italy and in the United States. *J. Food Sci.* **2006**, *71*, C81–C85. [CrossRef]
50. Pedreschi, F. Frying of Potatoes: Physical, Chemical, and Microstructural Changes. *Dry. Technol.* **2012**, *30*, 707–725. [CrossRef]
51. Kizito, K.F.; Abdel-Aal, M.H.; Ragab, M.H.; Youssef, M.M. Quality attributes of French fries as affected by different coatings, frozen storage and frying conditions. *J. Agric. Sci. Bot.* **2017**, *1*, 23–29.
52. Nourian, F.; Ramaswamy, H. Kinetics of quality change during cooking and frying of potatoes: Part ii. color. *J. Food Process Eng.* **2003**, *26*, 395–411. [CrossRef]
53. Pedreschi, F.; Moyano, P.; Kaack, K.; Granby, K. Color changes and acrylamide formation in fried potato slices. *Food Res. Int.* **2005**, *38*, 1–9. [CrossRef]
54. Deb, K.; Pratap, A.; Agarwal, S.; Meyarivan, T. A fast and elitist multiobjective genetic algorithm: NSGA-II. *IEEE Trans. Evol. Comput.* **2002**, *6*, 182–197. [CrossRef]
55. Mestdagh, F.; De Wilde, T.; Castelein, P.; Németh, O.; Van Peteghem, C.; De Meulenaer, B. Impact of the reducing sugars on the relationship between acrylamide and Maillard browning in French fries. *Eur. Food Res. Technol.* **2008**, *227*, 69–76. [CrossRef]
56. Johnson, A.M.; Porter, G.; Camire, M.E. Low-Acrylamide French Fry Acceptance: A Pilot Study. *J. Food Sci.* **2019**, *84*, 3717–3725. [CrossRef]
57. Ortiz-Martínez, V.M.; Martínez-Frutos, J.; Hontoria, E.; Hernández-Fernández, F.J.; Egea, J.A. Multiplicity of solutions in model-based multiobjective optimization of wastewater treatment plants. *Optim. Eng.* **2021**, *22*, 1–16. [CrossRef]
58. Farinu, A.; Baik, O.-D. Convective mass transfer coefficients in finite element simulations of deep fat frying of sweetpotato. *J. Food Eng.* **2008**, *89*, 187–194. [CrossRef]
59. Gholamibozanjani, G.; Leong, S.; Oey, I.; Bremer, P.; Silcock, P.; Farid, M. Heat and Mass Transfer Modeling to Predict Temperature Distribution during Potato Frying after Pre-Treatment with Pulsed Electric Field. *Foods* **2021**, *10*, 1679. [CrossRef] [PubMed]

foods

MDPI

Article

Dynamic Thermal Properties Estimation Using Sensitivity Coefficients for Rapid Heating Process

Anbuhkani Muniandy [1], Patnarin Benyathiar [2], Dharmendra K. Mishra [1,*] and Ferhan Ozadali [1,3]

1 Department of Food Science, Purdue University, 745 Agriculture Mall Dr, West Lafayette, IN 47907, USA; amunian@purdue.edu (A.M.); fozadali@purdue.edu (F.O.)
2 Department of Food Technology, Mahidol University, Kanchanaburi Campus, 199 Sangkraburi Road, Sai Yok, Kanchanaburi 71150, Thailand; patnarin.ben@mahidol.ac.th
3 Mead Johnson Nutrition, Reckitt Benckiser Health, 2400 W Lloyd Expy, Evansville, IN 47712, USA
* Correspondence: mishradh@purdue.edu; Tel.: +1-765-494-2594

Abstract: Thermal conductivity determination of food at temperatures > 100 °C still remains a challenge. The objective of this study was to determine the temperature-dependent thermal conductivity of food using rapid heating (TPCell). The experiments were designed based on scaled sensitivity coefficient (SSC), and the estimated thermal conductivity of potato puree was compared between the constant temperature heating at 121.10 °C (R12B10T1) and the rapid heating (R22B10T1). Temperature-dependent thermal conductivity models along with a constant conductivity were used for estimation. R22B10T1 experiment using the *k* model provided reliable measurements as compared to R12B10T1 with thermal conductivity values from 0.463 ± 0.011 W m^{-1} K^{-1} to 0.450 ± 0.016 W m^{-1} K^{-1} for 25–140 °C and root mean squares error (*RMSE*) of 1.441. In the R12B10T1 experiment, the analysis showed the correlation of residuals, which made the estimation less reliable. The thermal conductivity values were in the range of 0.444 ± 0.012 W m^{-1} K^{-1} to 0.510 ± 0.034 W m^{-1} K^{-1} for 20–120 °C estimated using the *k* model. Temperature-dependent models (linear and *k* models) provided a better estimate than the single parameter thermal conductivity determination with low *RMSE* for both types of experiments. SSC can provide insight in designing dynamic experiments for the determination of thermal conductivity coefficient.

Keywords: temperature-dependent thermal properties; scaled sensitivity coefficient; TPCell; parameter estimation; inverse problems

Citation: Muniandy, A.; Benyathiar, P.; Mishra, D.K.; Ozadali, F. Dynamic Thermal Properties Estimation Using Sensitivity Coefficients for Rapid Heating Process. *Foods* **2021**, *10*, 1954. https://doi.org/10.3390/foods10081954

Academic Editors: Carlos Vilas, Míriam R. García and Jose A. Egea

Received: 7 August 2021
Accepted: 19 August 2021
Published: 22 August 2021

1. Introduction

In food processing, experiments designed under dynamic heating conditions for estimation of thermal conductivity at elevated temperatures have received much attention recently due to the development and implementation of novel and innovative technologies. Given this, innovative product and process development in a very competitive market demands the development of challenging products, which will require the determination of their thermal properties under realistic processing conditions. The inverse problems technique is an effective tool which can be used to solve emerging challenges in food manufacturing [1–5]. Due to the lack of rapid methods, estimation of thermal properties is usually performed from experiments in a constant temperature environment [1]. The parameter estimation technique has been widely used in estimating the thermal properties of various food products [3,5–15]. It has also been used to estimate the fluid-to-particle heat transfer coefficient during aseptic processing of particulate foods [16] and heat flux during baking [17]. Constant temperature boundary condition can lead to prolonged exposure of heat to the sample. This can potentially degrade the product and then reliable estimates of thermal properties may not be obtained. Studies in the literature have used linear and non-linear models for the estimation of the thermal conductivity from the experimental temperature profile [1,2,18]. However, most of those studies worked on the slow heating experiments.

The objective of this study was to determine temperature-dependent thermal conductivity utilizing an experimental design based on scaled sensitivity coefficients (SSC). Parameter estimation was used to estimate constant and temperature-dependent thermal conductivity using experimental temperature profiles. The parameter SSC were studied to determine if the parameter can be estimated with relative accuracy associated with it [19,20]. A comparison of thermal conductivity estimation was presented using constant temperature boundary condition (R12B10T1, represents the traditional approach) vs. heat flux boundary condition (R22B10T1, represents the rapid heating method). The numbering systems (R12B10T1 and R22B10T1) used to describe the experiments were adopted from transient heat conduction solutions [21]. A high fat containing product to simulate soups that are high in fat content was chosen as a model food to compare the thermal properties between R12B10T1 and R22B10T1.

2. Materials and Methods

2.1. Sample Preparation

Potato puree, containing 22% (w/w) fat, was prepared with chicken broth, heavy cream, potato flakes, and butter. The ingredients in a vessel were heated on a hot plate at medium heat until reaching the temperature of 95 ± 2 °C and then the vessel was removed from the hot plate to cool down to room temperature before further analysis. The apparent viscosity of the sample was 16,735 cP measured by Brookfield AMETEK DVE Viscometer (Middleboro, MA, USA) at 6.27 s^{-1} with LV s64 spindle.

2.2. Mathematical Model for Transient Heat Conduction in Cylindrical Coordinate for Constant Temperature Boundary Condition (R12B10T1) Experiment

The predicted temperature profile for the R12B10T1 experiment was obtained based on the finite element numerical solution of 2D axisymmetric heat transfer equation in COMSOL (Burlington, MA, USA), as shown in Equation (1). The domain of the heat transfer included the thermocouple, sample, and stainless–steel cup. A predefined mesh size calibrated for heat transfer was used for the entire geometry with a total of 2734 elements. The minimum element size was 0.516 mm with an average of 0.915 mm. The total mesh area and element area ratio were 1783 mm^2 and 1.423 × 10^4, respectively.

$$\frac{1}{r}\frac{\partial}{\partial r}\left[kf_k(T,k)r\frac{\partial T}{\partial r}\right]+\frac{\partial}{\partial z}\left[kf_k(T,k)\frac{\partial T}{\partial z}\right]=Cf_c(T,k)\frac{\partial T}{\partial t}\ for\ R_A<r\le R_B,\ 0<z\le Z_A,\ t>0 \quad (1)$$

The boundary conditions were,

$$\frac{\partial T}{\partial r}(R_A,z,t)=T(t),\ \frac{\partial T}{\partial z}(r,0,t)=T(t),\ \frac{\partial T}{\partial z}(r,Z_A,t)=T(t) \quad (2)$$

The initial temperature was,

$$T(r,z,0)=T_o \quad (3)$$

For the R12B10T1 experiment, the sample was placed in a cylindrical stainless-steel 316L sample holder, which contained a thermocouple probe at the geometric center (Figure 1). Another thermocouple was placed on the external surface of the sample holder and secured with Kapton® polyimide tape (DuPont, Wilmington, DE, USA). The initial temperature (T_o) of the sample was ~20 °C. Prior to starting the experiment, the temperatures of the sample and sample holder were equilibrated for 10 min. The sample holder was pressurized to 30 psig and placed in a silicone oil bath that was set at 121.10 °C. The center and surface thermocouples were used to monitor temperature at the center and at the surface, respectively, using LabView (National Instruments, Austin, TX, USA) as the data acquisition software. Once the sample was placed in the oil bath, the experiment was performed until the center thermocouple reached 120 °C. To terminate the experiment, the sample holder was removed from the oil bath and cooled to room temperature before releasing the pressure. Triplicate analyses were executed for statistically verifiable data.

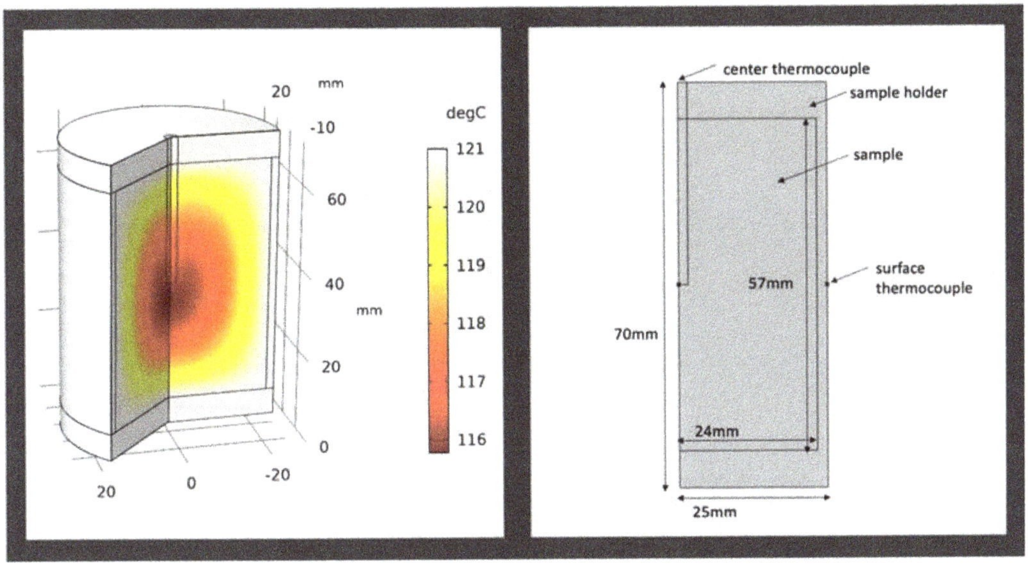

Figure 1. Simulation of R12B10T1 case using numerical solution. The solution shown was obtained at 40 min.

2.3. *Mathematical Model for Transient Heat Conduction in a Hollow Cylinder with Heat Flux on the Inside for Rapid Heating Condition (R22B10T1) Experiment*

Measurement of thermal conductivity by the TPCell device is based on R22B10T1 in a hollow cylinder with a heater located at the center [2]. The equations are shown below,

$$\frac{1}{r}\frac{\partial}{\partial r}\left[k_h r\frac{\partial T}{\partial r}\right] + \frac{\partial}{\partial z}\left[k_h\frac{\partial T}{\partial z}\right] + g_0 f(t) = C_h\frac{\partial T}{\partial t}\ for\ R_0 < r \leq R_{1,},\ 0 < z \leq Z_1,\ t > 0 \quad (4)$$

$$\frac{1}{r}\frac{\partial}{\partial r}\left[k f_k(T,k) r\frac{\partial T}{\partial r}\right] + \frac{\partial}{\partial z}\left[k_1 f_k(T,k)\frac{\partial T}{\partial z}\right] = C f_C(T,k)\frac{\partial T}{\partial t}\ for\ R_1 < r \leq R_2,\ 0 < z \leq Z_1,\ t > 0 \quad (5)$$

The insulation boundary conditions were used due to the short duration of experiment [2],

$$\frac{\partial T}{\partial r}(R_2, z, t) = 0,\ \frac{\partial T}{\partial z}(r, 0, t) = 0,\ \frac{\partial T}{\partial z}(r, Z_1, t) = 0 \quad (6)$$

The initial condition was,

$$T(r, z, 0) = T_0 \quad (7)$$

The thermal conductivity of the samples was measured using the TPCell by loading 275 mL of the sample into the cylindrical sample holder (Figure 2). The T_o of the sample was ~20 °C. The sample holder was sealed and pressurized up to 60 psig using air to achieve an elevated temperature of the sample. The heater was supplied with 20 W power for the duration of the experiment. Once the temperature of the heater reached 137.55 ± 0.42 °C, the power supply was cut off to stop the experiment. The resistance (R) of the heater was converted to temperature using a calibration equation, $T = 25.381R - 12{,}295$. Triplicate analyses were performed for statistical accuracy.

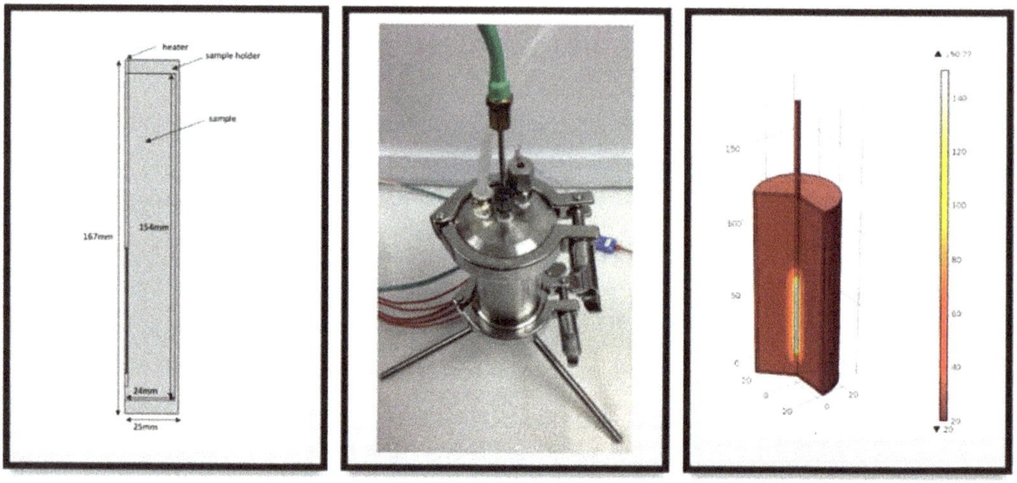

Figure 2. Schematic of R22B10T1 case: TPCell instrument with the center heater (**left**), and simulation of the center heater with product (**right**).

2.4. Parameter Estimation

Thermal conductivity was estimated using a sequential estimation method from the temperature profiles of R12B10T1 and R22B10T1 experiments. The functions used for thermal conductivity estimation for both experiments were single parameter, linear, and k model as shown below. The k model was a reparameterization of the linear model to improve the parameter identifiability.

(A) Single parameter model,

$$k = k_C \tag{8}$$

(B) Linear model with two parameters,

$$k(T) = a + b(T) \tag{9}$$

(C) k model with two parameters,

$$k(T) = k_1\left(\frac{T_2 - T}{T_2 - T_1}\right) + k_2\left(\frac{T - T_1}{T_2 - T_1}\right) \tag{10}$$

2.5. Scaled Sensitivity Coefficient and Sequential Estimation

Parameter identifiability was assessed by plotting the SSC to determine whether all the parameters in a model can be estimated uniquely and simultaneously with their relative errors. The SSC is also used in the optimal experimental design criteria where it maximizes the determinant of the sensitivity matrix. However, in this study, SSC was used to gain further understanding with regards to parameter correlation and identifiability. The sensitivity coefficient of thermal conductivity was derived by taking the first derivative of the temperature with respect to thermal conductivity. To perform a direct comparison, the sensitivity coefficient was scaled by multiplying with the parameter to obtain the SSC as shown in Equation (11).

$$X_i' = k_i \frac{\partial T}{\partial k_i} \tag{11}$$

The sum of SSC was calculated using Equation (12). All parameters in the model cannot be estimated uniquely and simultaneously if the sum of SSC is equal to zero [19,22].

$$k_1 \frac{\partial T}{\partial k_1}\bigg|_{k_2} + k_2 \frac{\partial T}{\partial k_2}\bigg|_{k1} = -(T - T_0) - \left[C_1 \frac{\partial T}{\partial C_1}\bigg|_{C_2} + C_2 \frac{\partial T}{\partial C_2}\bigg|_{C1}\right] \tag{12}$$

Based on the experimental temperature profile, the thermal conductivity was determined using sequential estimation. The sequential estimation procedure was developed in MATLAB® [19] based on the Gauss minimization method and required prior information of parameters. In this estimation procedure, the parameter estimates initially would have large fluctuations, but the estimates eventually attain a constant value once enough data have been added.

The mathematical form of non-linear sequential estimation is derived from maximum a posteriori (MAP) estimation. The minimization function in the Gauss method can be expressed as;

$$S = \left[Y - \hat{Y}(\beta)\right]' W \left[Y - \hat{Y}(\beta)\right] + [\mu - \beta]' U [\mu - \beta] \tag{13}$$

where Y is the experimental response variable and $\hat{Y}$ is the predicted response, μ is the prior information of parameter vector β, W is the inverse of covariance matrix of errors, and U is the inverse covariance matrix of parameters. β was solved and reported as the estimated thermal conductivity. The parameter estimates were reported along with its root mean square error (*RMSE*) and 95% confidence interval. The *RMSE* for the estimate was calculated based on Equation (14). The 95% confidence interval of parameter were calculated using MATLAB® built-in function nlparci (parameter, residual, sensitivity coefficient). Residuals were calculated by taking the difference between the experimental and predicted temperature at each time point. Standard statistical assumptions of uncorrelated errors, which are normally distributed with zero mean and constant variance, were verified for the residuals. Additional assumptions specific to the use of sequential estimation, which needs to be satisfied, are known as covariance matrix errors, no errors in independent variables, and the known prior of information of parameters.

$$RMSE = \sqrt{\frac{\sum_{i=1}^{n} \left(\hat{Y}_i - Y_i\right)^2}{n}} \tag{14}$$

3. Results and Discussion

Parameters of Equations (8–10) showed large SSC as illustrated in Figure 3. The plots were used to determine if the simultaneous estimation of parameters was possible. All parameters in the model can be estimated with a low error when the magnitude of the SSC is large and without linear dependency or correlation between the parameters [19]. To consider SSC to be large, it should be at least 10% of the temperature rise [1]. When SSC is small, the estimation may result in larger errors and hence larger confidence intervals of the parameter. Parameters are considered not correlated if their ratio was not constant [20]. Visually, the SSC curves would have the same pattern with the same or different magnitudes if the parameters are correlated.

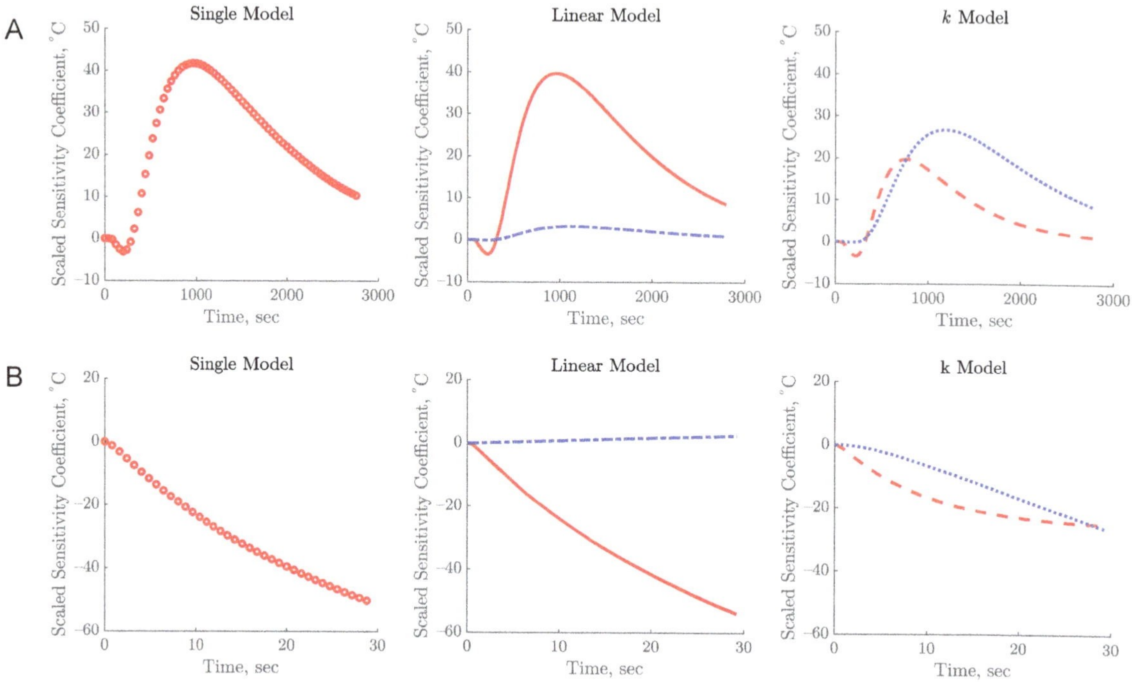

Figure 3. Scaled Sensitivity Coefficient (SSC) of thermal conductivity for single parameter model, linear model, and k model for R12B10T1 (**A**) and R22B10T1 (**B**) experiments. Legends: (o) k_c, (–) a, (–.) b, (– –) k_1, and (. . .) k_2.

For the R12B10T1 experiment, the SSC plots of the parameters for estimation of thermal conductivity were large and uncorrelated, as shown in Figure 3A. Based on the result from the single parameter, SSC was 41.68%, which is considered large. The SSCs of two parameters for estimation of thermal conductivity using the linear model were 39.65% and 3.17%, as compared to 15.78% and 22.18% for the k model and the sums of SSC for those models were not zero. The two parameters estimated using the linear and k models were not correlated as a result. This means both equations can be used to estimate the thermal conductivity with two parameters using the inverse problems methods. However, the SSC of parameter b in linear model was very small (3.17%) as compared to the parameter a (~39.65%), suggesting that it would be difficult to estimate b and probably would have large standard error. The magnitude SSC of both parameters k_1 and k_2 in the k model are evenly distributed as compared to the temperature rise. The SSC plots for the two parameters were quite identical to the plots reported previously [1]. This can be attributed to the identical nature of the R12B10T1 experiment and measurement of temperature at the geometric center of a cylindrical container.

Figure 3B shows the SSC plots of the parameters for estimation of thermal conductivity from the R22B10T1 experiment. The SSC for a single parameter was 44.6%. The values of SSCs for the two parameters estimated using the linear model were 53.90% and 2.40%, as compared to 21.25% and 22.08% for the k model. The parameter b in the linear model had the lowest SSC and hence it would be harder to estimate. The SSC of parameters estimated from this experiment also exhibited a large magnitude with no correlation between parameters. The sum of SSC for all parameters was not zero, which means the parameters can be estimated uniquely and simultaneously. In both experiments, the magnitude of the SSC was reduced in the linear and the k model when an additional parameter was added (Figure 3). This is because the magnitude of SSC with one parameter is now being shared by two parameters non-proportionally.

In order to estimate the parameters with low errors, the R12B10T1 experiment must be conducted for at least 20 min while the R22B10T1 experiment required only 30 s (Figure 3). The experimental time is optimal when the parameter SSC attains a maximum value. The optimal experimental time is further confirmed when a constant value is achieved by the sequential estimation (Figure 4). Further data acquisition beyond this optimal time may not add any substantial improvement to the estimated parameter [20]. The large difference in the experiment duration was due to the boundary conditions used in these experiments. The R12B10T1 experiment had constant temperature boundary on the walls of the sample holder, hence the temperature rise at the geometric center was relatively slow which led to longer experiment duration. In contrast, the R22B10T1 experiment utilizes a heat flux boundary at the center of the sample resulting in rapid temperature rise. The SSC indicates the magnitude of change in the temperature due to perturbation in the parameter [19]. Due to the different boundary condition used, the duration to reach the highest magnitude of SSC was different between these two experiments.

The sequential estimation of parameters based on the temperature profile obtained from the R12B10T1 experiment is shown in Figure 4. The predicted data from all three models showed a good fit with the experimental data. This estimation process requires appropriate prior information as the initial guess. During sequential estimation, the estimated parameter values keep changing as each datum is being added, with the goal of minimizing the sum of squares of the errors as illustrated in Figure 4A–C (center). The estimation was complete and reliable when parameter values attained a constant value and remained constant for the rest of the experimental time. When the parameter values do not attain a constant value toward the end of the experiment, it indicates that there might be some error in the model or in the experiment [2]. The final estimated values were reported along with their standard error and 95% confidence interval (Table 1).

Table 1. Estimation of the thermal conductivity in the single parameter model, linear model, and *k* model for R12B10T1 and R22B10T1 experiments.

Model		R12B10T1			R22B10T1		
		Rep 1	**Rep 2**	**Rep 3**	**Rep 1**	**Rep 2**	**Rep 3**
Single parameter model	k_c	0.457	0.490	0.504	0.464	0.459	0.458
	$RMSE$	0.582	0.663	0.838	1.400	1.540	1.485
	LCI_{k_c}	0.456	0.490	0.503	0.462	0.458	0.457
	UCI_{k_c}	0.457	0.491	0.505	0.465	0.460	0.459
Linear model	a	0.421	0.435	0.435	0.505	0.478	0.485
	$b \times 10^{-3}$	0.438	0.685	0.873	−0.515	−0.239	−0.224
	$RMSE$	0.485	0.436	0.525	1.285	1.538	2.275
	RE_a %	0.174	0.165	0.157	1.179	0.915	0.783
	RE_b %	2.008	1.292	0.985	−14.331	−22.417	−21.015
	LCI_a	0.419	0.433	0.433	0.491	0.467	0.471
	UCI_a	0.423	0.437	0.437	0.519	0.489	0.499
	$LCI_b \times 10^{-3}$	0.415	0.663	0.847	−0.692	−0.379	−0.394
	$UCI_b \times 10^{-3}$	0.463	0.706	0.898	−0.338	−0.099	−0.053
k model	k_1	0.430	0.449	0.452	0.484	0.469	0.450
	k_2	0.474	0.517	0.540	0.437	0.445	0.468
	$RMSE$	0.485	0.436	0.525	1.285	1.538	1.499
	RE_{k1} %	0.129	0.122	0.114	0.616	0.508	0.495
	RE_{k2} %	0.075	0.073	0.071	0.889	0.679	0.603
	LCI_{k_1}	0.428	0.447	0.451	0.477	0.463	0.444
	UCI_{k_1}	0.431	0.450	0.454	0.491	0.475	0.456
	LCI_{k_2}	0.473	0.516	0.538	0.427	0.437	0.461
	UCI_{k_2}	0.475	0.518	0.541	0.446	0.453	0.475

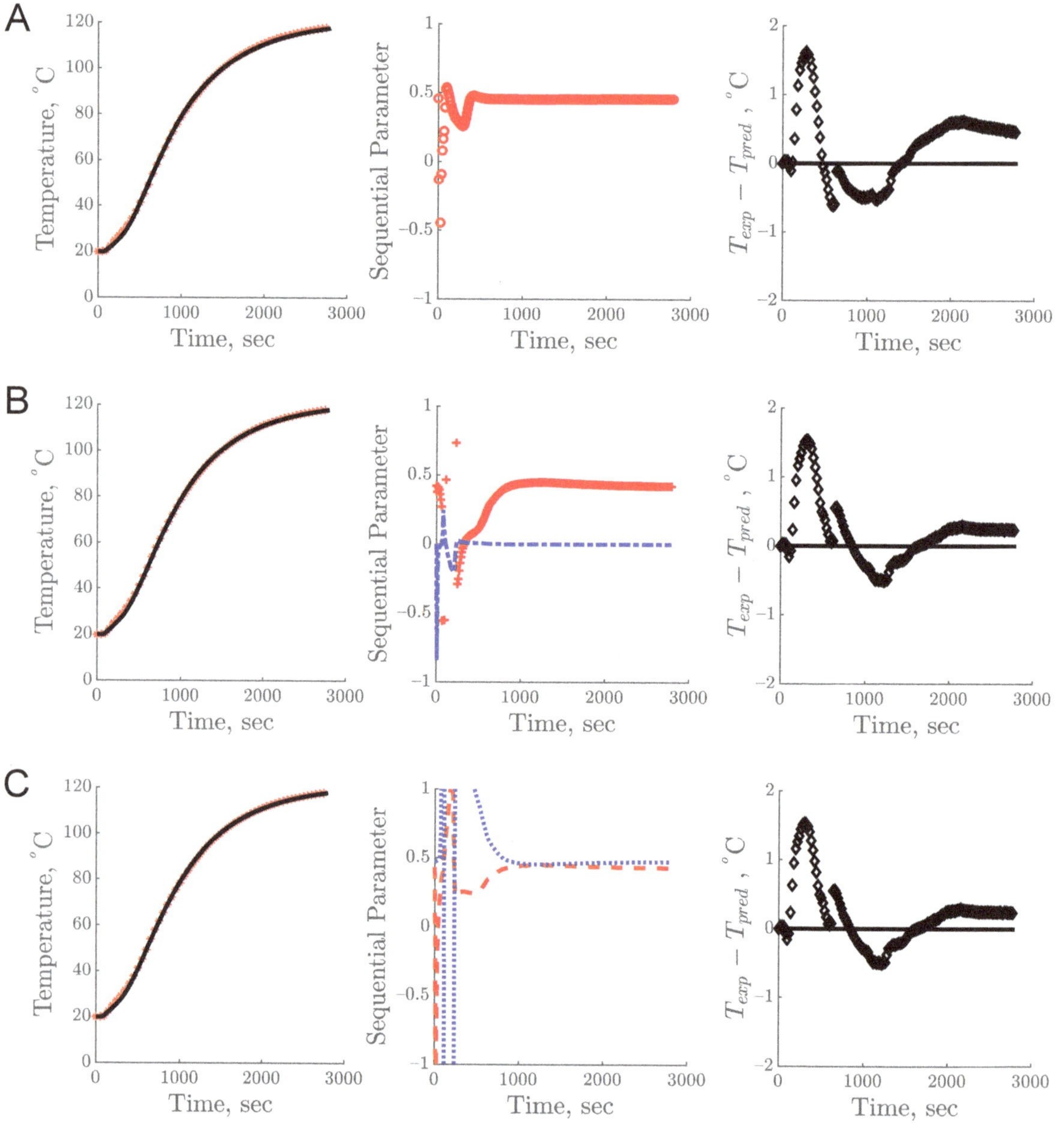

Figure 4. Experimental vs. predicted temperature profile over time (**left**), sequential estimation of parameters (**center**), and corresponding residual plot of experimental (T_{exp}) and predicted (T_{pred}) temperature (**right**) for R12B10T1 experiment: (**A**) single parameter model, (**B**) linear model, and (**C**) *k* model. Legends: (*) T_{exp}, (–) T_{pred}, (o) k_c, (+) a, (–.) b, (– –) k_1, (. . .) k_2, and (◇) residuals.

The residuals from all models in Figure 4 show a pattern, which is not desirable, and the mean value of the residuals was 0.27 for the constant model and 0.17 for both the linear and *k* model. These residuals were most likely due to a potential change in the sample during prolonged heating. A similar result was also reported from retort processing of cherry pomace [1]. Prolonged exposure of heat to a food product at a high temperature can lead to undesirable reactions within the food matrix including oxidation, separation of lipid and moisture, the formation of unwanted off-aroma/off-flavor compounds, browning, and degradation of nutrient and sensory quality attributes.

Overall, the estimation from R12B10T1 experiment had relatively low *RMSE* as shown in Table 1. The highest *RMSE* was observed for the single parameter model in all three replicates while the lowest *RMSE* was found for the linear and *k* model. Due to the parameter uncertainty with large relative error and large confidence interval of *b* in the linear model, the *k* model was chosen as the best thermal conductivity model.

For the single parameter model of R12B10T1, the thermal conductivity value of potato puree was constant at 0.484 ± 0.024 W m^{-1} K^{-1} as shown in Figure 5. An increase in thermal conductivity was observed in both the linear and *k* models. The *k* model showed the thermal conductivity values from 0.444 ± 0.012 W m^{-1} K^{-1} to 0.510 ± 0.034 W m^{-1} K^{-1} while those from the linear model were 0.447 ± 0.013 W m^{-1} K^{-1} to 0.523 ± 0.038 W m^{-1} K^{-1}. Thermal conductivity value of mashed potato in literature has been reported as 0.59 W m^{-1} K^{-1} [23] and blanched potato as 0.55 W m^{-1} K^{-1} at 20 °C [24]. Values reported in this study were slightly lower due to presence of high fat content which is known to decrease the thermal conductivity of foods [25,26]. The thermal conductivity value calculated from Choi-Okos model [27] based on the composition of the potato puree was 0.462 W m^{-1} K^{-1} at 25 °C which is well within the range reported in Figure 5.

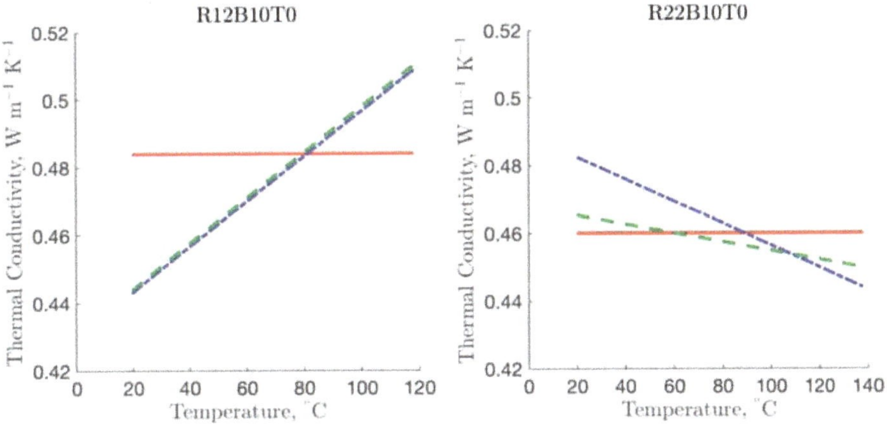

Figure 5. Thermal conductivity of potato puree estimated using the single parameter model, linear model, and *k* model for R12B10T1 and R22B10T1 experiments. Legends: (–) single parameter model, (–.) linear model, (– –) *k* model.

The sequential estimation of parameters based on the temperature profile obtained from the R22B10T1 experiment is shown in Figure 6. Based on the result, the model predicted temperature fits well with the experimental data. The sequential estimation from this experiment showed that parameter values remain unchanged toward the end of the experiment. The residuals for all models did not violate any standard statistical assumption. The average of residuals for linear and *k* model was −0.05. The mean value of the residuals from the R22B10T1 experiment were much smaller compared to the residuals in R12B10T1. This confirms that the parameter estimation from the R22B10T1 experiment was reliable. In this case, the R22B10T1 experiment was only 30 s as compared to 47 min for the R12B10T1 experiment. The parameter covariance matrix and correlation matrix for the linear and *k* model are presented in Table 2. The correlation coefficient of parameters in the linear model was quite high (0.99), which is not desirable when estimating multiple parameters. This was expected based on the SSC of parameters *a* and *b*.

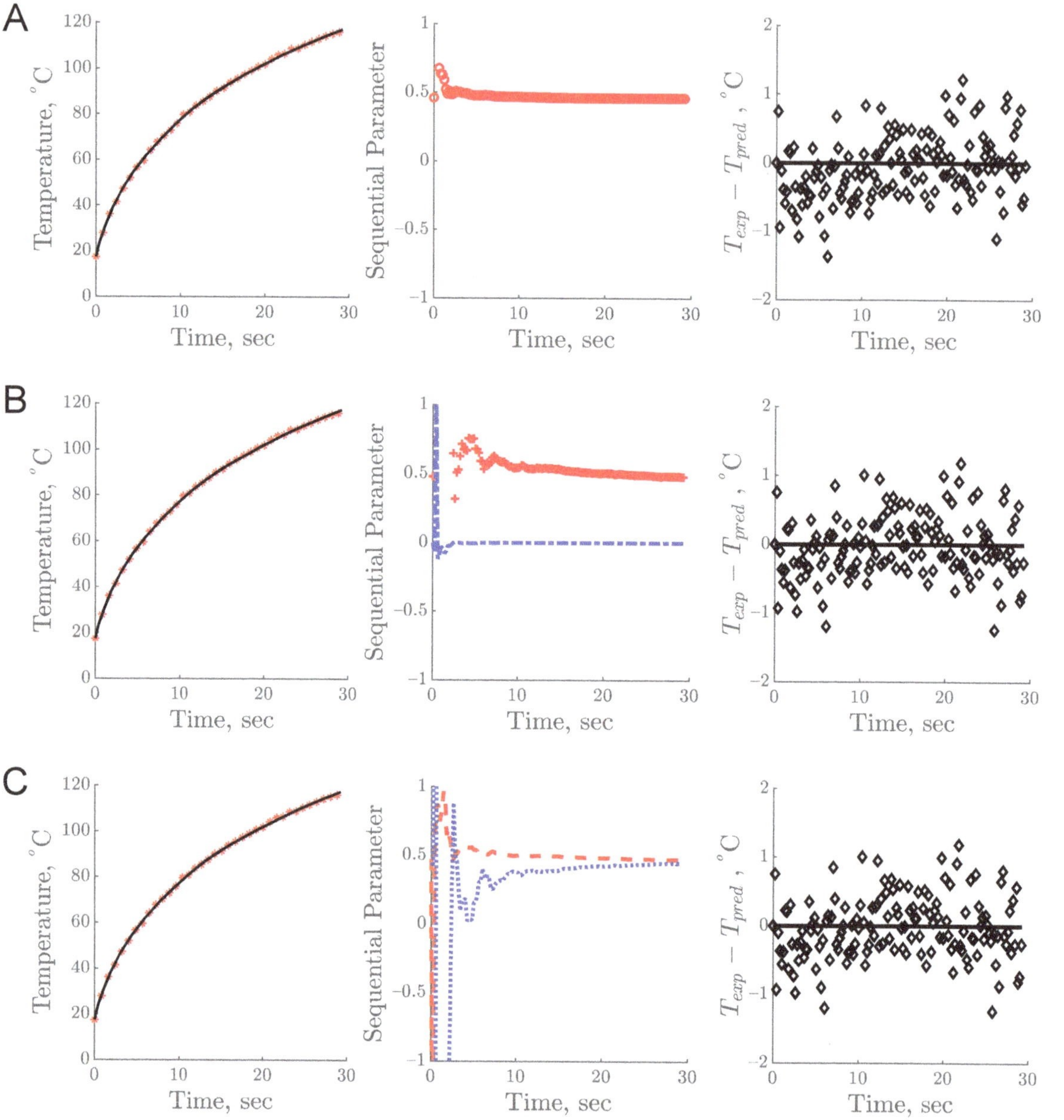

Figure 6. Experimental vs. predicted temperature profile over time (**left**), sequential estimation of parameters k_1 and k_2 (**center**) and corresponding residual plot (**right**) for R22B10T1 experiment: (**A**) single parameter model, (**B**) linear model, and (**C**) k model. Legends: (*) T_{exp}, (–) T_{pred}, (o) k_c, (+) a, (–.) b, (– –) k_1, (. . .) k_2, and (◇) residuals.

Table 2. Covariance and correlation matrices for the linear and k model for R12B10T1 and R22B10T1 experiments.

			Rep 1		Rep 2		Rep 3	
R12B10T2			a	b	a	b	a	b
Linear model	Covariance $\times 10^{-7}$	a	10.258	−0.122	8.014	−0.097	10.512	−0.130
		b	−0.122	0.002	−0.097	0.001	−0.130	0.002
	Correlation	a	1.000	−0.988	1.000	−0.985	1.000	−0.983
		b	−0.988	1.000	−0.985	1.000	−0.983	1.000
			k_1	k_2	k_1	k_2	k_1	k_2
k model	Covariance $\times 10^{-7}$	k_1	5.964	−3.288	4.612	−2.686	5.979	−3.698
		k_2	−3.288	2.412	−2.686	2.189	−3.698	3.279
	Correlation	k_1	1.000	−0.867	1.000	−0.845	1.000	−0.835
		k_2	−0.867	1.000	−0.845	1.000	−0.835	1.000
R22B10T2			a	b	a	b	a	b
Linear model	Covariance $\times 10^{-5}$	a	4.782	−0.059	3.697	−0.045	6.105	−0.075
		b	−0.059	0.001	−0.045	0.001	−0.075	0.001
	Correlation	a	1.000	−0.996	1.000	−0.996	1.000	−0.996
		b	−0.996	1.000	−0.996	1.000	−0.996	1.000
			k_1	k_2	k_1	k_2	k_1	k_2
k model	Covariance $\times 10^{-5}$	k_1	1.200	−1.493	1.096	−1.329	0.911	−1.102
		k_2	−1.493	2.037	−1.329	1.769	−1.102	1.464
	Correlation	k_1	1.000	−0.955	1.000	−0.955	1.000	−0.955
		k_2	−0.955	1.000	−0.955	1.000	−0.955	1.000

The RMSEs from R22B10T1 were higher than those from R12B10T1 due to the differences in accuracy of the temperature sensing elements. Based on the results from Table 1, the least *RMSE* values from the R22B10T1 were observed for the linear and k models. The *RMSE* from R22B10T1 of the linear and k models were close, which was not seen in the R12B10T1 experiment. Since the parameter b for the linear model exhibited large relative error and confidence interval (Table 1), it is not considered as the right model for the conductivity. Generally, thermal conductivity changes with temperature. Thus, the single parameter model is an average value over the temperature range [27]. Although it can be used for initial assessment, the k model would be appropriate and realistic.

Based on the results in Figure 5, the thermal conductivity result of potato puree using R22B10T1 showed that the values of the single parameter model remained constant over the temperatures at 0.460 ± 0.003 W m^{-1} K^{-1} while the average values of the linear model and k model decreased from 0.481 ± 0.010 W m^{-1} K^{-1} to 0.444 ± 0.010 W m^{-1} K^{-1} and 0.463 ± 0.011 W m^{-1} K^{-1} to 0.450 ± 0.016 W m^{-1} K^{-1}, respectively. The major difference using the R22B10T1 and R12B10T1 experiments was the variation of thermal conductivity with the increase in temperatures. The thermal conductivity values obtained from R22B10T1 showed a decrease in values with increasing temperature while an increasing trend was observed from the R12B10T1 experiment. While the thermal conductivity is known to increase with temperature, a slight decrease is evident in foods with high fat content [27]. Up to 20.4% decrease in thermal conductivity can occur in pure fat at temperatures between 25 °C to 140 °C [28]. In the current study, the decrease in thermal conductivity over the same temperature range was 8.3% and 2.8% for the linear and k model, respectively.

The temperature abuse during the experiment could negatively impact the reliability of estimated thermal properties. The separation of potato puree and potential changes in its matrix could occur due to prolonged exposure to the high temperature. This degradation phenomenon was observed from the R12B10T1 experiment. The increasing trend in the thermal conductivity values obtained from the R12B10T1 experiment might have an error due to the changes in the food matrix.

Even though the inverse problems were able to estimate the parameters of the linear model, the relative error and the confidence intervals were large for the parameter b.

In addition, the correlation coefficient of parameters of the linear model was higher (~0.99) as compared to the *k* model (~0.95). A study by da Silva et al. (2020) reported that simultaneous estimation of two parameters for determination of thermal diffusivity of coconut pulp was not possible due to the high correlation between the parameters, this is similar to what was observed in the current study with the linear model. The *k* model can be seen as a reparameterization of the linear model and had the same *RMSE* (Table 1). Reparameterization of the linear model to the *k* model improved the relative error, tightened the band width of the confidence intervals, and decreased the correlation between the parameters (Tables 1 and 2). Based on the SSC, sequential estimation, residual analysis, correlation coefficient, and *RMSE* values, the *k* function could be an appropriate model of temperature-dependent thermal conductivity for high fat-containing purees.

4. Conclusions

Innovations in the food industry toward rapid heating technologies such as ohmic and microwave heating requires thermal properties that are determined in realistic experimental conditions. Thermal property determination using rapid heating is suitable for novel applications in the food industry. The thermal conductivity of food was determined for both R12B10T1 (constant temperature boundary conditions) and R22B10T1 (heat flux boundary conditions) dynamic experiments using SSC and sequential estimation. The *k* model was sufficient in describing the dependence of thermal conductivity with temperature for both experiments. The linear model showed a large confidence interval of estimated parameters and high correlation between parameters. The thermal abuse created by the R12B10T1 experiment might have caused the higher conductivity measurements in both the single parameter model and temperature-dependent models due to the prolonged exposure at elevated temperatures. The new approach of rapid heating with TPCell, therefore, provides a quick and realistic measurement of the thermal conductivity in the food processing temperature range of 20–140 °C. This study will be beneficial to the food industry as a user-friendly tool for measuring thermal properties at elevated temperatures.

Author Contributions: Conceptualization, D.K.M.; methodology, D.K.M.; software, D.K.M. and A.M.; validation, A.M. and P.B.; formal analysis, A.M.; investigation, A.M.; resources, D.K.M. and F.O.; data curation, A.M. and D.K.M.; writing—original draft preparation, A.M.; writing—review and editing, D.K.M., P.B. and F.O.; visualization, P.B.; supervision, D.K.M.; project administration, D.K.M.; funding acquisition, D.K.M. All authors have read and agreed to the published version of the manuscript.

Funding: This work was supported by the NSF grant 109335 and USDA National Institute of Food and Agriculture, Hatch project 1023114.

Acknowledgments: The authors gratefully appreciate the guidance of James V. Beck on inverse problems and sequential parameter estimation.

Conflicts of Interest: The authors declare no conflict of interest.

Nomenclature

R12B10T1	transient heat conduction in cylindrical coordinate for constant temperature boundary condition
R22B10T1	transient heat conduction in a hollow cylinder with heat flux on the inside and insulated on the outside for rapid heating condition
RMSE	root meat square error
a, b	parameters in linear model
β	parameter
X_i'	scaled sensitivity coefficient, °C
μ	prior information of parameter vector, β, W m^{-1} K^{-1}
C_h	volumetric capacity of heater, J m^{-3} K^{-1}
C_{TC}	volumetric capacity of thermocouple, J m^{-3} K^{-1}
C_1	volumetric heat capacity of sample, J m^{-3} K^{-1} at T_1

C_2	volumetric heat capacity of sample, J m^{-3} K^{-1} at T_2
f_k	dimensionless thermal conductivity as a function of temperature
f_C	dimensionless heat capacity as a function of temperature
g_0	power, W m^{-3}
k	thermal conductivity of sample, W m^{-1} K^{-1}
k_h	thermal conductivity of heater, W m^{-1} K^{-1}
k_{TC}	thermal conductivity of thermocouple, W m^{-1} K^{-1}
k_C	thermal conductivity in single parameter model, W m^{-1} K^{-1}
k_1, k_2	parameters used in k model
LCI	95% lower confidence interval of parameters
i	index
n	number of responses
r	radial position, m
R	resistance of the heater, Ω
R_0	centerline of the cylinder, $R_0 = 0$
R_1	interface of heater and sample, m
R_2	wall of the cup, m
R_A	interface of thermocouple and sample, m
R_B	wall of the cup, m
RE	Relative error of parameter
T	temperature, °C
$T(t)$	temperature at t, °C
T_0	initial temperature of sample, °C
T_1	initial temperature for thermal properties, °C
T_2	final temperature for thermal properties, °C
t	time, s
U	inverse covariance matrix of parameters
UCI	95% upper confidence interval of parameters
W	inverse of covariance matrix of errors
Y	experimental response variable
$\hat{Y}$	predicted response variable
Z_A	total height of cylinder for (R12B10T1) experiment, mm
Z_1	total height of cylinder for (R22B10T1) experiment, mm
z	axial position, m

References

1. Greiby, I.; Mishra, D.K.; Dolan, K.D. Inverse method to sequentially estimate temperature-dependent thermal conductivity of cherry pomace during nonisothermal heating. *J. Food Eng.* **2014**, *127*, 16–23. [CrossRef]
2. Mishra, D.K.; Dolan, K.D.; Beck, J.V.; Ozadali, F. A novel instrument for rapid measurement of temperature-dependent thermal properties of conduction-heated food up to 140 °C. *J. Food Eng.* **2016**, *191*, 19–27. [CrossRef]
3. Muramatsu, Y.; Greiby, I.; Mishra, D.K.; Dolan, K.D. Rapid Inverse Method to Measure Thermal Diffusivity of Low-Moisture Foods. *J. Food Sci.* **2017**, *82*, 420–428. [CrossRef] [PubMed]
4. Mohamed, I.O. Simultaneous estimation of thermal conductivity and volumetric heat capacity for solid foods using sequential parameter estimation technique. *Food Res. Int.* **2009**, *42*, 231–236. [CrossRef]
5. Mohamed, I.O. Development of a simple and robust inverse method for determination of thermal diffusivity of solid foods. *J. Food Eng.* **2010**, *101*, 1–7. [CrossRef]
6. Martins, R.C.; Silva, C.L.M. Inverse problem methodology for thermal-physical properties estimation of frozen green beans. *J. Food Eng.* **2004**, *63*, 383–392. [CrossRef]
7. Zueco, J.; Alhama, F.; Gonazález Fernández, C.F. Inverse determination of the specific heat of foods. *J. Food Eng.* **2004**, *64*, 347–353. [CrossRef]
8. Da Silva, W.P.; E Silva, C.M.D.P.S.; De Souto, L.M.; Moreira, I.D.S.; Da Silva, E.C.O. Mathematical model for determining thermal properties of whole bananas with peel during the cooling process. *J. Food Eng.* **2018**, *227*, 11–17. [CrossRef]
9. Monteau, J.-Y. Estimation of thermal conductivity of sandwich bread using an inverse method. *J. Food Eng.* **2008**, *85*, 132–140. [CrossRef]
10. Ebrahimifakhar, A.; Yuill, D. Inverse estimation of thermophysical properties and initial moisture content of cereal grains during deep-bed grain drying. *Biosyst. Eng.* **2020**, *196*, 97–111. [CrossRef]
11. Mariani, V.C.; Barbosa De Lima, A.G.; Dos Santos Coelho, L. Apparent thermal diffusivity estimation of the banana during drying using inverse method. *J. Food Eng.* **2008**, *85*, 569–579. [CrossRef]

12. Gulati, T.; Datta, A. Enabling computer-aided food process engineering: Property estimation equations for transport phenomena-based models. *J. Food Eng.* **2013**, *116*, 483–504. [CrossRef]
13. Mariani, V.C.; Do Amarante Álvaro, C.C.D.; Dos Santos Coelho, L.D.S. Estimation of apparent thermal conductivity of carrot purée during freezing using inverse problem. *Int. J. Food Sci. Technol.* **2009**, *44*, 1292–1303. [CrossRef]
14. Mariani, V.C.; Dos Santos Coelho, L. Estimation of the apparent thermal diffusivity coefficient using an inverse technique. *Inverse Probl. Sci. Eng.* **2009**, *17*, 569–589. [CrossRef]
15. Ramsaroop, R.; Persad, P. Determination of the heat transfer coefficient and thermal conductivity for coconut kernels using an inverse method with a developed hemispherical shell model. *J. Food Eng.* **2012**, *110*, 141–157. [CrossRef]
16. Vidyarthi, S.K.; Mishra, D.K.; Dolan, K.D.; Muramatsu, Y. Inverse estimation of fluid-to-particle heat transfer coefficient in aseptic processing of particulate foods. *Biosyst. Eng.* **2020**, *198*, 210–222. [CrossRef]
17. Reddy, R.S.; Arepally, D.; Datta, A.K. Estimation of heat flux in bread baking by inverse problem. *J. Food Eng.* **2019**, *271*, 109774. [CrossRef]
18. Da Silva, W.P.; De Medeiros, M.S.; Gomes, J.P.; E Silva, C.M.D.P.S. Improvement of methodology for determining local thermal diffusivity and heating time of green coconut pulp during its pasteurization. *J. Food Eng.* **2020**, *285*, 110104. [CrossRef]
19. Beck, J.V.; Arnold, K.J. *Parameter Estimation in Engineering and Science*; John Wiley & Sons: New York, NY, USA, 1977.
20. Dolan, K.D.; Mishra, D.K. Parameter Estimation in Food Science. *Annu. Rev. Food Sci. Technol.* **2013**, *4*, 401–422. [CrossRef]
21. Beck, J.V.; Litkouhi, B. Heat conduction numbering system for basic geometries. *Int. J. Heat Mass Transf.* **1988**, *31*, 505–515. [CrossRef]
22. Mishra, D.K.; Dolan, K.D.; Beck, J.V.; Ozadali, F. Use of Scaled Sensitivity Coefficient Relations for Intrinsic Verification of Numerical Codes and Parameter Estimation for Heat Conduction. *J. Verif. Valid. Uncertain. Quantif.* **2017**, *2*, 031005. [CrossRef]
23. Mishra, D.K.; Dolan, K.D.; Beck, J.V.; Ozadali, F. Temperature-Dependent Dielectric and Thermal Properties of Whey Protein Gel and Mashed Potato. *Trans. ASABE* **2013**, *56*, 1457–1467. [CrossRef]
24. Krishna Kumar, P.K.; Bhunia, K.; Tang, J.; Rasco, B.A.; Takhar, P.S.; Sablani, S.S. Thermal transition and thermo-physical properties of potato (*Solanum tuberosum* L.) var. Russet brown. *J. Food Meas. Charact.* **2018**, *12*, 1572–1580. [CrossRef]
25. Marcotte, M.; Taherian, A.R.; Karimi, Y. Thermophysical properties of processed meat and poultry products. *J. Food Eng.* **2008**, *88*, 315–322. [CrossRef]
26. Tavman, I.H.; Tavman, S. Measurement of thermal conductivity of dairy products. *J. Food Eng.* **1999**, *41*, 109–114. [CrossRef]
27. Choi, Y.; Okos, M.R. Effect of temperature and composition on the thermal properties of food. In *Food Engineering and Process Applications 1*; Maguer, M.L., Jelen, P., Eds.; Elsevier: London, UK; New York, NY, USA, 1986; Volume 1, pp. 93–101.
28. Muniandy, A. *Application of Temperature-Dependent Thermal Properties in Food Thermal Process Simulation and Selection of Product Formulation*; Purdue University: West Lafayette, IN, USA, 2019.

foods

MDPI

Article

Effect of Cold- and Hot-Break Heat Treatments on the Physicochemical Characteristics of Currant Tomato (*Solanum pimpinellifolium*) Pulp and Paste

Kandi Sridhar [1,2], Hilal A. Makroo [1,3] and Brijesh Srivastava [1,*]

[1] Department of Food Engineering & Technology, Tezpur University, Tezpur 784 028, Assam, India; sridhar4647@gmail.com or sridhar.kandi@agrocampus-ouest.fr (K.S.); hilalmakroo@gmail.com or hilalmakroo@iust.ac.in (H.A.M.)

[2] UMR1253, Science et Technologie du Lait et de l'œuf, INRAE, L'Institut Agro Rennes-Angers, 65 Rue de Saint Brieuc, F-35042 Rennes, France

[3] Department of Food Technology, Islamic University of Science and Technology, Awantipora 192 122, Jammu & Kashmir, India

* Correspondence: brijesh@tezu.ernet.in; Tel.: +91-3712-27-5712

Abstract: Currant tomato (*Solanum pimpinellifolium*), an underutilized wild species of modern tomato, was investigated to determine the physicochemical properties and understand the effect of cold- and hot-break heat treatments on physicochemical characteristics. Moreover, a new Arrhenius-type equation was used to model the temperature-dependent viscosity of currant tomato pulp and paste. The currant tomato's porosity, surface area, and lycopene content were 40.96 ± 0.84%, 663.86 ± 65.09 mm^2, and 9.79 ± 1.88 mg/100 g, respectively. Cold- and hot-break heat treatments had a significant ($p < 0.05$) effect on tomato pulp and paste color change (0.09 to 0.26; 0.19 to 1.96), viscosity (0.06 to 0.02 Pa.s; 0.85 to 0.37 Pa.s), and lycopene content (9.70 to 9.07 mg/100 g; 9.60 to 9.37 mg/100 g), respectively. An Arrhenius-type equation described the temperature-dependent viscosity of currant tomato pulp and paste with activation energy (E_a) ranging from 7.54 to 11.72 kJ/mol and 8.62 to 8.97 kJ/mol, respectively. Principal component analysis (PCA) revealed a total of variance 99.93% in tomato pulp and paste as affected by the cold- and hot-break heat treatments. Overall, the findings may provide knowledge for design graders and process optimization to develop currant tomato-based products.

Keywords: underutilized wild species; mathematical modeling; lycopene; viscosity; thermal processing; color

Citation: Sridhar, K.; Makroo, H.A.; Srivastava, B. Effect of Cold- and Hot-Break Heat Treatments on the Physicochemical Characteristics of Currant Tomato (*Solanum pimpinellifolium*) Pulp and Paste. *Foods* **2022**, *11*, 1730. https://doi.org/10.3390/foods11121730

Academic Editors: Carlos Vilas, Míriam R. García and Jose A. Egea

Received: 18 May 2022
Accepted: 10 June 2022
Published: 13 June 2022

1. Introduction

Diets rich in fruits and vegetables have received much consumer attention due to their health-promoting properties and therefore play a vital role in human nutrition and food security. The domestication of major plant species, such as rice, wheat, and maize, provides almost 60% of total energy intake [1]. However, over the years, this approach has been dramatically narrowed, and other wild, semi-wild, and domesticated fruits and vegetable species are entirely neglected and underutilized. Underutilized crops represent a critical source of improved food production and a stable food supply for the projected human population. Moreover, many underutilized crops are considered sources of disease resistance, early maturation, resistance to drought, and soil erosion tolerance [2].

Currant tomato (*Solanum pimpinellifolium*), an underutilized wild species of modern tomato, is native to Western South America and grown (0.50 to 1 m in height) in well-drained and moist soils, with quite large bunches [3]. A study identified many disease-resistant genes in currant tomato varieties and good sources of lycopene with high antioxidant traits compared to commercial *Solanum lycopersicon* L. cultivars [4,5]. Currant tomato is an underutilized or undervalued fruit or vegetable crop but an important source of nutrient and bioactive compounds with beneficial properties [6]. A study by Delgadillo-Díaz et al. [7] investigated the physicochemical characteristics and biological activity of *S. lycopersicum* L. and *S. pimpinellifolium* L. in different cropping systems. Similarly, Bojarska et al. [8] investigated the physicochemical properties of different tomato cultivars, including small-sized tomatoes and concluded that the lower hardness and high biological activity of small-fruited tomatoes compared to commercial tomatoes. This indicates that small-sized tomatoes are an attractive source of health-promoting properties. One way to increase the utilization of currant tomatoes is by conducting extensive research regarding the physicochemical properties and its utilization as a raw material for industrial processing and manufacturing of currant tomato-based products. This signifies an additional option for producing lycopene-rich products and further increases food production by value addition and food sustainability. Generally, the physical properties of fruit are an important parameter for designing and modelling processing operations to create a specific model to obtain a high-standardized raw material for stabilizing commercial tomato-based products in the consumer market [9].

Generally, fresh tomato has a short storage life under ambient storage conditions, and thus tomato is processed in the form of pulp, paste, sauce, soup, and puree. Therefore, tomato processing industries consider tomato pulp or paste as the primary quality parameter of the final food product [10]. For example, tomato pulp or paste quality depends on processing conditions, such as temperature, which further influence the physicochemical characteristics [11]. Another study showed the inactivation of pectin-degrading enzymes (i.e., pectin methylesterase and polygalacturonase) in tomatoes through cold- and hot-break heat treatments improves the quality of the end product [12]. However, although these studies have demonstrated the effectiveness of industrial processes on the quality attributes of commercial/modern tomatoes, too little scientific attention has been devoted to the utilization of currant tomatoes and their high-quality end products. We hypothesize that a better understanding of physicochemical and processing methods could provide the basis to develop high-quality currant tomato-derived commercial products at an industrial scale.

Therefore, this study was aimed to determine the physicochemical properties (i.e., linear dimensions, seed and skin weights, juice yield, bulk and true densities, porosity, geometric mean diameter, surface area, sphericity, moisture, pH, total soluble solids, lycopene, and ascorbic acid) of currant tomato and to understand the effect of cold- and hot-break heat treatments on the physicochemical characteristics of currant tomato pulp and paste. Moreover, the change in viscosity of tomato pulp and paste the temperature was modelled by a new Arrhenius approach. This study contributes to the design and handling of currant tomato processing equipment. The findings from modelling and heat treatment can be used to standardize process parameters to maintain the high yield and quality of currant tomato-based processed products.

2. Materials and Methods

2.1. Sample Collection

Currant tomatoes (4 kg) were purchased from the Solmara local market (26° 41′ 25.10′′ N and 92° 48′ 2.10′′ E), Napaam, Assam, India during January, 2015. Unblemished and mature fresh samples were selected based on visual observation of degree of maturity, size, and color without any mechanical and/or pathological injuries (Figure 1). All the samples were transferred into low-density polyethylene (LDPE) food-grade sampling bags (250 × 200 mm) and then transported to the Department of Food Engineering & Technology, Tezpur University, Assam, India, for further experimental analysis (no longer than two days).

Figure 1. Pictorial representation of currant tomato fruits. Fresh unripe tomato fruits along with twiggy bushes (**A**) and fully ripened fresh currant tomato fruits (**B**).

2.2. Physicochemical Properties

Fresh fruit samples were thoroughly washed with tap water followed by rinsing with double-distilled water to remove all soil residues. Then, the fruit surface was blotted to dry with a bibulous paper and immediately analyzed to determine physicochemical properties.

The linear dimensions, such as major diameter (length), intermediate diameter (width), and minor diameter (thickness), were measured using a Vernier caliper with a sensitivity of ±0.02 mm. Skin, seed, and juice weights were measured using an electronic balance with an accuracy of ±0.01 g. The water displacement method was used to determine the volume of samples (n = 10) [13]. Briefly, randomly selected fruit samples were placed in a measuring cylinder containing a known water volume. The weight of water displaced by the sample was recorded, and the volume was calculated according to Equation (1).

$$\text{Volume}\left(\text{cm}^3\right) = \left[\frac{\text{Weight of displaced water (W)}}{\text{Weight density of water }(\rho)}\right] \quad (1)$$

Bulk density (ρ_b) was measured using the mass to volume relationship according to Li et al. [14], as shown in Equation (2). In contrast, true density (ρ_t) was calculated according to the liquid displacement method [15]. The porosity (ε) of the samples was determined from bulk and true densities using the relationship according to Equation (3).

$$\text{Bulk density}\left(\text{g/cm}^3\right) = \left[\frac{\text{Weight of sample}}{\text{Total volume of sample}}\right] \quad (2)$$

$$\text{Porosity }(\varepsilon) = \left[\left(\frac{\rho_t - \rho_b}{\rho_t}\right) \times 100\right] \quad (3)$$

where ρ_b = bulk density and ρ_t = true density.

Geometric mean diameter (GMD), surface area (S), and sphericity (φ) were calculated according to Equations (4)–(6):

$$\text{Geometric mean diameter (mm)} = \left[\sqrt[3]{\text{A} \times \text{B} \times \text{C}}\right] \quad (4)$$

$$\text{Surface area}\left(\text{mm}^2\right) = \left[\pi\,(\text{GMD})^2\right] \quad (5)$$

$$\text{Sphericity }(\%) = \left[\left(\frac{\text{GMD}}{\text{A}}\right) \times 100\right] \quad (6)$$

where A = major diameter (length), B = intermediate diameter (width), and C = minor diameter (thickness).

Moisture content was analyzed according to the Association of Official Analytical Chemists [16], and the results were expressed as %. The pH was determined by a digital pH-meter (pH 510, Eutech Instruments Pte Ltd., Singapore) using 5 mL of sample and calibrated with acidic and neutral buffer solutions (pH 4.0 and 7.0). Total soluble solids of the samples were analyzed by a traditional hand-held refractometer (Erma Inc., Tokyo, Japan), and the results were expressed as °Brix.

Lycopene content was determined by the method proposed by Fish et al. [17]. Briefly, samples (5 g) were thoroughly mixed with an acetonic solution of butylated hydroxytoluene (0.05%, 5 mL), ethanol (95%, 5 mL), and hexane (10 mL) over a shaker (Orbitek® LT, Chennai, India) for 15 min at 180 rpm and 4 °C. After shaking, distilled water (3 mL) was added to the reaction mixture and then it was kept for shaking at 180 rpm and 4 °C for 5 min. The sample was allowed to separate into phases at ambient temperature, and the absorbance of the hexane organic phase was measured at 503 nm using a UV/Vis spectrophotometer (Shimadzu 1700, Tokyo, Japan). Lycopene content of the sample was calculated using a molar extinction coefficient and a specific multiplication factor (31.20) according to Equation (7) and expressed as mg/100 g.

$$\text{Lycopene content (mg/100 g)} = \left[\left(\frac{31.20 \times \text{absorbance}}{\text{Weight of sample}}\right)\right] \tag{7}$$

Ascorbic acid content was measured according to the method described by Makroo et al. [10]. Briefly, a crushed sample (5 g) was diluted with distilled water to make the total volume 100 mL, and then metaphosphoric acid (20%, 25 μL) was added. The aliquot (10 mL) was titrated against 2, 6 dichloroindophenols (0.05%, w/v) until it turned to faint pink color for 20 s. Ascorbic acid standardization was carried out using the AOAC method and was calculated as shown in Equation (8). Ascorbic acid content was expressed as mg/100 g.

$$\text{Ascorbic acid}\left(\frac{\text{mg}}{100\text{ g}}\right) = \left[\left(\frac{\text{Titer value} \times \text{Dye factor} \times \text{Volume made up} \times 100}{\text{Amount of aliquot} \times \text{sample weight}}\right)\right] \tag{8}$$

2.3. Cold- and Hot-Break Heat Treatments

Fresh fruit samples were longitudinally cut into four non-identical pieces and then blended in a laboratory-scale mini-blender (Orpat® HHB-107E, Gujarat, India) over 5 min to make smooth pulp (5.50 ± 0.70 °Brix at 24 ± 2 °C). The pulp was preheated to a temperature of 55 and 65 °C for cold-break processing over 2 min and to 75, 85, and 95 °C for hot-break processing over 2 min, respectively [18]. The pulp mixture was sieved through a cheesecloth to remove the skin, seeds, and other solid residues. For tomato paste, freshly prepared pulp samples were concentrated to a viscous paste (25 ± 1 °Brix at 24 ± 2 °C) in a rotary vacuum evaporator (Eyela, NCB-1200, Tokyo, Japan) at 45 °C under controlled pressure. All the samples were cooled to room temperature and stored in sterilized food-grade reclosable pouches (100 × 150 mm) at 4 °C. Samples without cold- and hot-break heat treatments were treated as a control. The samples were coded as T55 and T65 (cold-break) and T75, T85, and T95 (hot-break). All the experiments were conducted in triplicate.

2.4. The Physicochemical Characteristics of Currant Tomato Pulp and Paste

2.4.1. Color Analysis

The color of samples was determined by a colorimeter (UltraScan VIS, Hunter Associates Laboratory, Inc., Reston, VA, USA) in terms of CIELAB color parameters, L (light to dark), a (green to red), and b (blue to yellow). The colorimeter was standardized using a black glossy ceramic plate as a reference to the measurements. The total color difference (ΔE) was calculated according to Equation (9).

$$\text{Color difference } (\Delta E) = \left[\Delta L^2 + \Delta a^2 + \Delta b^2\right]^{\frac{1}{2}} \tag{9}$$

where ΔL = difference in lightness, Δa = difference in intensity of red color, and Δb = difference in intensity of yellow color.

2.4.2. Apparent Viscosity

Apparent viscosity measurements were determined using a Rapid Visco Analyser (Newport Scientific Pty. Ltd., Warriewood NSW, Australia). Studies by Makroo et al. [12] and Kapoor and Metzger [19] reported the determination of apparent viscosity by a Rapid Visco Analyser. Briefly, samples (25 mL) were subjected to holding temperatures of 30, 45, 60, 75, and 90 °C and a stirring speed of 100 rpm for 3 min after attaining the required temperature. The apparent viscosity of samples was expressed as Pa.s. Lycopene and total soluble solids for pulp and paste were determined according to the above-mentioned methods [12,17]. Total titratable acidity was determined by titration of samples against NaOH (0.10 N) containing phenolphthalein (0.50%) indicator to an endpoint of faint pink color for 1 min. The results were reported based on % citric acid [12].

2.5. Modeling of the Temperature-Dependent Apparent Viscosity of Currant Tomato Pulp and Paste

The effect of temperature on the apparent viscosity of pulp and paste is usually expressed by the linear Arrhenius relationship [20], as shown in Equation (10).

$$\ln(\eta) = \ln(A_s) + \frac{E_a}{R}\left(\frac{1}{T}\right) \tag{10}$$

where η = apparent viscosity (mPa.s), R (gas constant) = 8.314 J/mol K, E_a = activation energy of flow (J/mol), A_s = the pre-exponential (entropic) factor of the Arrhenius equation for the liquid system, and T = absolute temperature (K).

The plot of the logarithm of apparent shear viscosity (ln (η)) vs. reciprocal of absolute temperature (1/T) over different studied temperature ranges was plotted to construct a straight line ($R^2 \geq 0.98$) and intercept [20] as shown in Figure 2. The slope of the straight line is equal to E_a/R and the intercept on the ordinate is equal to ln A_s. In addition to E_a/R and ln A_s, we determined the Arrhenius temperature (T_A, K), which was deduced from the intercept according to Messaâdi et al. [20] as shown in Equation (11).

$$T_A = \left[\frac{-E_a}{R\ln(A_s)}\right] \tag{11}$$

The viscosity-temperature dependence can be simplified based on Equations (10) and (11), as shown in Equation (12).

$$\ln(\eta) = \left[\frac{E_a}{R}\left(\frac{1}{T} - \frac{1}{T_A}\right)\right] \tag{12}$$

For homogenous dimensions, the Arrhenius activation temperature (T*, K) was determined based on Equation (13), as shown in Figure 2.

$$T^* = \left[\frac{E_a}{R}\right] \tag{13}$$

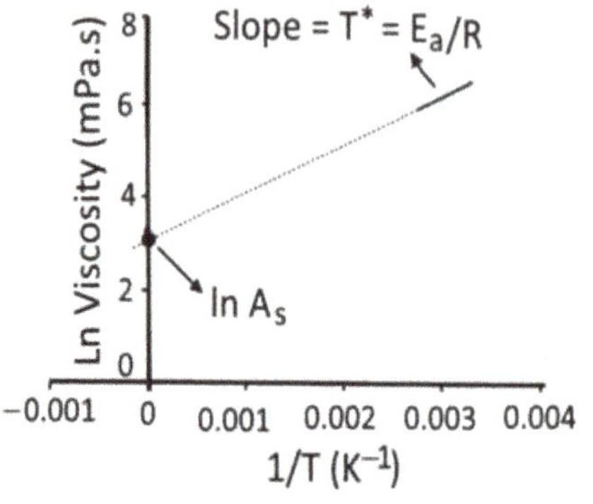

Figure 2. Graphical representation for determination of the Arrhenius parameters: activation energy (E_a/R), the logarithm of pre-exponential factor ($\ln A_s$), and Arrhenius activation temperature (T*, K). R = universal gas constant [20]. Figure 2 was reconstructed using experimental data of the present study, which was based on Messaâdi et al. [20] and is available under the Creative Commons Attribution License.

2.6. Principal Component Analysis (PCA)

PCA, an unsupervised multivariate analysis, is used to provide an exploratory grouping of the samples by transferring a set of correlated variables into a new set of linearly uncorrelated variables (i.e., principal components) based on Eigenvalue > 1. We performed the PCA (a Kaiser–Meyer–Olkin value of 0.80, $p < 0.05$) to visualize the differences and similarities between the samples as affected by cold- and hot-break heat treatments using Origin® 2019b version 9.65 (OriginLab Corporation, Northampton, MA, USA). The findings were illustrated by a loading plot containing first two principal components.

2.7. Statistical Analysis

The experimental data were reported as the mean ± standard deviation (SD) of ≥three independent determinations. The difference in experimental data was statistically assessed by one-way analysis of variance (ANOVA) using IBM® SPSS® version 16.0 (IBM Ltd., Chicago, IL, USA), and the differences were considered significant at $p < 0.05$ by Duncan's multiple range test. All graphs were constructed using Microsoft Excel® version 2021 (Microsoft Co., Ltd., Redmond, WA, USA). Bivariate correlation analysis was performed among the defined Arrhenius parameters using IBM® SPSS® version 16.0.

3. Results and Discussion

3.1. Physicochemical Properties

The physicochemical properties of the currant tomato are presented in Table 1. The average major and minor diameters ranged from 13 to 16.01 mm and 14.02 to 18.03 mm, respectively, with an average intermediate diameter of 14.14 mm at a mean moisture content of 90.86%. The presence of high moisture indicated that the preservation technologies might be required to extend the shelf-life over storage. The fruit contained 40% juice, while 60% contributed to the presence of seeds (34%) and skin (23.30%). The bulk and true density of fruit varied from 0.48 to 0.97 (g/cm^3) with a porosity of 40.10 to 42.15%. The density of fruits is a fundamental material property and a function of moisture content, which could be used to design materials handling equipment for drying and storage [15]. The high porosity indicated that the drying of currant tomato could be faster with low power requirements. The geometric mean diameter derived from major, minor, and intermediate diameters ranged from 13.67 to 15.93 mm, with a mean value of 14.12 mm.

Table 1. Descriptive statistics for the physicochemical characteristics of currant tomato [1].

Parameter	Minimum	Maximum	Mean	SD
Major diameter (mm)	13	16.01	14.31	0.82
Intermediate diameter (mm)	13.01	17.01	14.14	1.08
Minor diameter (mm)	14.02	18.03	14.92	1.22
Skin weight (%)	22.05	25.42	23.30	0.02
Seed weight (%)	32.00	35.25	34.00	0.01
Juice (%)	38.96	42.78	40.00	0.01
Bulk density (g/cm^3)	0.48	0.59	0.54	0.04
True density (g/cm^3)	0.88	0.97	0.92	0.03
Porosity (%)	40.10	42.15	40.96	0.84
Geometric mean diameter (mm)	13.67	15.93	14.12	0.70
Surface area (mm^2)	585.91	797.12	663.86	65.09
Sphericity (%)	0.98	1.04	1.01	0.08
Moisture (%)	88.1	93.63	90.86	1.81
pH	4.30	4.60	4.32	0.12
Total soluble solids (°Brix)	5	7	5.50	0.70
Lycopene (mg/100 g)	8.21	11.45	9.79	1.88
Ascorbic acid (mg/100 g)	38.15	43.06	39.93	1.88

[1] The results were expressed as the mean ± standard deviation ($n \geq 3$). SD = standard deviation of the mean determined from ≥3 independent determinations.

The surface area and sphericity values varied from 585.91 to 797.12 mm^2 and 0.98 to 1.04%, respectively. The surface area could determine the wax and packaging material to be applied to the fruit. At the same time, sphericity may be related to the diameter of the fruit and useful for hopper design and pricking machines to handle the fruit [21]. The pH of the fruit was found to be acidic (4.32), with a total soluble solids content ranging from 5 to 7 °Brix. These results agreed with the earlier findings on the pH (4.29) and total soluble solids (5 to 8 °Brix) content of commercial *S. lycopersicon* L. [10,22]. The lower pH was probably due to the presence of organic acids, including citric and maleic acids, that contributed to the tomato's acidic nature. Lycopene and ascorbic acid contents ranged from 8.21 to 11.45 mg/100 g and 38.15 to 43.06 mg/100 g, respectively, which indicated that the currant tomato is a good source of lycopene and ascorbic acid contents. Similar results were reported for lycopene content in tomato-based products [23]. In brief, the information on the physicochemical properties of any food material may provide knowledge on designing and optimizing process equipment and predicting the behavior of food material. Thus, these findings can assist in developing handling equipment for currant tomatoes.

3.2. *Effect of Cold- and Hot-Break Heat Treatments on the Physicochemical Characteristics of Currant Tomato Pulp*

3.2.1. Color

The color parameters of currant tomato pulp as a function of heat treatments indicated a significant difference ($p < 0.05$), as shown in Table 2. The values of the lightness (L), green/red (a), and blue/yellow (b) decreased significantly ($p < 0.05$) at all heat treatments. The variations in L, a, and b values might be due to the consequences of heat treatments that altered the color parameters of tomato pulp [24]. An earlier study by Ganje et al. [25] concluded the variations in color parameters of tomato pulp as affected by thermal processing, which could be related to the degradation of pigments during heat treatments. For example, lycopene in tomatoes may be degraded to *cis* form during heat treatments, resulting in a color change [10]. No significant ($p > 0.01$) change was observed in the a/b value, ranging from 0.99 to 1.04. Application of statistical analysis revealed that heat treatments had a significant ($p < 0.05$) effect on the ΔE of tomato pulp, which further supported the visual appearance of tomato pulp after heat treatments (Figure 3A). Therefore, the findings from this study recommended considering the optimization of heat treatments of tomato juice to maximize the retention of the color attributes of the currant tomato pulp for the consumer's acceptance.

Table 2. Effect of heat treatments on the color parameters of currant tomato pulp and paste [1].

Treatment	L	a	b	a/b	ΔE
Pulp					
Control	29.45 ± 0.54 d	7.61 ± 1.08 e	7.32 ± 0.62 c	1.04 ± 0.01 a	0.00 ± 0.00 a
T55	29.40 ± 0.97 d	7.53 ± 2.28 d	7.31 ± 0.83 c	1.03 ± 0.05 a	0.09 ± 0.01 b
T65	29.33 ± 1.04 d	7.47 ± 2.11 c	7.34 ± 0.14 c	1.02 ± 0.01 a	0.09 ± 0.01 b
T75	29.11 ± 2.01 c	7.40 ± 0.08 c	7.21 ± 0.22 b	1.03 ± 0.07 a	0.26 ± 0.04 c
T85	28.92 ± 3.33 b	7.27 ± 1.29 b	7.25 ± 0.28 b	1.01 ± 0.02 a	0.23 ± 0.04 c
T95	28.78 ± 0.07 a	7.13 ± 1.58 a	7.17 ± 0.09 a	0.99 ± 0.08 a	0.21 ± 0.05 c
Paste					
Control	25.54 ± 0.71 a	9.88 ± 0.05 c	6.54 ± 0.08 e	1.51 ± 0.01 a	0.00 ± 0.00 a
T55	27.15 ± 0.85 c	8.91 ± 0.03 a	5.97 ± 0.01 b	1.49 ± 0.09 a	1.96 ± 0.09 f
T65	26.94 ± 0.34 b	9.12 ± 0.06 a	6.12 ± 0.09 c	1.49 ± 0.08 a	0.33 ± 0.01 d
T75	26.99 ± 0.68 b	9.27 ± 0.01 b	6.24 ± 0.01 d	1.48 ± 0.05 a	0.19 ± 0.01 b
T85	27.26 ± 0.55 c	9.21 ± 0.04 b	6.19 ± 0.05 c	1.48 ± 0.01 a	0.28 ± 0.01 c
T95	27.58 ± 0.26 d	8.99 ± 0.05 a	5.89 ± 0.06 a	1.52 ± 0.09 a	0.49 ± 0.01 e

[1] The results were expressed as the mean ± standard deviation (n = 3). Mean values within a column with different lowercase superscripts (a–f) were significantly different ($p < 0.05$, Duncan's multiple range test). T55 and T95 represent coded temperature values for cold- and hot-break heat treatments.

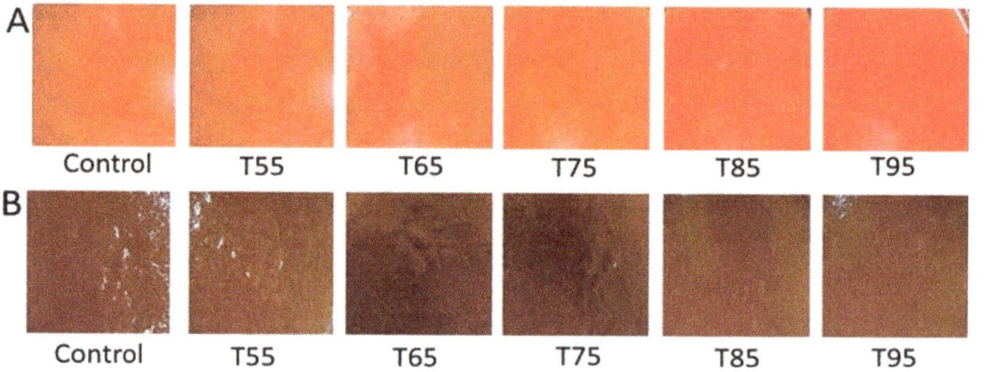

Figure 3. Effect of heat treatments on currant tomato. Pulp (**A**) and paste (**B**). In Figure 3 (**A**,**B**), T55 to T95 represent coded temperature values for cold- and hot-break heat treatments, respectively.

3.2.2. Apparent Viscosity, Lycopene Content, Total Soluble Solids, and Total Titratable Acidity

Figure 4 shows the apparent viscosity, lycopene content, total soluble solids, and total titratable acidity of tomato pulp as a function of heat treatments. The apparent viscosity of tomato pulp differed significantly ($p < 0.05$) among the heat treatments and followed no definite trend. However, T85 and T95 showed a decreased apparent viscosity (average of 0.03 Pa.s) compared to control (0.05 Pa.s), indicating the effect of heat treatments on currant tomato pulp. The variations in apparent viscosity could be associated with the alignment of the heterogeneous suspended particles that contributed to a strong particle aggregation [18]. Another study by Augusto et al. [26] explained the contribution of particle size in tomato pulp and suggested the homogenization of tomato products to maintain consistency. Heating treatment had a significant impact on the pectin integrity and viscosity of tomato pulp [27]. The heating treatment further showed the conformational changes in pectin solubility, which may change the viscosity of tomato pulp [28]. Another study by Verlent et al. [29] showed the variations in tomato viscosity as affected by heat treatments due to the action of polygalacturonase on pectin, which may be partially active to depolymerization of pectin. Based on the studies mentioned above, we assume that the complexity and heterogeneity of the dispersions may contribute to the variations in the rheological properties of the tomato pulp.

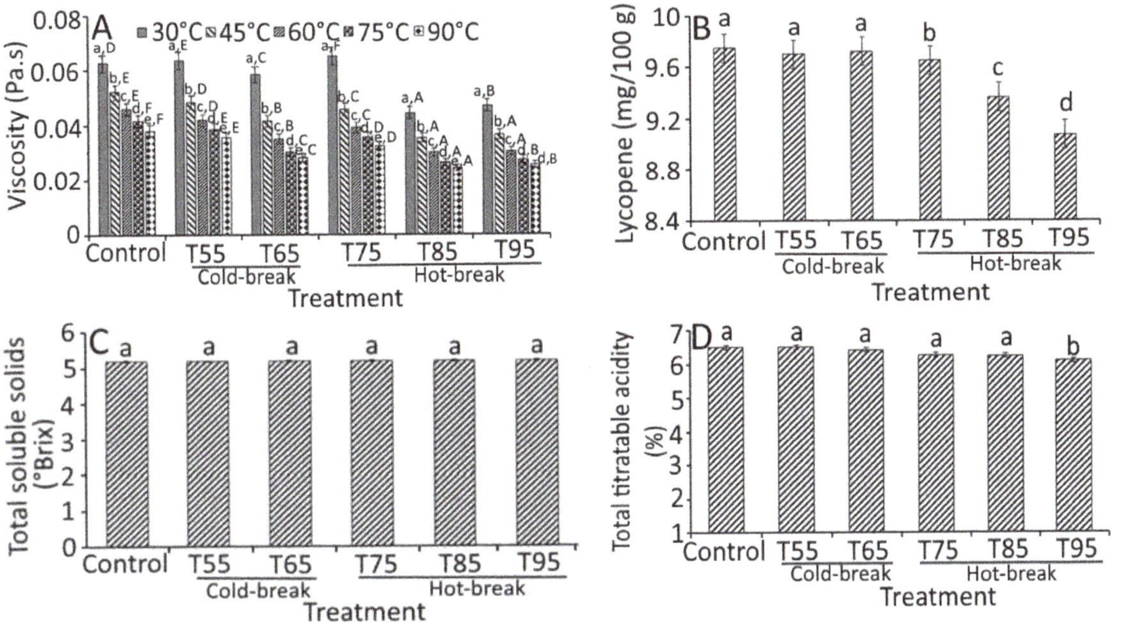

Figure 4. Effect of heat treatments on the quality characteristics of currant tomato pulp. Apparent viscosity (**A**), lycopene (**B**), total soluble solids (**C**), and total titratable acidity (**D**). The error bar represents the standard deviation from the mean of independent triplicate results. Different lowercase letters (a–e) within the treatment indicate a significant difference ($p < 0.05$) based on Duncan's multiple range test. In Figure 4A, different uppercase letters (A–F) among the holding temperatures indicate a significant difference ($p < 0.05$) based on Duncan's multiple range test. T55 and T95 represent coded temperature values for cold- and hot-break heat treatments.

On the other hand, viscosity holding temperatures (30, 45, 60, 75, and 90 °C) showed a significant ($p < 0.05$) effect on the apparent viscosity of tomato pulp (Figure 4A). At 45 °C, heat-treated samples collectively showed a 0.90-fold lower apparent viscosity than the control sample at 30 °C. Moreover, a similar decreased trend was observed for viscosity at different holding temperatures, which was the lowest at 90 °C. At 90 °C, all treated samples demonstrated a 0.95-fold lower apparent viscosity than the control sample (0.04 Pa.s). These findings were likely related to the rise in temperature and heat treatments [18].

A significant difference ($p < 0.05$) in lycopene content and total titratable acidity was observed based on the heat treatments except for the TSS content, where heat treatments showed (Figure 4B–D) insignificant change ($p > 0.05$). Similarly, Boubidi and Boutebba [30] reported the insignificant change in TSS content of tomato products as affected by the change in temperature, which may relate to the short time treatments. For lycopene content, no significant change was observed in T55, T65, and T75 compared to control. However, a slight but significant decrease in lycopene content was observed in T85 (9.36 mg/100 g) and T95 (9.07 mg/100 g). At a higher temperature (130 °C), Miki and Akatsu [31] reported the loss of lycopene content (0.17 mg/100 g), which could be related to the higher degradation of *trans* lycopene to *cis* form [25]. High loss of lycopene at high holding temperature has also been reported due to dissolved air in the pulp that may destroy lycopene [32]. Therefore, the degradation of tomato pulp during thermal processing can be reduced by de-aerating the pulp before thermal treatment and reducing the duration of treatment.

The total titratable acidity was decreased significantly from 6.50 (control) to 6.10% (T95), which was found to be 0.40% lower than control. We propose that reducing total titratable acidity could be the consequences of inactivated pectin enzyme activity and solubilization of acids in an aqueous medium. A study by Boubidi and Boutebba [30] highlighted the interdependence of natural acids by heat treatments. Compared to the natural acids, heat treatments significantly affected the parameters like apparent viscosity, lycopene content, and total titratable acidity.

3.3. *Effect of Cold- and Hot-Break Heat Treatments on the Physicochemical Characteristics of Currant Tomato Paste*

3.3.1. Color

The color characteristics of the currant tomato paste are summarized in Table 2. The L value significantly ($p < 0.05$) ranged from 25.54 to 27.58, while the 'a' value varied from 8.99 to 9.88 in all heat treatments. Similarly, there was a significant change ($p < 0.05$) in color 'b' value after heat treatments, ranging from 5.89 to 6.24. Moreover, the a/b value had an insignificant effect ($p > 0.05$) that ranged from 1.48 to 1.52. As shown in Table 2, heat treatments had a significant decrease in ΔE of currant tomato paste, which varied from 0.19 to 1.96. According to the findings from the color characteristics of currant tomato paste, it can be concluded that the tomato paste followed a similar trend as the color characteristics of the currant tomato pulp. This indicated that the heat treatments had a similar influence on currant tomato pulp and paste color characteristics. Our visual scoring of currant tomato pastes also evidenced the change of color attributes (Figure 3B). Similar findings were documented by Shatta et al. [10] that showed significant changes in color parameters (L, a, b, a/b, and ΔE) of tomato paste after thermal processing. Another study by Ganje et al. [25] highlighted the change in the color characteristics of tomato paste overheat processing, which could be the consequences of lycopene thermal degradation from *trans* to *cis* form and isomerization. Considering the color change characteristics of currant tomato paste, application of heat treatments on tomato products should be carefully monitored to receive high consumer perception on thermally processed currant tomato-based products.

3.3.2. Apparent Viscosity, Lycopene Content, Total Soluble Solids, and Total Titratable Acidity

The apparent viscosity, lycopene, total soluble solids, and total titratable acidity of currant tomato paste as a function of heat treatments are depicted in Figure 5A–D. The findings reported that the apparent viscosity, lycopene, and total titratable acidity of currant tomato paste differed significantly ($p < 0.05$), but an insignificant ($p > 0.05$) change was observed in the total soluble solids content of currant tomato paste (Figure 5). The apparent viscosity values (0.47 to 0.85 Pa.s) of currant tomato paste demonstrated an increased trend for heat treatments (T55 to T85) and then decreased to 0.75 Pa.s (T95). Higher apparent viscosity for currant tomato paste could be ascribed to the solid concentration of the samples, indicating that solids content plays a vital role in the consistency of tomato paste [22]. For heat treatments, the highest apparent viscosity was observed for T85 at a holding temperature of 30 °C (0.85 Pa.s), whereas the lowest was recorded for control (0.26 Pa.s) at a holding temperature of 90 °C (Figure 5A). This indicated that the heat treatments influenced the viscosity of currant tomato paste. For example, T85 reported the maximum apparent viscosity at all holding temperatures (30 to 90 °C) compared to control, T55, T65, and T75. This could be related to the action of pectin enzymes due to their strong intra- and inter-molecular bonding of pectin. At a higher temperature (T95), significantly lower apparent viscosity was observed than at T75 and T85. A possible explanation for this might be related to the breakdown of pectin structural units of the sample. Other researchers have suggested the low viscosity of tomato paste during heat treatment [22].

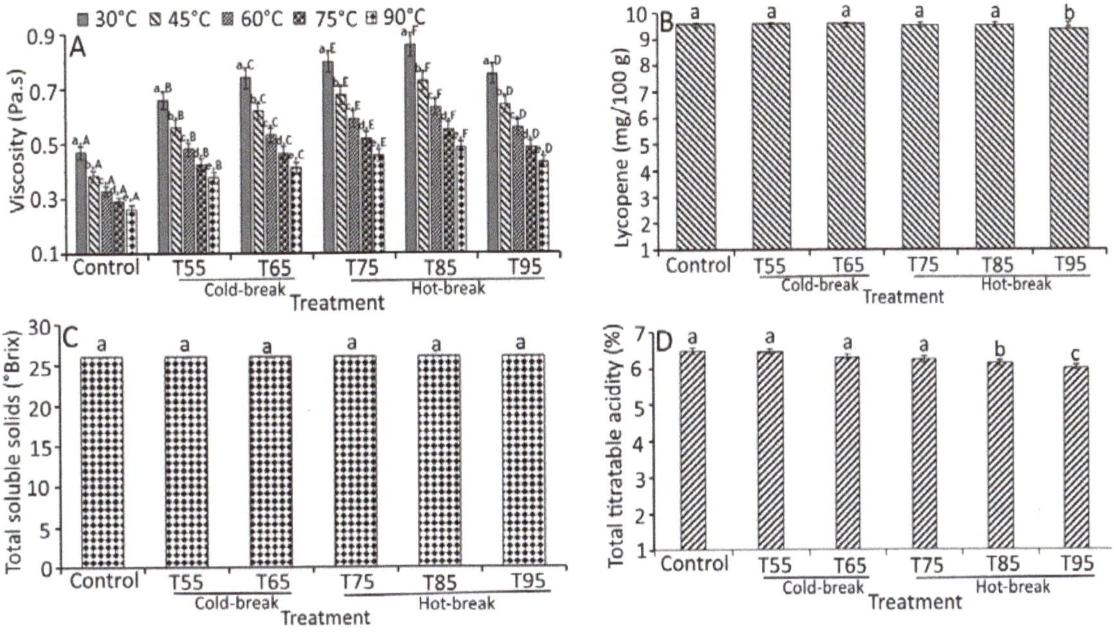

Figure 5. Effect of heat treatments on the quality characteristics of currant tomato paste. Apparent viscosity (**A**), lycopene (**B**), total soluble solids (**C**), and total titratable acidity (**D**). The error bar represents the standard deviation from the mean of independent triplicate results. Different lowercase letters (a–e) within the treatment indicate a significant difference ($p < 0.05$) based on Duncan's multiple range test. In Figure 5A, different uppercase letters (A–F) among the holding temperatures indicate a significant difference ($p < 0.05$) based on Duncan's multiple range test. T55 and T95 represent coded temperature values for cold- and hot-break heat treatments.

On the other hand, we compared the influence of holding temperature on apparent viscosity of currant tomato paste (Figure 5A). The apparent viscosity of currant tomato paste significantly decreased ($p < 0.05$) in control and heat-treated samples over the rise in holding temperature from 30 to 90 °C. The heat-treated samples showed a 1.62-fold higher apparent viscosity than the control sample at a holding temperature of 30 °C, which followed a similar tendency at all holding temperatures (45 to 90 °C). These differences can be partially elucidated by the proximity of holding temperature and heat treatments.

Lycopene content and total titratable acidity significantly varied and ranged from 9.75 to 9.07 mg/100 g and 6.5 to 6.10%, respectively, for all treatments. Still, no significant change ($p > 0.05$) was observed in total soluble solids (5.20 °Brix). Similar findings were reported by Kaur et al. [33] that showed a slight variation in lycopene content during different processing methods. At higher heat treatments (T75 to T95), lycopene content was found to be lower (~9.36 mg/100 g) than in control (9.75 mg/100 g), indicating that the lycopene was stable at a lower temperature (T55 and T65) and then degraded at higher temperature (T75 to T95). This was similar to the lycopene degradation findings of Shi et al. [34], where lycopene was stable at lower temperatures. Then, a significant decrease was observed over a temperature rise. The maximum decrease in total titratable acidity of the currant tomato paste was obtained at T95, which was 0.82-fold lower than the control (6.50%). This may be due to the precipitation of salts existing in the paste [35]. Based on our findings, it can be concluded that the heat treatments may influence the viscosity and quality attributes of currant tomato paste. Hence, based on requirements, heat treatments and holding temperature may be altered for a subsequent application in the development of currant tomato products.

3.4. Modeling of the Temperature-Dependent Apparent Viscosity of Currant Tomato Pulp and Paste

In general, the effect of temperature on the apparent viscosity of currant tomato pulp and paste is explained according to the Arrhenius equation (Equation (10)). The Arrhenius model constants were evaluated in terms of natural log (ln) of viscosity vs. inverse of absolute temperature (1/T, K^{-1}) for pulp and paste (Figure 6). The activation energy values for pulp and paste were observed in the range of 7.54 to 11.72 kJ/mol and 8.62 to 8.97 kJ/mol, respectively. The coefficient of determination (R^2) for pulp and paste was >0.98, which explained the strong linear relationship between inverse of temperature (1/T) and apparent viscosity of pulp and paste. These findings agreed with earlier findings, which reported activation energy values from 8.60 to 14.08 kJ/mol for tomato concentrates [36].

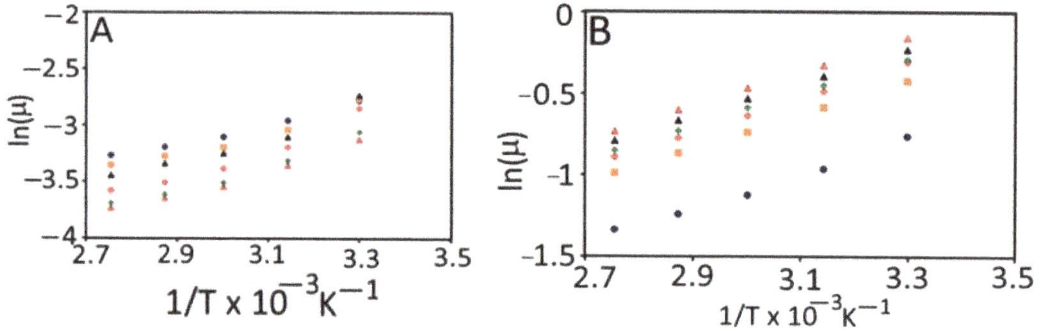

Figure 6. Arrhenius plot for the viscosity of currant tomato pulp (**A**) and paste (**B**). Control (●), T55 (■), T65 (♦), T75 (▲), T85 (▲), and T95 (●). T55 and T95 represent coded temperature values for cold- and hot-break heat treatments.

The Arrhenius temperature (T_A, K) and the Arrhenius activation temperature (T^*, K) for pulp and paste are shown in Table 3. The T_A and T^* showed interdependence and ranged from 313.49 to 6106 K and 907.77 to 1410.75 K for both pulp and paste, respectively. We further performed the correlation analysis among the defined Arrhenius parameters (T_A, ln A_s, and E_a) for currant tomato pulp and paste (Table 4). ln A_s and E_a exhibited a negative correlation with T_A, while ln A_s showed a positively weak correlation with E_a for tomato pulp. The similar correlation trend was reported for paste, where T_A showed a strong negative correlation (−0.90) with ln A_s; however, T_A for paste demonstrated a strong positive correlation (0.79) with E_a compared to pulp. This indicated the interdependence of Arrhenius parameters, which may highly influence the determination of E_a for food products. A study by Messaâdi et al. [20] reported the linear and non-linear correlations among the Arrhenius parameters. Thus, these results concluded that the currant tomato pulp and paste obey the linear Arrhenius behavior. An Arrhenius equation could describe the temperature dependence of viscosity of currant tomato pulp and paste. Hence, we recommend using a new Arrhenius-type equation with two and/or three parameters and thus useful for determining the nature of fluids.

Table 3. Arrhenius parameters for temperature-dependent apparent viscosity of currant tomato pulp and paste [1].

Treatment	T* (K)	ln A_s	E_a (kJ/mol)	T_A (K)
Pulp				
Control	907.77	1.12	7.54	810
T55	1071.83	0.57	8.91	1877
T65	1410.75	0.62	11.72	2264
T75	1242.12	0.01	10.32	1035
T85	1117.59	0.07	9.29	1490
T95	1092.95	0.18	9.08	6106
Paste				
Control	1076.27	2.58	8.94	417.18
T55	1040.19	3.05	8.64	340.61
T65	1079.61	3.03	8.97	355.46
T75	1028.20	3.28	8.54	313.49
T85	1080.03	3.19	8.97	338.53
T95	1030.50	3.20	8.62	324.11

[1] T* = Arrhenius activation temperature, A_s = the pre-exponential (entropic) factor of the Arrhenius equation for the liquid system, E_a = activation energy of flow, and T_A = Arrhenius temperature. T55 and T95 represent coded temperature values for cold- and hot-break heat treatments.

Table 4. Correlation among the defined Arrhenius parameters of currant tomato pulp and paste [1].

Sample	Arrhenius Parameters	T_A	ln A_s	E_a
Pulp	T_A	1	−0.01	−0.29
	ln A_s		1	0.32
	E_a			1
Paste	T_A	1	−0.90 *	0.79
	ln A_s		1	−0.46
	E_a			1

[1] T_A = Arrhenius temperature, A_s = the pre-exponential (entropic) factor of the Arrhenius equation for the liquid system, and E_a = activation energy of the flow. * Correlation is significant at the 0.05 level (2-tailed). T55 and T95 represent coded temperature values for cold- and hot-break heat treatments.

3.5. Principal Component Analysis (PCA)

The PCA of tomato samples as affected by cold- and hot-break heat treatments resulted a total variance of 99.93%, in which first two principal components (PCs 1 and 2) represented a major variance of 98.88% and 1.05%, respectively with an Eigenvalue >1, indicating close interdependence of the samples. Figure 7 illustrates the clear separation of samples within the quadrants 1 and 4 (Q1 and Q4), indicating the effect of cold- and hot-break heat treatments that had different influences on tomato pulp and paste. Moreover, both pulp and paste samples formed well-separated groups as a function of cold- and hot-break heat treatments. For example, pulp samples treated with T55, T65, and T75 formed a group 1, while group 2 was formed by the pulp samples treated with T85 and T95. However, paste samples treated with cold- and hot-break heat treatments (T55, T65, T75, T85, and T95) collectively formed group 4 (Figure 7). Likewise, untreated samples (control pulp and paste) were clearly separated as group 3. The formation of different groups could be related to the similarities (due to less angle) among the samples treated with cold- and hot-break heat treatments. Thus, samples with less angle showed a high similarity and were close to each other among the samples. Based on these observations, the findings concluded that the tomato pulp and paste characteristics might have been influenced by the cold- and hot-break heat treatments and could be considered while preparing heat-treated tomato-based processed products.

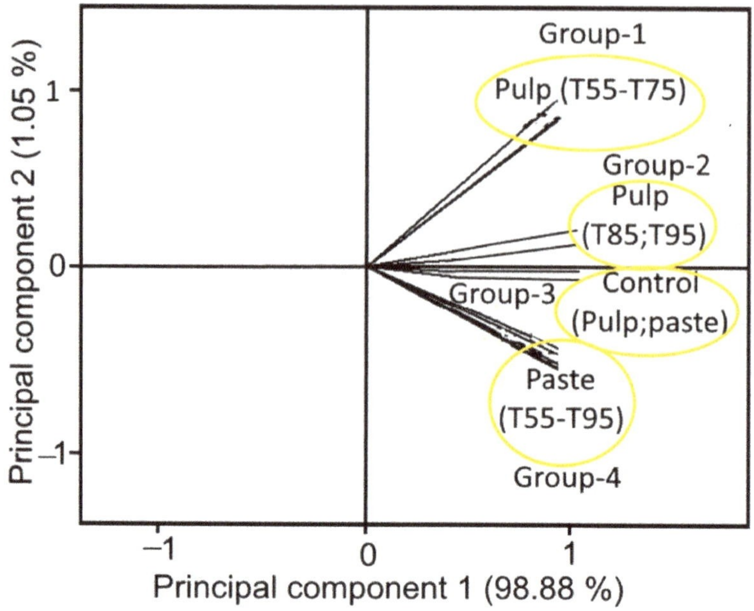

Figure 7. Loading plot of the first two principal components of tomato samples as affected by cold- and hot-break heat treatments. T55 and T95 represent coded temperature values for cold- and hot-break heat treatments.

4. Conclusions

The results showed the physicochemical properties of currant tomato pulp and paste. Currant tomato pulp and paste were significantly affected by the heat treatments. Color, lycopene content, and viscosity of tomato pulp and paste were significantly influenced by the cold- and hot-break heat treatments. However, total soluble solids, and total titratable acidity showed no significant differences after the cold- and hot-break heat treatments. The activation energy values ranged from 7.54 to 11.72 kJ/mol and 8.62 to 8.97 kJ/mol, respectively, for pulp and paste. The results further indicated that the linear Arrhenius-type equation successfully described the temperature-dependent apparent viscosity of currant tomato pulp and paste. PCA analysis revealed the differences in tomato pulp and paste characteristics (a total variance of 99.93%) as affected by the cold- and hot-break heat treatments. This is one of the first attempts to provide the basic physicochemical information on currant tomatoes that could be used in the design of currant tomatoes cleaning, handling, and separation machines. Data from heat treatments and modeling can be used to control and optimize process parameters in developing high-quality currant tomato-based food products.

Author Contributions: Conceptualization, B.S. and K.S.; methodology, K.S.; software, K.S.; validation, B.S., K.S. and H.A.M.; formal analysis, B.S. and K.S.; investigation, K.S.; resources, B.S.; data curation, K.S.; writing—original draft preparation, K.S.; writing—review and editing, B.S., K.S. and H.A.M.; visualization, B.S.; supervision, B.S.; project administration, B.S.; funding acquisition, B.S. All authors have read and agreed to the published version of the manuscript.

Funding: This research received no external funding.

Institutional Review Board Statement: Not applicable.

Informed Consent Statement: Not applicable.

Data Availability Statement: The data that supports the findings of this study are available within the manuscript.

Acknowledgments: The authors sincerely acknowledge the support of the staff at the Department of Food Engineering & Technology, Tezpur University—India, for their technical assistance. The authors also acknowledge DST-FIST, NEQIP-AICTE, and UGC-SAP for providing financial support for infrastructure development of the Department of Food Engineering & Technology, Tezpur University—India.

Conflicts of Interest: The authors declare no conflict of interest.

References

1. Katoch, R. Underutilized crops: An overview. In *Ricebean: Exploiting the Nutritional Potential of an Underutilized Legume*; Katoch, A., Ed.; Springer: Singapore, 2020; pp. 1–27.
2. Padulosi, S.; Cawthorn, D.-M.; Meldrum, G.; Flore, R.; Halloran, A.; Mattei, F. Leveraging neglected and underutilized plant, fungi, and animal species for more nutrition-sensitive and sustainable food systems. In *Encyclopedia of Food Security and Sustainability*; Ferranti, P., Berry, E.M., Anderson, J.R., Eds.; Elsevier: Oxford, UK, 2019; pp. 361–370.
3. Iriti, M.; Belli, L.; Nali, C.; Lorenzini, G.; Gerosa, G.; Faoro, F. Ozone sensitivity of currant tomato (*Lycopersicon pimpinellifolium*), a potential bioindicator species. *Environ. Pollut.* **2006**, *141*, 275–282. [CrossRef] [PubMed]
4. Walgenbach, J.F. Integrated pest management strategies for field-grown tomatoes. In *Sustainable Management of Arthropod Pests of Tomato*; Wakil, W., Brust, G.E., Perring, T.M., Eds.; Academic Press: San Diego, CA, USA, 2018; pp. 323–339.
5. Top, O.; Bar, C.; Ökmen, B.; Özer, D.Y.; Rusçuklu, D.; Tamer, N.; Frary, A.; Doğanlar, S. Exploration of three *Solanum* species for improvement of antioxidant traits in tomato. *HortScience* **2014**, *49*, 1003–1009. [CrossRef]
6. Murthy, H.N.; Bapat, V.A. (Eds.) Importance of underutilized fruits and nuts. In *Bioactive Compounds in Underutilized Fruits and Nuts*; Springer International Publishing: Cham, Switzerland, 2020; pp. 3–19.
7. Delgadillo-Díaz, M.; Gullian-Klanian, M.; Sosa-Moguel, O.; Sauri-Duch, E.; Cuevas-Glory, L.F. Evaluation of physico-chemical characteristics, antioxidant compounds and antioxidant capacity in creole tomatoes (*Solanum lycopersicum* L. and *S. pimpinellifolium* L.) in an aquaponic system or organic soil. *Int. J. Veg. Sci.* **2019**, *25*, 124–137. [CrossRef]
8. Bojarska, J.E.; Piłat, B.; Majewska, K.M.; Sobiechowska, D.A.; Narwojsz, A. Selected physical parameters and chemical compounds of different types of tomatoes. *Czech J. Food Sci.* **2020**, *38*, 28–35. [CrossRef]
9. Mirzaee, E.; Rafiee, S.; Keyhani, A.; Djom-eh, Z.E.; Kheiralipour, K. Mass modeling of two varieties of apricot (*Prunus armenaica* L.) with some physical characteristics. *J. Plant Omics* **2008**, *1*, 37–43.
10. Shatta, A.; Youssef, K.; Al Sanabani, A.; El Samahy, S. Impact of processing steps on physicochemical and rheological properties of tomato paste (cold-break). *MOJ Food Process. Technol.* **2017**, *5*, 00122. [CrossRef]
11. Farahnaky, A.; Majdinasab, M.; Majzoobi, M.; Mesbahi, G. A comparative study of physicochemical and rheological properties of Iranian tomato pastes. *Int.J. Food Eng.* **2010**, *6*, 1–16. [CrossRef]
12. Makroo, H.A.; Rastogi, N.K.; Srivastava, B. Enzyme inactivation of tomato juice by ohmic heating and its effects on physico-chemical characteristics of concentrated tomato paste. *J. Food Process Eng.* **2017**, *40*, e12464. [CrossRef]
13. Aydın, C.; Özcan, M.M. Determination of nutritional and physical properties of myrtle (*Myrtus communis* L.) fruits growing wild in Turkey. *J. Food Eng.* **2007**, *79*, 453–458. [CrossRef]
14. Li, Z.; Li, P.; Liu, J. Physical and mechanical properties of tomato fruits as related to robot's harvesting. *J. Food Eng.* **2011**, *103*, 170–178. [CrossRef]
15. Boukouvalas, C.J.; Krokida, M.K.; Maroulis, Z.B.; Marinos-Kouris, D. Density, and porosity: Literature data compilation for foodstuffs. *Int. J. Food Prop.* **2006**, *9*, 715–746. [CrossRef]
16. AOAC. *Official Methods of Analysis of AOAC International*; AOAC International: Rockville, MD, USA, 2016; ISBN 978-0-935584-87-5.
17. Fish, W.W.; Perkins-Veazie, P.; Collins, J.K. A Quantitative assay for lycopene that utilizes reduced volumes of organic solvents. *J. Food Comp. Anal.* **2002**, *15*, 309–317. [CrossRef]
18. Kubo, M.T.K.; Rojas, M.L.; Miano, A.C.; Augusto, P.E.D. Rheological properties of tomato products. In *Tomato Chemistry, Industrial Processing and Product Development*; The Royal Society of Chemistry: London, UK, 2019; pp. 1–25.
19. Kapoor, R.; Metzger, L.E. Small-scale manufacture of process cheese using a rapid visco analyzer. *J. Dairy Sci.* **2005**, *88*, 3382–3391. [CrossRef]
20. Messaâdi, A.; Dhouibi, N.; Hamda, H.; Belgacem, F.B.M.; Adbelkader, Y.H.; Ouerfelli, N.; Hamzaoui, A.H. A new equation relating the viscosity Arrhenius temperature and the activation energy for some Newtonian classical solvents. *J. Chem.* **2015**, *2015*, 163262. [CrossRef]
21. Goyal, R.K.; Kingsly, A.R.P.; Kumar, P.; Walia, H. Physical and mechanical properties of aonla fruits. *J. Food Eng.* **2007**, *82*, 595–599. [CrossRef]
22. Wu, B.; Patel, B.K.; Fei, X.; Jones, O.; Campanella, O.H.; Reuhs, B.L. Variations in physical-chemical properties of tomato suspensions from industrial processing. *LWT* **2018**, *93*, 281–286. [CrossRef]
23. Vioque, M.; de la Cruz-Ares, S.; Gómez, R. Preliminary investigation on the physicochemical and functional properties of commercial *Salmorejo* found in Spanish supermarkets. *Foods* **2021**, *10*, 1146. [CrossRef]

24. Kaur, D.; Sogi, D.S.; Abas Wani, A. Degradation kinetics of lycopene and visual color in tomato peel isolated from pomace. *Int. J. Food Prop.* **2006**, *9*, 781–789. [CrossRef]
25. Ganje, M.; Jafari, S.M.; Farzaneh, V.; Malekjani, N. Kinetics modelling of color deterioration during thermal processing of tomato paste with the use of response surface methodology. *Heat Mass Transf.* **2018**, *54*, 3663–3671. [CrossRef]
26. Augusto, P.E.D.; Ibarz, A.; Cristianini, M. Effect of high-pressure homogenization (HPH) on the rheological properties of tomato juice: Time-dependent and steady-state shear. *J. Food Eng.* **2012**, *111*, 570–579. [CrossRef]
27. Gao, R.; Wu, Z.; Ma, Q.; Lu, Z.; Ye, F.; Zhao, G. Effects of breaking methods on the viscosity, rheological properties and nutritional value of tomato paste. *Foods* **2021**, *10*, 2395. [CrossRef] [PubMed]
28. Yeom, H.W.; Ramirez, M.; Conte, J. Effects of storage and heating on serum viscosity of commercial hot break tomato paste. *J. Food Sci. Eng.* **2018**, *1*, 14–19.
29. Verlent, I.; Hendrickx, M.; Rovere, P.; Moldenaers, P.; Loey, A.V. Rheological properties of tomato-based products after thermal and high-pressure treatment. *J. Food Sci.* **2006**, *71*, S243–S248. [CrossRef]
30. Boubidi, F.; Boutebba, A. Effects of heat treatments on quality parameters and the natural antioxidants of triple concentrated tomato paste. *J. Ann. Food Sci. Technol.* **2013**, *14*, 5–12.
31. Miki, N.; Akatsu, K. Effects of heating sterilization on the color of tomato juice. *Nippon. Shokuhin Kogyo Gakkaishi* **1970**, *17*, 175–181. [CrossRef]
32. Shi, J. Lycopene in tomatoes: Chemical and physical properties affected by food processing. *Crit. Rev. Food Sci. Nutr.* **2000**, *20*, 293–334.
33. Kaur, C.; George, B.; Deepa, N.; Jaggi, S.; Kapoor, H.C. Viscosity and quality of tomato juice as affected by processing methods. *J. Food Qual.* **2007**, *30*, 864–877. [CrossRef]
34. Shi, J.; Le Maguer, M.; Bryan, M.; Kakuda, Y. Kinetics of lycopene degradation in tomato puree by heat and light irradiation. *J. Food Proc. Eng.* **2003**, *25*, 485–498. [CrossRef]
35. Popescu, E.C.; Iordan, M. Effect of thermal processing on bioactive compounds of tomato paste. *J. Ann. Food Sci. Technol.* **2012**, *13*, 185–189.
36. Dak, M.; Verma, R.C.; Jaaffrey, S. Rheological properties of tomato concentrate. *Int. J. Food Eng.* **2008**, *4*, 1–19. [CrossRef]

MDPI
St. Alban-Anlage 66
4052 Basel
Switzerland
Tel. +41 61 683 77 34
Fax +41 61 302 89 18
www.mdpi.com

Foods Editorial Office
E-mail: foods@mdpi.com
www.mdpi.com/journal/foods

www.ingramcontent.com/pod-product-compliance
Lightning Source LLC
Chambersburg PA
CBHW040806250926
51850CB00007BA/1095
* 9 7 8 3 0 3 6 5 6 6 9 2 4 *